FIREFLY

WORLD FACTBOOK

SECOND EDITION
KEITH LYE

Area 2,988,885 sq mi [7,741,229 sq km]
Population 19,913,000
Capital (population) Canberra (309,000)
Government Federal constitutional monarchy
Ethnic groups Caucasian 92%. Asian 7%. Aboriginal 1%
Languages English

FIREFLY BOOKS

A FIREFLY BOOK

Published by Firefly Books Ltd. 2006

Copyright © 2006 Philip's

Cartography by Philip's
Text: Keith Lye

First printing

Publisher Cataloging-in-Publication Data (U.S.)

Lye, Keith.
    World factbook / Keith Lye.—2nd ed.
[400] p. : col. ill. , maps ; cm.
Includes index.
Summary: Illustrated reference guide to every country, territory, principality and dependency in the world, summarizing basic statistics, climate, vegetation, history, politics and economy.
ISBN-13: 978-1-55407-201-9 (pbk.)
ISBN-10: 1-55407-201-8 (pbk.)
1. Geography -- Handbooks, manuals, etc. 2. History -- Handbooks, manuals, etc. 3. Economics -- Handbooks, manuals, etc. 4. Political science -- Handbooks, manuals, etc. I. Title.
909 dc22   G123.L94 2006

Library and Archives Canada Cataloguing in Publication

Lye, Keith
    World factbook / Keith Lye. — 2nd ed.

Previous ed. published under title: Firefly world factbook.
Includes index.
ISBN-13: 978-1-55407-201-9
ISBN-10: 1-55407-201-8

    1. Geography—Handbooks, manuals, etc.
2. History—Handbooks, manuals, etc.   3. Economics—Handbooks, manuals, etc.   4. Political science--Handbooks, manuals, etc.   I. Title.

Published in the United States by
Firefly Books (U.S.) Inc.
P.O. Box 1338, Ellicott Station
Buffalo, New York 14205

Published in Canada by
Firefly Books Ltd
66 Leek Crescent
Richmond Hill, Ontario L4B 1H1

Published in Great Britain by Philip's,
a division of Octopus Publishing Group Ltd,
2–4 Heron Quays, London E14 4JP

Printed in China

EDITOR Daisy Leitch

EXECUTIVE ART EDITOR Mike Brown

DESIGNER Will Butler

PRODUCTION Åsa Sonden

FRONT COVER : Hong Kong Harbour, CORBIS; Peru, Philip's; Flag of Denmark, Philip's; Climate of Kabul, Philip's; Sand Dune and Trees, CORBIS.
BACK COVER : Flags of South Africa, Switzerland, Suriname, Tuvalu, Sri Lanka, and Iraq, Philip's; Maps of Costa Rica and Ivory Coast, Philip's.

Introduction

Firefly World Factbook contains articles and maps for all the countries of the world, arranged in alphabetical order. A simple locator map shows the position of the country within its region, and the national flag is illustrated with an explanation of its origin. The smallest countries or islands, such as the islands of the Atlantic Ocean, are grouped together within their region. Other small countries are found alongside their larger neighbors; for example, Andorra and Gibraltar are included in the entry for Spain.

In addition to the main article, each country features a separate piece about the capital city; a statistics box detailing area, population, capital city, government, ethnic groups, languages, religions, currency and website; and a climate graph.

The country maps are physically colored and are usually a column width in size, but a country's shape and area have sometimes merited a larger map. Where neighboring countries are in dispute, the maps show the de facto boundary between nations; that is, the boundary that exists in the real world rather than the boundary that a particular country may wish to be shown on a map.

Each map contains a number of symbols indicating the major towns and cities, principal roads and railways, with the main border crossings. Major international airports and the highest point in most countries are also shown. A scale bar indicates the distances between places, and the lines of latitude and longitude show the country's global position.

The spellings of the placenames on the maps match the forms that are used in everyday English; for example, Rome (not Roma), Munich (not München), Moscow (not Moskva). Only the more common accented characters of the Western European languages have been used.

KEY

■● City or town	✈ Major airport	⌒ Railroad
✹ Capital city	△ Highest point in country	✳ Road
	↩ International boundary	River

Contents

WORLD STATISTICS

Country/Territory	Area (1,000 sq km)	Area (1,000 sq mi)	Population (1,000s)	Capital City	Annual Income US$
Afghanistan	652	252	28,514	Kabul	700
Albania	28.7	11.1	3,545	Tirana	4,400
Algeria	2,382	920	32,129	Algiers	5,400
American Samoa (US)	0.20	0.08	58	Pago Pago	8,000
Andorra	0.47	0.18	68	Andorra La Vella	19,000
Angola	1,247	481	10,979	Luanda	1,700
Anguilla (UK)	0.10	0.04	13	The Valley	8,600
Antigua & Barbuda	0.44	0.17	68	St John's	11,000
Argentina	2,780	1,074	39,145	Buenos Aires	10,500
Armenia	29.8	11.5	2,991	Yerevan	3,600
Aruba (Netherlands)	0.19	0.07	71	Oranjestad	28,000
Australia	7,741	2,989	19,913	Canberra	26,900
Austria	83.9	32.4	8,175	Vienna	27,900
Azerbaijan	86.6	33.4	7,868	Baku	3,700
Azores (Portugal)	2.2	0.86	243	Ponta Delgada	15,000
Bahamas	13.9	5.4	297	Nassau	15,300
Bahrain	0.69	0.27	678	Manama	15,100
Bangladesh	144	55.6	141,340	Dhaka	1,800
Barbados	0.43	0.17	277	Bridgetown	15,000
Belarus	208	80.2	10,311	Minsk	8,700
Belgium	30.5	11.8	10,348	Brussels	29,200
Belize	23.0	8.9	273	Belmopan	4,900
Benin	113	43.5	7,250	Porto-Novo	1,100
Bermuda (UK)	0.05	0.02	65	Hamilton	35,200
Bhutan	47.0	18.1	2,186	Thimphu	1,300
Bolivia	1,099	424	8,724	La Paz/Sucre	2,500
Bosnia-Herzegovina	51.2	19.8	4,008	Sarajevo	1,900
Botswana	582	225	1,562	Gaborone	8,500
Brazil	8,514	3,287	184,101	Brasília	7,600
Brunei	5.8	2.2	358	Bandar Seri Begawan	18,600
Bulgaria	111	42.8	7,518	Sofia	6,500
Burkina Faso	274	106	13,575	Ouagadougou	1,100
Burma (= Myanmar)	677	261	42,720	Rangoon	1,700
Burundi	27.8	10.7	6,231	Bujumbura	500
Cambodia	181	69.9	13,363	Phnom Penh	1,600
Cameroon	475	184	16,064	Yaoundé	1,700
Canada	9,971	3,850	32,508	Ottawa	29,300
Canary Is. (Spain)	7.4	2.9	1,673	Las Palmas/Santa Cruz	19,900
Cape Verde Is.	4.0	1.6	415	Praia	1,400
Cayman Is. (UK)	0.26	0.10	47	George Town	35,000
Central African Republic	623	241	3,742	Bangui	1,200
Chad	1,284	496	9,539	Ndjaména	1,000
Chile	757	292	15,824	Santiago	10,100
China	9,597	3,705	1,298,848	Beijing	4,700
Colombia	1,139	440	42,311	Bogotá	6,100
Comoros	2.2	0.84	652	Moroni	700
Congo	342	132	2,998	Brazzaville	900
Congo (Dem. Rep. of the)	2,345	905	58,318	Kinshasa	600
Cook Is. (NZ)	0.29	0.11	21	Avarua	5,000
Costa Rica	51.1	19.7	3,957	San José	8,300
Croatia	56.5	21.8	4,497	Zagreb	9,800
Cuba	111	42.8	11,309	Havana	2,700
Cyprus	9.3	3.6	776	Nicosia	13,200
Czech Republic	78.9	30.5	10,246	Prague	15,300

Listed above are the principal countries and territories of the world. If a territory is not completely independent, then the country it is associated with is named. The area figures give the total area of land, inland water and ice. The population figures are 2004 estimates. The annual income is the Gross Domestic Product per capita in US dollars. [Gross Domestic Product per capita has been measured using the purchasing power parity method. This enables comparisons to be made between countries through their purchasing power (in US dollars), showing real price levels of goods and services rather than using currency exchange rates.] The figures are the latest available, usually 2002 estimates. *OPT 5 Occupied Palestinian Territory; N/A 5 Not available.

Country / Territory	Area (1,000 sq km)	Area (1,000 sq mi)	Population (1,000s)	Capital City	Annual Income US$
Denmark	43.1	16.6	5,413	Copenhagen	28,900
Djibouti	23.2	9.0	467	Djibouti	1,300
Dominica	0.75	0.29	70	Roseau	5,400
Dominican Republic	48.5	18.7	8,834	Santo Domingo	6,300
East Timor	14.9	5.7	1,019	Dili	500
Ecuador	284	109	13,213	Quito	3,200
Egypt	1,001	387	76,117	Cairo	4,000
El Salvador	21.0	8.1	6,588	San Salvador	4,600
Equatorial Guinea	28.1	10.8	523	Malabo	2,700
Eritrea	118	45.4	4,447	Asmara	700
Estonia	45.1	17.4	1,342	Tallinn	11,000
Ethiopia	1,104	426	67,851	Addis Ababa	700
Faeroe Is. (Denmark)	1.4	0.54	47	Tórshavn	22,000
Fiji	18.3	7.1	881	Suva	5,600
Finland	338	131	5,215	Helsinki	25,800
France	552	213	60,424	Paris	26,000
French Guiana (France)	90.0	34.7	191	Cayenne	14,400
French Polynesia (France)	4.0	1.5	266	Papeete	5,000
Gabon	268	103	1,355	Libreville	6,500
Gambia, The	11.3	4.4	1,547	Banjul	1,800
Gaza Strip (OPT)*	0.36	0.14	1,325	–	600
Georgia	69.7	26.9	4,694	Tbilisi	3,200
Germany	357	138	82,425	Berlin	26,200
Ghana	239	92.1	20,757	Accra	2,000
Gibraltar (UK)	0.006	0.002	28	Gibraltar Town	17,500
Greece	132	50.9	10,648	Athens	19,100
Greenland (Denmark)	2,166	836	56	Nuuk (Godthåb)	20,000
Grenada	0.34	0.13	89	St George's	5,000
Guadeloupe (France)	1.7	0.66	440	Basse-Terre	9,000
Guam (US)	0.55	0.21	166	Agana	21,000
Guatemala	109	42.0	14,281	Guatemala City	3,900
Guinea	246	94.9	9,246	Conakry	2,100
Guinea-Bissau	36.1	13.9	1,388	Bissau	700
Guyana	215	83.0	706	Georgetown	3,800
Haiti	27.8	10.7	7,656	Port-au-Prince	1,400
Honduras	112	43.3	6,824	Tegucigalpa	2,500
Hong Kong (China)	1.1	0.42	6,855	–	27,200
Hungary	93.0	35.9	10,032	Budapest	13,300
Iceland	103	39.8	294	Reykjavik	30,200
India	3,287	1,269	1,065,071	New Delhi	2,600
Indonesia	1,905	735	238,453	Jakarta	3,100
Iran	1,648	636	69,019	Tehran	6,800
Iraq	438	169	25,375	Baghdad	2,400
Ireland	70.3	27.1	3,970	Dublin	29,300
Israel	20.6	8.0	6,199	Jerusalem	19,500
Italy	301	116	58,057	Rome	25,100
Ivory Coast (= Côte d'Ivoire)	322	125	17,328	Yamoussoukro	1,400
Jamaica	11.0	4.2	2,713	Kingston	3,800
Japan	378	146	127,333	Tokyo	28,700
Jordan	89.3	34.5	5,611	Amman	4,300
Kazakhstan	2,725	1,052	15,144	Astana	7,200
Kenya	580	224	32,022	Nairobi	1,100
Kiribati	0.73	0.28	101	Tarawa	800
Korea, North	121	46.5	22,698	Pyŏngyang	1,000
Korea, South	99.3	38.3	48,598	Seoul	19,600
Kuwait	17.8	6.9	2,258	Kuwait City	17,500
Kyrgyzstan	200	77.2	5,081	Bishkek	2,900

WORLD STATISTICS

Country / Territory	Area (1,000 sq km)	Area (1,000 sq mi)	Population (1,000s)	Capital City	Annual Income US$
Laos	237	91.4	6,068	Vientiane	1,800
Latvia	64.6	24.9	2,306	Riga	8,900
Lebanon	10.4	4.0	3,777	Beirut	4,800
Lesotho	30.4	11.7	1,865	Maseru	2,700
Liberia	111	43.0	3,391	Monrovia	1,000
Libya	1,760	679	5,632	Tripoli	6,200
Liechtenstein	0.16	0.06	33	Vaduz	25,000
Lithuania	65.2	25.2	3,608	Vilnius	8,400
Luxembourg	2.6	1.0	463	Luxembourg	48,900
Macau (China)	0.02	0.007	445	–	18,500
Macedonia	25.7	9.9	2,071	Skopje	5,100
Madagascar	587	227	17,502	Antananarivo	800
Madeira (Portugal)	0.79	0.31	253	Funchal	22,700
Malawi	118	45.7	11,907	Lilongwe	600
Malaysia	330	127	23,522	Kuala Lumpur/Putrajaya	8,800
Maldives	0.30	0.12	339	Malé	3,900
Mali	1,240	479	11,957	Bamako	900
Malta	0.32	0.12	397	Valletta	17,200
Marshall Is.	0.18	0.07	58	Majuro	1,600
Martinique (France)	1.1	0.43	426	Fort-de-France	10,700
Mauritania	1,026	396	2,999	Nouakchott	1,700
Mauritius	2.0	0.79	1,220	Port Louis	10,100
Mayotte (France)	0.37	0.14	173	Mamoundzou	600
Mexico	1,958	756	104,960	Mexico City	8,900
Micronesia, Fed. States of	0.70	0.27	108	Palikir	2,000
Moldova	33.9	13.1	4,446	Chişinău	2,600
Monaco	0.001	0.0004	32	Monaco	27,000
Mongolia	1,567	605	2,751	Ulan Bator	1,900
Montserrat (UK)	0.10	0.04	9	Plymouth	3,400
Morocco	447	172	32,209	Rabat	3,900
Mozambique	802	309	18,812	Maputo	1,100
Namibia	824	318	1,954	Windhoek	6,900
Nauru	0.02	0.008	13	Yaren District	5,000
Nepal	147	56.8	27,071	Katmandu	1,400
Netherlands	41.5	16.0	16,318	Amsterdam/The Hague	27,200
Netherlands Antilles (Neths)	0.80	0.31	216	Willemstad	11,400
New Caledonia (France)	18.6	7.2	214	Nouméa	14,000
New Zealand	271	104	3,994	Wellington	20,100
Nicaragua	130	50.2	5,360	Managua	2,200
Niger	1,267	489	11,361	Niamey	800
Nigeria	924	357	137,253	Abuja	900
Northern Mariana Is. (US)	0.46	0.18	78	Saipan	12,500
Norway	324	125	4,575	Oslo	33,000
Oman	310	119	2,903	Muscat	8,300
Pakistan	796	307	159,196	Islamabad	2,000
Palau	0.46	0.18	20	Koror	9,000
Panama	75.5	29.2	3,000	Panamá	6,200
Papua New Guinea	463	179	5,420	Port Moresby	2,100
Paraguay	407	157	6,191	Asunción	4,300
Peru	1,285	496	27,544	Lima	5,000
Philippines	300	116	86,242	Manila	4,600
Poland	323	125	38,626	Warsaw	9,700
Portugal	88.8	34.3	10,524	Lisbon	19,400
Puerto Rico (US)	8.9	3.4	3,898	San Juan	11,100
Qatar	11.0	4.4	522	Doha	20,100
Réunion (France)	2.5	0.97	766	St-Denis	5,600
Romania	238	92.0	22,356	Bucharest	7,600
Russia	17,075	6,593	143,782	Moscow	9,700
Rwanda	26.3	10.2	7,954	Kigali	1,200

Country/Territory	Area (1,000 sq km)	Area (1,000 sq mi)	Population (1,000s)	Capital City	Annual Income US$
St Kitts & Nevis	0.26	0.10	39	Basseterre	8,800
St Lucia	0.54	0.21	162	Castries	5,400
St Vincent & Grenadines	0.39	0.15	117	Kingstown	2,900
Samoa	2.8	1.1	178	Apia	5,600
San Marino	0.06	0.02	27	San Marino	34,600
São Tomé & Príncipe	0.96	0.37	187	São Tomé	1,200
Saudi Arabia	2,150	830	25,796	Riyadh	11,400
Senegal	197	76.0	10,852	Dakar	1,500
Serbia & Montenegro	102	39.4	10,826	Belgrade	2,200
Seychelles	0.46	0.18	81	Victoria	7,800
Sierra Leone	71.7	27.7	5,884	Freetown	500
Singapore	0.68	0.26	4,354	Singapore City	25,200
Slovak Republic	49.0	18.9	5,424	Bratislava	12,400
Slovenia	20.3	7.8	2,011	Ljubljana	19,200
Solomon Is.	28.9	11.2	524	Honiara	1,700
Somalia	638	246	8,305	Mogadishu	600
South Africa	1,221	471	42,719	Cape Town/Pretoria/ Bloemfontein	10,000
Spain	498	192	40,281	Madrid	21,200
Sri Lanka	65.6	25.3	19,905	Colombo	3,700
Sudan	2,506	967	39,148	Khartoum	1,400
Suriname	163	63.0	437	Paramaribo	3,400
Swaziland	17.4	6.7	1,169	Mbabane	4,800
Sweden	450	174	8,986	Stockholm	26,000
Switzerland	41.3	15.9	7,451	Bern	32,000
Syria	185	71.5	18,017	Damascus	3,700
Taiwan	36.0	13.9	22,750	Taipei	18,000
Tajikistan	143	55.3	7,012	Dushanbe	1,300
Tanzania	945	365	36,588	Dodoma	600
Thailand	513	198	64,866	Bangkok	7,000
Togo	56.8	21.9	5,557	Lomé	1,400
Tonga	0.65	0.25	110	Nuku'alofa	2,200
Trinidad & Tobago	5.1	2.0	1,097	Port of Spain	10,000
Tunisia	164	63.2	9,975	Tunis	6,800
Turkey	775	299	68,894	Ankara	7,300
Turkmenistan	488	188	4,863	Ashkhabad	6,700
Turks & Caicos Is. (UK)	0.43	0.17	19	Cockburn Town	9,600
Tuvalu	0.03	0.01	11	Fongafale	1,100
Uganda	241	93.1	26,405	Kampala	1,200
Ukraine	604	233	47,732	Kiev	4,500
United Arab Emirates	83.6	32.3	2,524	Abu Dhabi	22,100
United Kingdom	242	93.4	60,271	London	25,500
United States of America	9,629	3,718	293,028	Washington, DC	36,300
Uruguay	175	67.6	3,399	Montevideo	7,900
Uzbekistan	447	173	26,410	Tashkent	2,600
Vanuatu	12.2	4.7	203	Port-Vila	2,900
Vatican City	0.0004	0.0002	1	Vatican City	N/A
Venezuela	912	352	25,017	Caracas	5,400
Vietnam	332	128	82,690	Hanoi	2,300
Virgin Is. (UK)	0.15	0.06	22	Road Town	16,000
Virgin Is. (US)	0.35	0.13	125	Charlotte Amalie	19,000
Wallis & Futuna Is. (France)	0.20	0.08	16	Mata-Utu	2,000
West Bank (OPT)*	5.9	2.3	2,311	–	800
Western Sahara	266	103	267	El Aaiún	N/A
Yemen	528	204	20,025	San'a	800
Zambia	753	291	10,462	Lusaka	800
Zimbabwe	391	151	12,672	Harare	2,100

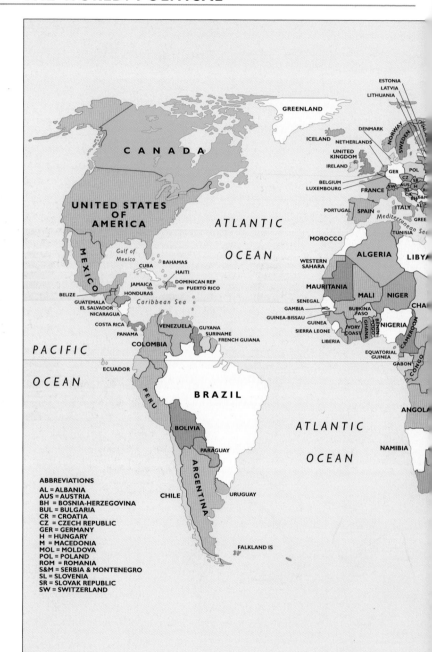

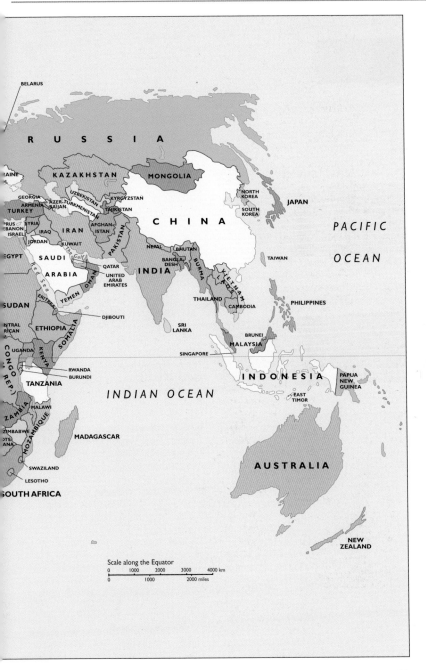

AFGHANISTAN

Introduced in January 2002, this flag replaces that of the Mujaheddin ("holy warriors"), who defeated Afghanistan's socialist government but lost power at the end of 2001. The flag is the 19th different design used by the country since 1901.

The Islamic Republic of Afghanistan is a landlocked country bordered by Turkmenistan, Uzbekistan, Tajikistan, China, Pakistan, and Iran. The main regions are the northern plains, the central highlands, and the southwestern lowlands.

The central highlands, comprising most of the Hindu Kush and its foothills, with peaks rising to more than 21,000 ft [6,400 m], cover nearly three-quarters of the land. Many Afghans live in the deep valleys of the highlands. The River Kabul flows east to the Khyber Pass border with Pakistan.

Much of the southwest is desert, while the northern plains contain most of the country's limited agricultural land. Grasslands cover much of the north, while the vegetation in the dry south is sparse.

Trees are rare in both regions. But forests of such coniferous trees as pine and fir grow on the higher mountain slopes, with cedars lower down. Alder, ash, juniper, oak, and walnut grow in the mountain valleys.

Area 251,772 sq mi [652,090 sq km]
Population 28,514,000
Capital (population) Kabul (1,565,000)
Government Transitional regime
Ethnic groups Pashtun (Pathan) 44%, Tajik 25%, Hazara 10%, Uzbek 8%, others 8%
Languages Pashtu, Dari/Persian (both official), Uzbek
Religions Islam (Sunni Muslim 84%, Shiite Muslim 15%), others
Currency Afghani = 100 puls
Website www.afghan-web.com

CLIMATE

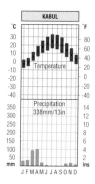

The height of the land and the country's remote position have a great effect on the climate. In winter, northerly winds bring cold, snowy weather in the mountains, but summers are hot and dry. The rainfall decreases to the south with temperatures higher throughout the year.

HISTORY

In ancient times, the area was invaded by Aryans, Persians, Greeks, and Macedonians, and warrior armies from central Asia. Arab armies introduced Islam in the late 7th century. It has always occupied a strategic position, because the Khyber Pass was both the gateway to India and the back door to Russia.

Its modern history began in 1747, when local tribes united for the first time, though a civil war was fought between 1819 and 1835 as factions struggled for power. In 1839, British troops invaded Afghanistan, in an attempt to reduce Russian influence. Over the next 80 years, Britain fought three Anglo-Afghan wars to maintain control over the region. The British finally withdrew in 1921, when Afghanistan became independent.

POLITICS

In 1964, Afghanistan adopted a democratic constitution, but the country's ruler, King Zahir, and the legislature failed to agree on reforms. Muhammad Daoud Khan, the king's cousin, seized power in 1973 and abolished the monarchy. He ruled as president until 1978, when he was killed during a left-wing coup. The new regime's socialist policies conflicted with Islam and provoked a rebellion.

On December 25, 1987, Soviet troops invaded Afghanistan to support the left-wing regime. The Soviet occupation led to a protracted civil war. Various Muslim groups united behind the banner of the Mujaheddin ("holy warriors") to wage a guerrilla campaign, financed by the United States and aided by Pakistan. Soviet forces withdrew in 1989.

By 1992, the Mujaheddin had overthrown the government. The fundamentalist Muslim Taliban ("students") became the dominant group and, by 2000, the Taliban regime controlled 90% of the land.

In October 2001, the Taliban regime refused to hand over Saudi-born Osama bin Laden, the man suspected of masterminding the attacks on New York City and Washington D.C. on September 11, 2001. This led to international action being taken against Afghanistan, with the United States to the fore. The objective was to destroy both bin Laden's terrorist organization, al Qaida, and the Taliban. In November, the Taliban regime collapsed and a coalition government was set up, led by Hamid Karzai, who was sworn into office in 2002. Later that year, more than 1,000 people died in an earthquake in northern Afghanistan.

Despite ongoing conflict, a draft constitution was approved in January 2004. The first democratic elections for president were held in October 2004, won by Hamid Karzai. September 2005 saw parliamentary and provincial elections.

ECONOMY

Afghanistan is one of the world's poorest countries. About 60% of the people are farmers, many of whom are semi-nomadic herders. Wheat is the chief crop. Natural gas is produced, together with some coal, copper, gold, precious stones and salt. There are few factories. Exports include karakul skins (which are used to make hats and jackets), carpets, dried fruit, and nuts.

KABUL

Capital of Afghanistan situated on the River Kabul in the eastern part of the country. It is strategically located in a high mountain valley in the Hindu Kush. The city was taken by Genghis Khan in the 13th century. Later it became part of the Mogul Empire, from 1526 to 1738. It has been the capital since 1776 and was occupied by the British during the Afghan Wars in the 19th century. Following the Soviet invasion in 1979, Kabul was the scene of bitter fighting. Unrest continued into the mid-1990s as rival Muslim groups fought for control. Industries include textiles, leather goods, and furniture.

ALBANIA

Albania's official name, Shqiperia, means "Land of the Eagle," and the black double eagle was the emblem of the 15th-century hero Skanderbeg. A star placed above the eagle in 1946 was removed in 1992 when a non-Communist government was formed.

The Republic of Albania lies in the Balkan Peninsula. It faces the Adriatic Sea in the west and is bordered by Serbia and Montenegro, Macedonia, and Greece. About 70% of the land is mountainous, with the highest point, Korab, reaching 9,068 ft [2,764 m] on the Macedonian border. Most Albanians live in the west on the coastal lowlands—the main farming region. Albania lies in an earthquake zone and severe earthquakes occur occasionally.

Area 11,100 sq mi [28,748 sq km]
Population 3,545,000
Capital (population) Tirana (300,000)
Government Multiparty republic
Ethnic groups Albanian 95%, Greek 3%, Macedonian, Vlachs, Gypsy
Languages Albanian (official)
Religions Many people say they are nonbelievers; of the believers, 70% follow Islam and 30% follow Christianity (Orthodox 20%, Roman Catholic 10%)
Currency Lek = 100 qindars
Website www.parlament.al

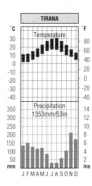

CLIMATE

The coastal areas of Albania have a typical Mediterranean climate, with fairly dry, sunny summers and cool, moist winters. The mountains have a severe climate, with heavy winter snow.

HISTORY

Albania was originally part of a region called Illyria. In 167 BC, it became part of the Roman Empire. When the Roman Empire broke up in AD 395, much of Albania became part of the Byzantine Empire. The country was subsequently conquered by Goths, Bulgarians, Slavs, and Normans, although southern Albania remained part of the Byzantine Empire until 1204.

Much of Albania became part of the Serbian Empire in the 14th century and in the 15th century, a leader named Skanderbeg, now regarded as a national hero, successfully led the Albanians against the invading Ottoman Turks. But after his death in 1468, the Turks took over the country. Albania was part of the Ottoman Empire until 1912, when Albania declared its independence.

Italy invaded Albania in 1939, but German forces took over the country in 1943. At the end of World War II, an Albanian People's Republic was formed under the Communist leaders who had led the partisans against the Germans. Pursuing a modernization program on rigid Stalinist lines, the regime of Enver Hoxha at various times associated politically and economically with Yugoslavia (to 1948), the Soviet Union (1948–61) and China (1961–77), before following a fiercely independent policy. After Hoxha died in 1985, his successor, Ramiz Alia, continued the dictator's austere policies, but by the end of the decade, even Albania was affected by the sweeping changes in Eastern Europe.

POLITICS

In 1990, the more progressive wing of the Communist Party, led by Ramiz Alia,

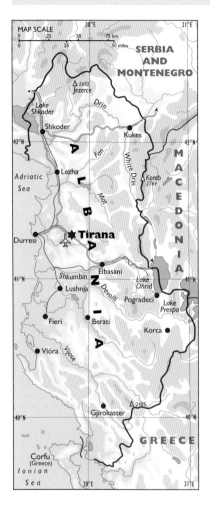

tee took over, but collapsed within six months.

Elections in 1992 finally brought to an end the last Communist regime in Europe when the non-Communist Democratic Party won power. In 1997, amid a financial crisis caused by the collapse of fraudulent pyramid-selling schemes, fresh elections took place. The socialist-led government that took power was reelected in 2001. The stability of the region was threatened when Albanian-speaking Kosovars and Macedonians, many favoring the creation of a Greater Albania, fought with government forces in northwestern Macedonia.

ECONOMY

Albania is Europe's poorest country. Agriculture employs 62% of the population. Major crops include fruits, corn, olives, potatoes, sugarbeet, vegetables and wheat. Livestock farming, and the fishing industry are also important.

Private ownership has been encouraged since 1991, but change has been slow. Albania has some minerals, and chromite, copper, and nickel are exported. There is also some petroleum, lignite, and hydroelectricity.

won the struggle for power. The new government instituted a wide program of reform, including the legalization of religion, the encouragement of foreign investment, the introduction of a free market for peasants' produce, and the establishment of pluralist democracy. The Communists comfortably retained their majority in 1991 elections, but the government was brought down two months later by a general strike. An interim coalition "national salvation" commit-

TIRANA (TIRANË)

Capital of Albania situated on the banks of the River Ishm in central Albania. Tirana was founded in the early 17th century by the Ottoman Turks and became the capital in 1920. In 1946 the communists came to power and the industrial sector of the city was developed. Industries include metal goods, agricultural machinery, and textiles. Places of interest include Skanderbeg Square with the main historical buildings. Also the 18th-century Haxhi Ethem Bey Mosque; Art Gallery (Albanian); National Museum of History.

ALGERIA

The star and crescent and the color green on Algeria's flag are traditional symbols of the Islamic religion. The liberation movement which fought for independence from French rule from 1954 used this flag. It became the national flag when Algeria became independent in 1962.

The People's Democratic Republic of Algeria is Africa's second largest country after Sudan. Most Algerians live in the north, on the fertile coastal plains and hill country. South of this region lie high plateaus and ranges of the Atlas Mountains. Four-fifths of Algeria is in the Sahara, the world's largest desert.

Area 919,590 sq mi [2,381,741 sq km]
Population 32,129,000
Capital (population) Algiers (Alger, 1,722,000)
Government Socialist republic
Ethnic groups Arab-Berber 99%
Languages Arabic and Berber (both official), French
Religions Sunni Muslim 99%
Currency Algerian dinar = 100 centimes
Website www.algeria-un.org

CLIMATE

The coast has a Mediterranean climate, with warm and dry summers and mild and moist winters. The northern highlands have warmer summers and colder winters. The arid Sahara is hot by day and cool by night. Annual rainfall is less than 8 in [200 mm].

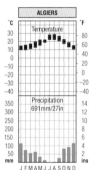

HISTORY

In early times, the region came under such rulers as the Phoenicians, Carthaginians, Romans, and Vandals. Arabs invaded the area in the AD 600s, converting the local Berbers to Islam and introducing Arabic. Intermarriage has made it difficult to distinguish Arabs from Berbers by ancestry, though Berber dialects are still spoken. A law, effective from July 1998 making Arabic the only language allowed in public life, met with much opposition in Berber-speaking areas.

POLITICS

Algeria experienced French colonial rule and colonization by settlers, finally achieving independence in 1962, following years of bitter warfare between nationalist guerrillas and French armed forces. After independence, the socialist FLN (National Liberation Party) formed a one-party government. Opposition parties were permitted in 1989.

ALGIERS

Capital and largest city of Algeria, north Africa's chief port on the Mediterranean. Founded by the Phoenicians, it has been ruled by Romans, Berber Arabs, Turks, and Muslim Barbary pirates. In 1830 the French made Algiers the capital of the colony of Algeria. In World War II it was the headquarters of the Allies and seat of the French provisional government. During the 1950s and 1960s it was a focus for the violent struggle for independence. The old city is based round a 16th-century Turkish citadel. The 11th-century Sidi Abderrahman Mosque is a major destination for pilgrims.

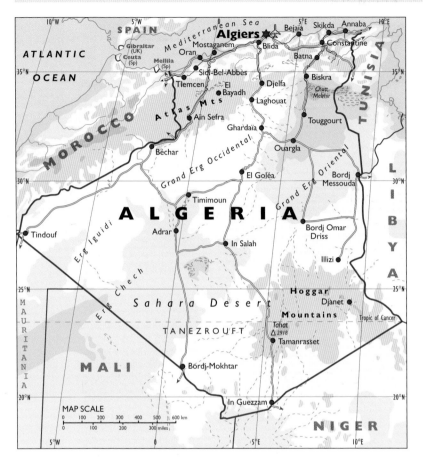

In 1991, a Muslim party, the FIS (Islamic Salvation Front) won an election. The FLN canceled the election results and declared a state of emergency. Terrorist activities rose and, between 1991 and 1999, about 100,000 people were killed. A proposal to ban political parties based on religion was approved in a referendum in 1996. In 1999, Abdelaziz Boutflika, the candidate thought to be favored by the army, was elected president. The scale of the violence fell. In 2005, the government agreed to the demands of the Berber community, including official recognition of the Berber language. In September an amnesty for Islamist guerrillas was approved in a referendum.

ECONOMY

Algeria is a developing country, whose main income is from its two main natural resources, petroleum and natural gas. Its natural gas reserves are among the world's largest. Petroleum and natural gas account for around two-thirds of the country's total revenues and more than 90% of the exports. Algeria's petroleum refining capacity is the biggest in Africa. Agriculture employs about 16% of the population.

17

ANGOLA

The flag is based on the flag of the MPLA (the Popular Movement for the Liberation of Angola) during the independence struggle. The emblem includes a star symbolizing socialism, one half of a gearwheel to represent industry, and a machete symbolizing agriculture.

The Republic of Angola is a large country, more than twice the size of France, on the southwestern coast of Africa. The majority of the country is part of the plateau that forms most of southern Africa, with a narrow coastal plain in the west.

Angola has many rivers. In the northeast, several rivers flow northward to become tributaries of the River Congo, while in the south, some rivers, including the Cubango (Okavango) and the Cuanda, flow southeastward into inland drainage basins in the interior of Africa.

Area 481,351 sq mi [1,246,700 sq km]
Population 10.979,000
Capital Luanda (2,250,000)
Government Multiparty republic
Ethnic groups Ovimbundu 37%, Kimbundu 25%, Bakongo 13%, others 25%
Languages Portuguese (official), many others
Religions Traditional beliefs 47%, Roman Catholic 38%, Protestant 15%
Currency Kwanza = 100 lwei
Website www.angola.org

CLIMATE

Angola has a tropical climate, with temperatures of over 68°F [20°C] all year round, though the upland areas are cooler. The coastal regions are dry, increasingly so to the south of Luanda, but the rainfall increases to the north and east. The rainy season is between November and April. Tropical forests flourish in the north, but the vegetation along the coast is sparse, with semidesert in the south.

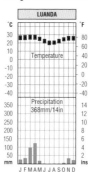

HISTORY

Bantu-speaking peoples from the north settled in Angola around 2,000 years ago. In the late 15th century, Portuguese navigators, seeking a route to Asia around Africa, explored the coast and, in the early 16th century, the Portuguese set up bases.

Angola became important as a source of slaves for Brazil, Portugal's huge colony in South America. After the decline of the slave trade, Portuguese settlers began to develop the land. The Portuguese population increased gently in the 20th century.

In the 1950s, local nationalists began to demand independence. In 1956, the MPLA (Popular Movement for the Liberation of Angola) was founded with support from the Mbundu and mestizos (people of African and European descent). The MPLA led a revolt in Luanda in 1961, but it was put down by Portuguese troops.

Other opposition groups developed. In the north, the Kongo set up the FNLA (Front for the Liberation of Angola), while, in 1966, southern peoples, including many Ovimbundu, formed UNITA (National Union for the Total Independence of Angola).

POLITICS

The Portuguese agreed to grant Angola independence in 1975, after which rival

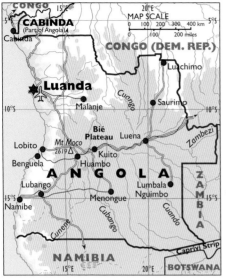

army and rebels signed a ceasefire to end conflict. Angola then started the lengthy process of rebuilding its devastated infrastructure with thousands of refugees to be resettled and landmines to be cleared.

ECONOMY

Angola is a developing country, where 70% of the people are poor farmers, although agriculture contributes only about 9% of the gross domestic product. The main food crops include cassava, corn, sweet potatoes, and beans, while bananas, coffee, palm products, seed cotton and sugarcane are grown for export. Cattle are the leading livestock, but sheep and goats are raised in drier areas.

Despite the poverty of most of its people and its low per capita GNP, Angola has much economic potential. It has petroleum reserves near Luanda and in the enclave of Cabinda, which is separated from Angola by a strip of land belonging to the Democratic Republic of Congo. Petroleum and mineral fuels are the leading exports.

Other resources include diamonds (the second most important export), copper, and manganese. Angola also has a growing industrial sector. Manufactures include cement, chemicals, processed food and textiles.

nationalist forces began a struggle for power. A long-running civil war developed between the government forces, who received aid from the Soviet Union and Cuba, the FNLA in the north, and UNITA in the south. As the war developed, both the FNLA and UNITA turned to the West for support, while UNITA received support from South Africa. FNLA guerrilla activity ended in 1984, but UNITA took control of large areas. Economic progress was hampered not only by the vast spending on defense and security, but also by the MPLA government's austere Marxist policies.

In 1991, a peace accord was agreed and multiparty elections were held, in which the MPLA, which had renounced Marxism-Leninism, won a majority with Jose Eduardo Dos Santos, president since 1979, retaining power. But UNITA's leaders rejected the election result and civil war resumed in 1994. In 1997, the government invited UNITA leader, Jonas Savimbi, to join a coalition but he refused.

Savimbi was killed in action in February 2002, raising hopes of peace and the

LUANDA

Capital, chief port, and largest city of Angola, on the Atlantic coast. Luanda was first settled by the Portuguese in 1575. Its economy thrived on the shipment of more than 3 million slaves to Brazil until the abolition of slavery in the 19th century. Today, it exports crops from the province of Luanda.

ANTARCTICA

Antarctica is the fifth-largest continent (larger than Europe or Australasia), covering almost 10% of the world's total land area. Surrounding the South Pole, it is bordered by the Antarctic Ocean and the southern sections of the Atlantic, Pacific, and Indian Oceans. Almost entirely within the Antarctic Circle, it is of great strategic and scientific interest.

No people live there permanently, though scientists frequently stay for short periods to conduct research and exploration. Seven nations lay claim to sectors of it. Covered by an ice-sheet with an average thickness of c. 5,900 ft (1,800 m), it contains c. 90% of the world's ice and more than 70% of its fresh water and plays a crucial role in the circulation of the atmosphere and ocean, and hence in determining the planetary climate.

LAND
Resembling an open fan, with the Antarctic Peninsula as a handle, the continent is a snowy desert covering approximately 5.5 million sq mi [14.2 million sq km]. The land is a high plateau with an average elevation of 6,000 ft [1,800 m] and rising to 16,863 ft [5,140 m] in the Vinson Massif. Mountain ranges occur near the coasts.

The interior, or South Polar Plateau, lies beneath c. 6,500 ft [2,000 m] of snow, accumulated over tens of thousands of years. Mineral deposits exist in the mountains, but their recovery is not practicable. Coal may be plentiful, but the value of known deposits of copper, nickel, gold, and iron will not repay the expense, both financial and environmental, of their extraction and export.

SEAS AND GLACIERS
Antarctic rivers are frozen, inching toward the sea, and instead of lakes there are large bodies of ice along the coasts. The great Beardmore Glacier creeps down from the South Polar Plateau, and eventually becomes part of the Ross Ice Shelf. The southernmost part of the

Area 5,405,430 sq mi [108,108 sq mi ice-free, 5,297,322 sq mi ice-covered]
14 million sq km [280,000 sq km ice-free, 13.72 million sq km ice-covered]
Government The Antarctic Treaty
Websites www.antarctica.ac.uk; www.aad.gov.au

Atlantic is the portion of the Antarctic Ocean known as the Weddell Sea.

CLIMATE AND VEGETATION
The coldest, windiest, highest (on average), and driest continent. During summer, more solar radiation reaches the surface at the South Pole than is received at the Equator in an equivalent period.

Mostly uninhabitable, Antarctica remains cold all year, with only a few coastal areas being free from snow or ice in summer (December to February). On most of the continent the temperature remains below freezing, and in August it has been recorded at nearly -130°F [-90°C]. Precipitation is generally 7–15in [18–38cm] of snow a year, but it melts at a slower rate, allowing a build-up over the centuries.

Mosses manage to survive on rocks along the outer rim of the continent. Certain algae grow on the snow, and others appear in pools of fresh water when melting occurs.

CASEY — Temperature / Precipitation 224mm/8in

HISTORY
Antarctic islands were sighted first in the 18th century, and in 1820 Nathaniel Palmer reached the Antarctic Peninsula. Between 1838 and 1840, US explorer Charles Wilkes discovered enough of the coast to prove that a continent existed, and the English explorer James Clark Ross made coastal maps. Toward the end of the 19th century, exploration of the

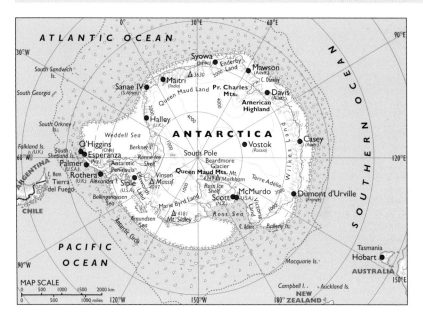

interior developed into a race for the South Pole. Roald Amundsen reached the Pole on December 14, 1911, a month before Captain Robert Scott. The airplane brought a new era of exploration, and Richard E. Byrd became the best-known of the airborne polar explorers.

ANTARCTIC TREATY

The Antarctic Treaty in 1961 set aside the area for peaceful uses only, guaranteeing freedom of scientific investigation, banning waste disposal and nuclear testing, and suspending the issue of territorial rights. By 1990 the original 12 signatories had grown to 25, with a further 15 nations granted observer status in subsequent deliberations. But the Treaty itself was threatened by wrangles between different countries, government agencies and international pressure groups.

Finally in July 1991, the belated agreement of the UK and the USA assured unanimity on a new accord to ban all mineral exploration for a further 50 years. This can only be rescinded if all the present signatories, plus a majority of any future adherents, agree. While the treaty has always lacked a formal mechanism for enforcement, it is firmly underwritten by public concern generated by the efforts of environmental pressure groups such as Greenpeace, which has been foremost in the campaign to have Antarctica declared as a "World Park."

The continent appears to be under threat from global warming. Some scientists believe this was the cause of the breakup of ice shelves along the Antarctic peninsula. Rising temperatures have also disturbed the breeding patterns of the Adelie penguins.

ANTARCTIC CIRCLE

Southernmost of the Earth's parallels, 66.5° south of the Equator. At this latitude the sun neither sets on the day of summer solstice (December 22) nor rises on the day of winter solstice (June 21).

The "celeste" (sky blue) and white stripes were the symbols of independence around the city of Buenos Aires, where an independent government was set up in 1810. It became the national flag in 1816. The gold May Sun was added two years later.

The Argentine Republic is the largest of South America's Spanish-speaking countries. Its western boundary lies in the Andes, with basins, ridges and peaks of more than 19,685 ft [6,000 m] in the north. South of latitude 27°S, the ridges merge into a single high cordillera, with Aconcagua, at 22,849 ft [6,962 m], the tallest mountain in the western hemisphere.

In the south, the Andes are lower, with glaciers and volcanoes. Eastern Argentina is a series of alluvial plains, from the Andean foothills to the sea. The Gran Chaco in the north slopes down to the Paraná River, from the high desert of the Andean foothills to lowland swamp forest. Between the Paraná and Uruguay rivers is Mesopotamia, a fertile region. Further south are the damp and fertile pampa grasslands. Thereafter, the pampa gives way to the dry, windswept plateaus of Patagonia toward Tierra del Fuego.

Area 1,073,512 sq mi [2,780,400 sq km]
Population 39,145,000
Capital (population) Buenos Aires (2,965,000)
Government Federal republic
Ethnic groups European 97%, Mestizo, Amerindian
Languages Spanish (official)
Religions Roman Catholic 92%, Protestant 2%, Jewish 2%, others
Currency Peso = 10,000 australs
Website www.sectur.gov.ar

CLIMATE

The climate varies from subtropical in the north to temperate in the south. Rainfall is abundant in the northeast, but is lower to the west and south. Patagonia is a dry region, crossed by rivers that rise in the Andes.

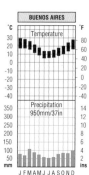

HISTORY

Spanish explorers first reached the coast in 1516, landing on the shores of the Rio de la Plata. They were soon followed by others in search of gold and silver. Early prosperity, based on stock raising and farming, combined with stable government, was boosted from 1870 by a massive influx of European immigrants, particularly Italians and Spaniards, for whom Argentina was a viable alternative to the United States. They settled lands recently cleared of Native Americans, often organized by huge land companies.

Development of a good railroad network to the ports, plus steamship services to Europe, and, from 1877, refrigerated vessels, helped to create the strong meat, wool, and wheat economy that carried Argentina into the 20th century. Before the Great Depression in the 1930s, Argentina was one of the world's more prosperous nations.

POLITICS

The collapse in the economy during the Great Depression led to a military coup in 1930. This started a long period of military intervention in the politics of the country.

From 1976, the "dirty war," saw the torture, wrongful imprisonment, and murder ("disappearance") of up to

Islands, the first since 1982. This meant that Argentines were allowed to visit the Falkland Islands and erect a memorial to their war dead, with Argentina agreeing to allow flights from the Falkland Islands to Chile.

In December 2001, violent protests broke out when the government introduced severe austerity measures, with the peso devalued and policies aimed at restoring the economy announced. The economy finally began to grow again in 2003 and 2004.

ECONOMY

An "upper-middle-income" developing country and one of the richest in South America in terms of natural resources, especially its fertile farmland. The economic base is mainly agricultural. Chief products are beef, corn and wheat. Sheep are raised in drier parts of the country, while other crops include citrus fruits, cotton, flax, grapes, potatoes, sorghum, sugarcane, sunflower seeds and tea.

Oil fields in Patagonia and the Piedmont make Argentina almost self-sufficient in petroleum and natural gas, these are a valuable export.

15,000 people by the military with 2 million people fleeing the country. In 1982, the government, blamed for the poor state of the economy, launched an invasion of the Falkland Islands (Islas Malvinas), which they had claimed since 1820. Britain regained the islands by sending an expeditionary force. After losing the conflict Argentina's President Galtieri resigned. Constitutional government was restored in 1983 though the army remained influential.

In 1999, Argentina and Britain signed an agreement concerning the Falkland

BUENOS AIRES

Capital of Argentina, on the estuary of the Río de la Plata, 150 mi (240 km) from the Atlantic Ocean. Originally founded by Spain in 1536, it was refounded in 1580 after being destroyed by the indigenous population. It became a separate federal district and capital of Argentina in 1880. Buenos Aires later developed as a commercial center for beef, grain, and dairy products. It is the seat of the National University (1821). Industries: meat processing, flour milling, textiles, metalworks, and automobile assembly.

Armenia's flag was first used between 1918 and 1922, when the country was an independent republic. It was readopted on August 24, 1990. The red represents the blood shed in the past, the blue the land of Armenia, and the orange the courage of the people.

The Republic of Armenia is a landlocked country in southwestern Asia. Mostly consisting of a rugged plateau, criss-crossed by long faults. Movements along the faults cause earth tremors and occasionally major earthquakes. Armenia's highest point is Mount Aragats, at 13,149 ft [4,090 m] above sea level. The lowest land is in the northwest, where the capital Yerevan is situated. The largest lake is Ozero (Lake) Sevan.

The vegetation in Armenia ranges from semidesert to grassy steppe, forest, mountain pastures and treeless tundra at the highest levels. Oak forests are found in the southeast, with beech being the most common tree in the forests of the northeast. Originally it was a much larger kingdom centered on Mount Ararat incorporating present-day northeast Turkey and parts of north west Iran.

Area 11,506 sq mi [29,800 sq km]
Population 2,991,000
Capital (population) Yerevan (1,249,000)
Government Multiparty republic
Ethnic groups Armenian 93%, Russian 2%, Azeri 1%, others (mostly Kurds) 4%
Languages Armenian (official)
Religions Armenian Apostolic 94%
Currency Dram = 100 couma
Website www.armeniaforeignministry.com

CLIMATE

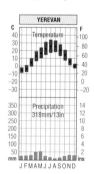

The height of the land, which averages 4,920 ft [1,500 m] gives rise to severe winters and cool summers. The highest peaks are snow-capped, but the total yearly rainfall is low, between and 8 and 31 in [200 and 800 mm].

HISTORY

Armenia was an advanced ancient kingdom, considered to be one of the original sites of iron and bronze smelting. A nation was established in the 6th century BC and Alexander the Great expelled the Persians in 330 BC. In 69 BC Armenia was incorporated into the Roman Empire. In AD 303, Armenia became the first country to adopt Christianity as its state religion. From 886 to 1046 Armenia was an independent kingom. From the 11th to 15th centuries the Mongols were the greatest power in the region. By the 16th century Armenia was controlled by the Ottoman Empire. Despite religious discrimination, the Armenians generally prospered under Turkish rule. Eastern Armenia was the battleground between the rival Ottoman and Persian empires. In 1828 Russia acquired Persian Armenia and (with many promises of religious tolerance) many Armenians moved to the Russian-controlled area. In Turkish Armenia, British promises of protection encouraged nationalist movements. The Turkish response was uncompromising, killing about 200,000 in 1896 alone. In the Russian sector, a process of Russification was enforced.

During World War I, Armenia was the battleground for the Turkish and Russian armies. Armenians were accused of aiding the Russians, and Turkish atrocities

republic. In 1922, it became, with Azerbaijan and Georgia, part of the Transcaucasian Republic within the Soviet Union. But the three territories became separate Soviet Socialist Republics in 1936. Earthquakes in 1984 and 1988 killed more than 80,000 people and destroyed many cities.

After the breakup of the Soviet Union in 1991, Armenia became an independent republic and joined the Commonwealth of Independent States (CIS).

POLITICS

Armenia has long disputed the status of Nagorno-Karabakh, an area enclosed by Azerbaijan where the majority of the people are Armenians. In 1992, Armenia occupied the territory between its eastern border and Nagorno-Karabakh. A ceasefire in 1994 left Armenia in control of about 20% of Azerbaijan's land area. With Azerbaijan and its ally Turkey blockading its borders, Armenia became increasingly dependent on Iran and Georgia for access to the outside world.

In 1998 Robert Kocharian former leader of Nagorno-Karabakh, became president. In 1999, gunmen stormed parliament and killed the prime minister.

intensified. More than 600,000 Armenians were killed by Turkish troops and 1.75 million were deported to Syria and Palestine. The Armenian Autonomous Republic was set up in the area held by Russia in 1918, but the western part of historic Armenia remained in Turkey, and the northwest was held by Iran. In 1920, Armenia became a Communist

ECONOMY

The World Bank classifies Armenia as a "lower-middle-income" economy. Conflict with Azerbaijan in the early 1990s and the earthquakes have damaged the economy, but since 1992 the government has encouraged free enterprise.

Poverty, corruption, and political assassinations contributed to Armenia losing 20% of its population in the 1990s. The country is highly industrialized with production dominated by mining and chemicals. Copper is the chief metal, but gold, lead, and zinc are also mined. Agriculture is the second-largest sector, with cotton, tobacco, fruit and rice the main products.

YEREVAN

Capital of Armenia, on the River Razdan, southern Caucasus. One of the world's oldest cities, it was capital of Armenia from as early as the 7th century (though under Persian control). A crucial crossroads for caravan routes between India and Transcaucasia, it is the site of a 16th-century Turkish fortress. It is a traditional winemaking center. Industries include chemicals, plastics, cables, tyres, metals, and vodka.

ATLANTIC ISLANDS

AZORES

Area 868 sq mi [2,247 sq km]
Population 243,000
Capital (population) Ponta Delgada (21,000)
Government Autonomous region of Portugal
Ethnic groups Azorean
Languages Portuguese (official)
Religion Roman Catholic
Currency Euro = 100 cents
Website www.drtacores.pt

The Azores is a group of nine large and several small islands rising from the Mid-Atlantic Ridge in the North Atlantic Ocean. They are mostly mountainous, of relatively recent volcanic origin and lie about 745 mi [1,200 km] west of Lisbon. They have been Portuguese since the mid-15th century. From 1938 until 1978, they were governed as three districts of Portugal, becoming an autonomous region in 1976. Farming and fishing are the main occupations.

CANARY ISLANDS

Area 2,875 sq mi [7,447 sq km]
Population 1,672,689
Capital Santa Cruz (Tenerife), Las Palmas (Gran Canaria)
Government Constitutional monarchy
Ethnic groups Spanish
Languages Spanish (official)
Religion Roman Catholic
Currency Euro = 100 cents
Website www.icanarias.com

The Canary Islands are seven large islands and many small volcanic islands situated off southern Morocco. The climate is subtropical, dry at sea level, wetter in the mountains. Claimed by Portugal in 1341, they were ceded to Spain in 1479. Two Spanish provinces since 1927. Tourism is a major occupation. Farming and fishing are important.

BERMUDA

Area 21 sq mi [53 sq km]
Population 65,000
Capital (population) Hamilton (1,000)
Government Parliamentary British overseas territory with internal self-government
Ethnic groups Black 55%, White 34%, others
Languages English (official), Portuguese
Religion Anglican 23%, Roman Catholic 15%, African Methodist Episcopal 11%, others
Currency Bermudian dollar = 100 cents
Website www.bermudatourism.com

Bermuda comprises some 150 small islands, the coral caps of ancient volcanoes rising from the floor of the North Atlantic Ocean. Uninhabited when discovered in 1503 by the Spaniard Juan Mermúdez, the islands were taken over by the British over a century later, with slaves brought from Virginia. Bermuda is Britain's oldest overseas territory, but has a long tradition of self-government. Tourism is the mainstay of the economy, but the islands are a tax haven for overseas companies.

CAPE VERDE

Area 1,557 sq mi [4,033 sq km]
Population 415,000
Capital (population) Praia (95,000)
Government Multiparty republic
Ethnic groups Creole (mulatto) 71%, African 28%
Languages Portuguese and Crioulo
Religion Roman Catholic and Protestant
Currency Cape Verdean escudo = 100 centavos
Website www.caboverde.com

The Republic of Cape Verde consists of ten large and five small islands, divided into the Barlavento (windward) and Sotavento (leeward) groups. They are volcanic and mainly mountainous, with steep cliffs and rocky headlands. The highest point is on the island of Fogo, an

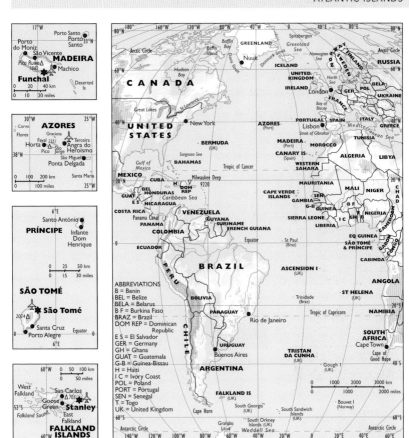

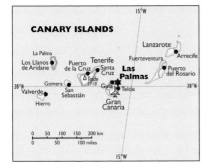

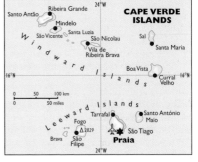

active volcano standing at 9,281 ft [2,829 m]). The climate is tropical, being hot for most of the year and mainly dry at sea level. The higher ground is cooler.

Portuguese since the 15th century, Verde included Portuguese Guinea (now

27

Guinea-Bissau) until 1879, when the mainland territory was separated. It was populated with slaves from Africa, and used chiefly as a provisioning station and assembly point for slaves in the trade from West Africa. In 1991, the ruling party was soundly trounced in the country's first multiparty elections by a newly legalized opposition party, the Movement for Democracy (MPD). The former ruling African Independence Party (PAICV) regained power in 2001.

Bananas, beans, coffee, fruit, peanuts, corn, and sugarcane are grown on the wetter, higher ground, when they are not ruined by endemic droughts. Cape Verde's exports comprise fish and fish preparations, and bananas. Economic problems, include high unemployment levels and the arrival of thousands of Angolan refugees. Cape Verde became fully independent in 1975.

FALKLAND ISLANDS

Area 12,173 sq km [4,700 sq mi]
Population 2,967
Capital (population) Stanley (2,000)
Government Overseas British territory
Ethnic groups British
Languages English (official)
Religions Anglican
Currency Falkland pound = 100 pence
Website www.falklandislands.com

The Falkland Islands (Islas Malvinas) lie 480 km [300 mi] to the east of Argentina and consist of two main islands, and more than 200 small ones.

Discovered in 1592 by the English navigator John Davis, the Falklands were first occupied nearly 200 years later by the French (East) and the British (West). The French interest, bought by Spain in 1770, was assumed by Argentina in 1806. The British, who had withdrawn in 1774, returned in 1832. They dispossessed the Argentinian settlers and

founded a settlement of their own, one that became a colony in 1892. In 1982, Argentinian forces invaded the islands, but two months later, the United Kingdom regained possession. In 1999, a formal agreement between Britain and Argentina permitted Argentinians to visit the islands. The economy is dominated by sheep-farming.

MADEIRA

Area 794 sq km [307 sq mi]
Population 253,482
Capital (population) Funchal (5,618)
Government Autonomous region of Portugal
Ethnic groups Portuguese
Languages Portuguese (official)
Religions Roman Catholic
Currency Euro = 100 cents
Website www.madeiratourism.org

Madeira is the largest of the group of volcanic islands lying 350 mi [550 km] west of the Moroccan coast. Porto Santo, the uninhabited Islas Selvagens (not shown on the map) and the Desertas complete the group. The island of Madeira makes up more than 90% of the total area.

With a warm climate and fertile soils, the Madeira Islands are known for their rich exotic plant life. The abundance of species is all the more surprising because rainfall is confined to the winter months. The present name, meaning "wood," was given by the Portuguese when they first saw the forested islands in 1419. The forests were largely destroyed and a farming industry was established. Spain held the islands between 1580 and 1640, while Britain occupied the islands twice early in the 19th century.

Major crops include bananas, corn, mangoes, oranges, and sugarcane. Grapes are grown to make Madeira wine. Fishing is important, as also is tourism.

ST HELENA

Area 47 sq mi [122 sq km]
Population 5,000
Capital (population) Jamestown
Government British overseas territory
Ethnic groups Britsih
Languages English (official)
Religion Anglican
Currency Pound sterling = 100 pence
Website www.sainthelena.gov.sh

South Atlantic island, 1,190 mi [1,920 km] from the coast of west Africa Discovered by the Portuguese in 1502, it was captured by the Dutch in 1633 and passed to the British East India Company in 1659. It became a British crown colony in 1834 and is chiefly known as the place of Napoleon I's exile. It is now a UK dependent territory and administrative center for the islands of Ascension and Tristan da Cunha. It services ships and exports fish and handcrafts.

ST PIERRE & MIQUELON

Area 93 sq mi [242 sq km]
Population 7,012
Capital (population) St. Pierre (5,618)
Government Self-governing territorial collectivity of France
Ethnic groups Basques and Bretons
Languages French (official)
Religions Roman Catholic
Currency Euro = 100 cents
Website www2.st-pierre-et-miquelon.info

A group of eight small islands in the Gulf of St. Lawrence, southwest of Newfoundland, Canada. Miquelon is the largest island. The group was claimed for France in 1535, and since 1985 has been a "territorial collectivity," sending delegates to the French parliament. Fishing is the most important activity, and has led to disputes with Canada.

SÃO TOMÉ & PRÍNCIPE

Area 372 sq mi [964 sq km]
Population 187,410
Capital (population) São Tomé (5,618)
Government Republic
Ethnic groups Mestico, Angolares, Forros, Servicais, Tongas, Europeans (primarily Portuguese)
Languages Portuguese (official)
Religions Roman Catholic
Currency Dobra = 100 céntimos
Website www.saotome.st

In the Gulf of Guinea, 200 mi [300 km] off the west coast of Africa. São Tome is the largest of two volcanic and mountainous islands, the vegetation is mainly tropical rainforest. They were discovered in 1471. The Portuguese established plantations in the late 18th century. The islands became independent in 1975. Cocoa, coffee, bananas, and coconuts are grown on plantations, and their export is the republic's major source of income.

TRISTAN DA CUNHA

Area 38 sq mi [98 sq km]
Population 275
Capital (population) Edinburgh (275)
Government British overseas territory
Ethnic groups British
Languages English (official)
Religions Anglican
Currency Pound sterling = 100 pence
Website www.tristandc.com

Tristan da Cunha is one of a group of four islands in the south Atlantic Ocean, between South Africa and South America. The group was discovered in 1506 by the Portuguese and annexed by Britain in 1816. In 1961, Tristan, the only inhabitable island, suffered a volcanic eruption that caused a temporary evacuation. It is administered from St. Helena.

AUSTRALIA

The national flag (above left) was adopted in 1901. It includes the British Union Flag, revealing Australia's historic links with Britain. In 1995, the Australian government put the flag used by the country's Aboriginal people (above right) on the same footing as the national flag.

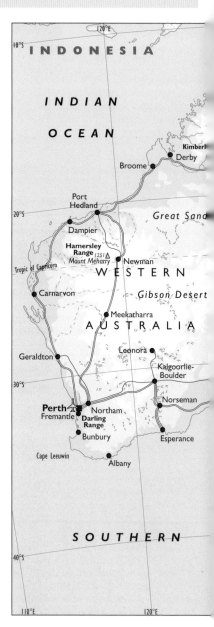

Australia is the world's sixth-largest country. The huge Western Plateau makes up 66% of its land area, and is mainly flat and dry. Off the coast of northeast Queensland lies the Great Barrier Reef. The Great Dividing Range extends down the entire east coast and into Victoria. The mountains of Tasmania are a southerly extension of the range. The highlands separate the east coastal plains from the Central Lowlands and include Australia's highest peak, Mount Kosciuszko, in New South Wales. The capital, Canberra, lies in the foothills. The southeast lowlands are drained by the Murray and Darling, Australia's two longest rivers. Lake Eyre is the continent's largest lake. It lies on the edge of the Simpson Desert and is a dry salt flat for most of the year. Alice Springs lies in the heart of the continent, close to Ayers Rock (Uluru).

Much of the Western Plateau is desert, although areas of grass and low shrubs are found on its margins. The grasslands of the Central Lowlands are used to raise livestock. The north has areas of savanna and rainforest. In dry areas, acacias are common. Eucalyptus grows in wetter regions.

CLIMATE

Only 10% of Australia has an average annual rainfall greater than 39 in [1,000 mm]. These areas include some of the tropical north (where Darwin is situated), the northeast coast, and the south-

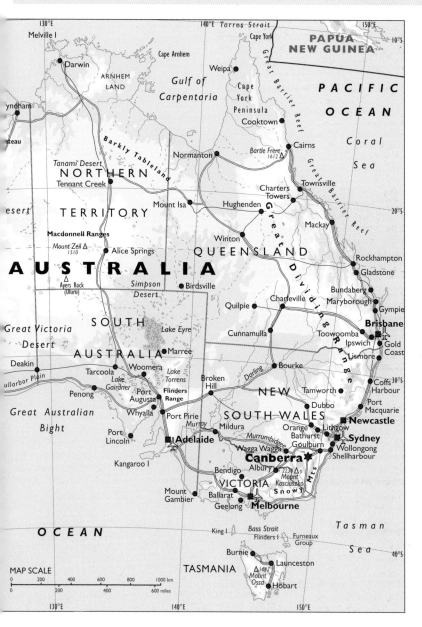

east. The coasts are usually warm and many parts of the south and southwest, including Perth, enjoy a Mediterranean climate of dry summers and moist winters. The interior is dry and many rivers are only seasonal.

31

AUSTRALIA

HISTORY

Native Australians (Aborigines) entered the continent from southeast Asia more than 50,000 years ago. They settled throughout the country and remained isolated from the rest of the world until the first European explorers, the Dutch, arrived in the 17th century. The Dutch

Area 2,988,885 sq mi [7,741,229 sq km]
Population 19,913,000
Capital (population) Canberra (309,000)
Government Federal constitutional monarchy
Ethnic groups Caucasian 92%. Asian 7%. Aboriginal 1%
Languages English (official)
Religions Roman Catholic 26%, Anglican 26%, other Christian 24%, non-Christian 24%
Currency Australian dollar = 100 cents
Website www.australia.gov.au

AYERS ROCK (ULURU)

Outcrop of rock, 280 mi [448 km] southwest of Alice Springs, Northern Territory, Australia. Named after the South Australian politician Sir Henry Ayers (1821–97), it remained undiscovered by Europeans until 1872. It stands 1,142 ft [348 m] high, and is the second largest single rock in the world—the distance around its base is c. 6 mi [10 km]. The rock, caves of which are decorated with ancient paintings, is of great religious significance to Native Australians.

did not settle. In 1770 British explorer Captain James Cook reached Botany Bay and claimed the east coast for Great Britain. In 1788 Britain built its first settlement (for convicts) on the site of present-day Sydney. The first free settlers arrived three years later.

In the 19th century, the economy developed rapidly, based on mining and sheep-rearing. At this time the continent was divided into colonies, which later were to become states. In 1901 the states of Queensland, Victoria, Tasmania, New South Wales, South Australia and Western Australia, federated to create the Commonwealth of Australia. In 1911 Northern Territory joined the federation. A range of progressive social welfare policies were adopted, such as payments for the elderly, in 1909. The federal capital was established in 1927 at Canberra, Australian Capital Territory. Australia fought as a member of the Allies in both world wars. The Battle of the Coral Sea in 1942 prevented a full-scale attack on the continent.

POLITICS

Post-1945 Australia steadily realigned itself with its Asian neighbors. Robert Menzies, Australia's longest-serving prime minister, oversaw many economic and social reforms and dispatched Australian troops to the Vietnam War. In 1977 Prime Minister Gough Whitlam was removed from office by the British governor general. He was succeeded by Malcolm Fraser. In 1983 elections, the Labor Party defeated Fraser's Liberal

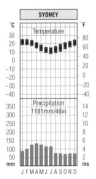

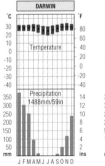

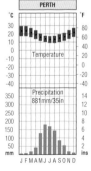

Party, and Bob Hawke became prime minister. His swhrewd handling of industrial disputes and economic recession helped him win a record four terms in office. In 1991 Hawke was forced to

CANBERRA

Capital of Australia on the River Molonglo, Australian Capital Territory, southeastern Australia. Settled in the early 1820s, it was chosen in 1908 as the new site for Australia's capital (succeeding Melbourne). The transfer of all governmental agencies was not completed until after World War II. Canberra has the Australian National University (1946), Royal Australian Mint (1965), Royal Military College and Stromlo Observatory. The new Parliament House was opened in 1988. Other important buildings include the National Library, National Museum, and National Gallery. Tidbinbilla Nature Reserve is located just outside the city.

resign as leader and was succeeded by Paul Keating. Backed by a series of opinion polls, Keating proposed that Australia should become a republic by 2001.

Keating won the 1993 general election and persevered with his free market reforms. In 1996 elections, Keating was defeated by a coalition led by John Howard. In 1998 Howard narrowly secured a second term in office. In a referendum of 1999 Australia voted against becoming a republic. In 2000 Sydney hosted the 28th Summer Olympic Games, nicknamed the "Friendly Games."

The historic maltreatment of Native Australians remains a contentious political issue. In 1993 the government passed the Native Title Act which restored to Native Australians land rights over their traditional hunting and sacred areas. In

January 2002, eastern Australia suffered devastating bush fires.

Howard secured a fourth term in 2004 elections and vowed to stay leader of the Liberal Party for as long as the members want him.

In recent years Australia has brokered peace deals in the more troubled spots of the Pacific such as Papua New Guinea and the Solomon Islands.

ECONOMY

Australia is a prosperous country. Originally an agrarian economy, although crops grow on only 6% of the land. The country remains a major producer and exporter of farm products, particularly cattle, wheat, and wool. Grapes grown for winemaking are also important. Australia is rich in natural resources and is a major producer of minerals, such as bauxite, coal, copper, diamonds, gold, iron ore, manganese, nickel, silver, tin, tungsten, and zinc. Some petroleum and natural gas is also produced.

The majority of Australia's imports are manufactured products. They include machinery and other capital goods required by factories. The country has a highly developed manufacturing sector; the major products include consumer goods, notably foodstuffs and household articles. Tourism is a vital industry.

GREAT BARRIER REEF

World's largest coral reef, in the Coral Sea off the northeast coast of Queensland. It was first explored by James Cook in 1770. It forms a natural breakwater and is up to 2,600 ft [800 m] wide. The reef is separated from the mainland by a shallow lagoon, 7–15 mi [11–24 km] wide. It is a world heritage site. The reef is 1,250 mi [2,000 km] in length, with an area of approximately 80,000 sq mi [207,000 sq km].

AUSTRIA

According to legend, the colors on Austria's flag date back to a battle in 1191, during the Third Crusade, when an Austrian duke's tunic was stained with blood, except under his swordbelt, where it remained white. The flag was officially adopted in 1918.

The Republic of Austria is a landlocked country in the heart of Europe. About three-quarters of the land is mountainous. Northern Austria contains the valley of the River Danube, and the Vienna Basin. This is Austria's main farming region.

Southern Austria contains ranges of the Alps, which rise to their highest point of 12,457 ft [3,797 m] at Grossglockner.

Area 32,378 sq mi [83,859 sq km]
Population 8,175,000
Capital (population) Vienna (1,560,000)
Government Federal republic
Ethnic groups Austrian 90%, Croatian, Slovene, others
Languages German (official)
Religions Roman Catholic 78%, Protestant 5%, Islam and others 17%
Currency Euro = 100 cents
Website www.austria.info

CLIMATE

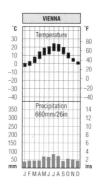

The climate is influenced both by westerly and easterly winds. The moist westerly winds bring rain and snow. They also moderate the temperatures. Dry easterly winds bring very cold weather during the winter, and hot weather during the summer.

The Habsburg ruler of the Holy Roman Empire became Emperor Francis I of Austria. In 1867, Austria and Hungary set up the powerful dual monarchy of Austria-Hungary.

Austria-Hungary was allied to Germany in World War I, but the defeated empire collapsed in 1918. Austria's present boundaries derive from the Versailles Treaty, signed in France in June 1919. In 1933, the Christian Socialist Chancellor Engelbert Dollfuss ended parliamentary democracy and ruled as a dictator. He was assassinated in 1934 due to his opposition to the Austrian Nazi Party's aim of uniting Austria and Germany.

The *Anschluss* (union with Germany) was achieved by the German invasion in March 1938. Austria became a province of the Third Reich called Ostmark until the defeat of the Axis powers in 1945.

HISTORY

Following the collapse of the Roman Empire, of which Austria, south of the Danube, formed a part, the area was invaded and settled by waves of Asian, Germanic, and Slav peoples. In the late 8th century, Austria came under the rule of Charlemagne, but in the 10th century, the area was overrun by Magyars.

In 955, the German king Otto I brought Austria under his rule, and in 962 it became part of what later became known as the Holy Roman Empire. German emperors ruled the area until 1806, when the Holy Roman Empire broke up.

POLITICS

After World War II, Austria was occupied by the Allies, Britain, France, and the United States and it paid reparations for a 10-year period. After agreeing to be permanently neutral, Austria became an independent federal republic in 1955.

In 1994, two-thirds of the people voted in favor of joining the European Union and the country became a member in 1995. Austria became a center of controversy in 1999, when the extreme right-wing Freedom Party, led by Jörg Haider, who had described Nazi Germany's employment policies as "sound," came second in national elections. In February 2000, a coalition government was formed consisting of equal numbers of ministers from the conservative People's Party, which had come third in the elections, and the Freedom Party. However, the Freedom Party suffered a setback in 2001 when its vote fell in city elections in Vienna.

ECONOMY

Austria is a prosperous country with plenty of hydroelectric power, some petroleum and natural gas, and reserves of lignite. The country's leading economic activity is manufacturing metals and metal products, including iron and steel, vehicles, machinery, machine tools, and ships. Craft industries, making such things as fine glassware, jewelry, and porcelain are also important. Dairy and livestock farming are the leading agricultural activities. Major crops include barley, potatoes, rye, sugarbeet, and wheat.

VIENNA (WIEN)

Capital of Austria, on the River Danube. Vienna became important under the Romans, but after their withdrawal in the 5th century it fell to a succession of invaders from eastern Europe. The first Habsburg ruler was installed in 1276. It was the seat of the Holy Roman Empire from 1558 to 1806. Occupied by the French during the Napoleonic Wars, it was later chosen as the site of the Congress of Vienna. As the capital of the Austro-Hungarian Empire, it was the cultural and social center of 19th-century Europe under Emperor Franz Joseph. After World War II, it was occupied (1945–55) by joint Soviet-Western forces. Historical buildings include the 12th-century St. Stephen's Cathedral, the Schönbrunn, and the Hofburg.

35

AZERBAIJAN

Azerbaijan's flag was adopted in 1991. Light blue is a traditional Turkic color. At the center of the red stripe is a white crescent and star, traditional symbols of Islam. The points of the star represent the eight groups of people in Azerbaijan.

The Republic of Azerbaijan lies in eastern Transcaucasia, bordering the Caspian Sea to the east. The Caucasus Mountains are in the north and include Azerbaijan's highest peak, Mount Bazar-Dyuzi, at 14,698 ft [4,480 m]. Another highland region including the Little Caucasus Mountains and part of the rugged Armenian plateau, lies in the southwest.

Between these regions lies a broad plain drained by the River Kura, its eastern part (south of the capital Baku) lies below sea level. Azerbaijan also includes the Nakhichevan Autonomous Republic on the Iran frontier, an area cut off from the rest of Azerbaijan by Armenian territory.

Forests grow on the mountains, while the lowlands comprise grassy steppe or semidesert.

Area 33,436 sq mi [86,600 sq km]
Population 7,868,000
Capital (population) Baku (1,792,000)
Government Federal multiparty republic
Ethnic groups Azeri 90%, Dagestani 3%, Russian, Armenian, others
Languages Azerbaijani (official),Russian, Armenian
Religions Islam 93%, Russian Orthodox 2%, Armenian Orthodox
Currency Azerbaijani manat = 100 gopik
Website www.president.az

migrated to the area from the east by the 9th century.

Azerbaijan was ruled by the Monguls between the 13th and 15th centuries and then by the Persian Safavid dynasty. By the early 19th century it was under Russian control.

CLIMATE

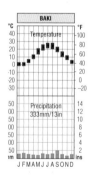

Azerbaijan has hot summers and cool winters. The plains have low rainfall ranging from c. 5 to 15 in [130–380 mm] a year. The uplands have much higher rainfall as does the subtropical southeast coast.

HISTORY

In ancient times, the area now called Azerbaijan was invaded many times. Arab armies introduced Islam in 642, but most modern Azerbaijanis are descendants of Persians and Turkic peoples who

CAUCASUS

Mountain region in southeast Europe, Russia, Georgia, Armenia, and Azerbaijan, extending southeast from the mouth of the River Kuban on the Black Sea to the Apscheron Peninsula in the Caspian Sea. It stretches for 750 mi [1,210 km].The system includes two major regions: north Caucasia (steppes) and Transcaucasia. It forms a barrier between Asia and Europe. There are deposits of petroleum, manganese, and iron, and cotton, fruit, and cereal crops are grown. The highest peak is Mount Elbrus 18,493 ft [5,637 m].

of people are Christian Armenians. A ceasefire in 1994 left Armenia in control of about 20% of Azerbaijan's land area. Talks held in 2001 in an attempt to resolve the dispute proved fruitless and sporadic fighting continues.

In 1998 Aliev was reelected president. In 2001 Azerbaijan joined the Council of Europe. In 2003, Aliev's son Ilham Aliev became president. His government was reelected in 2005, despite charges of fraud.

ECONOMY

In the mid-1990s, the World Bank classified Azerbaijan as a "lower-middle-income" economy. Yet, by the late 1990s, the petroleum reserves in the Baku area on the Caspian Sea, and in the sea itself, held great promise. Petroleum extraction and manufacturing, including petroleum refining and the production of chemicals, machinery, and textiles, are now the most valuable sources of revenue.

Large areas of land are irrigated and crops include cotton, fruit, grains, tea, tobacco, and vegetables. Fishing is still important, although the Caspian Sea is becoming increasingly polluted. Private enterprise is now encouraged.

After the Russian Revolution of 1917, attempts were made to form a Transcaucasian Federation made up of Armenia, Azerbaijan and Georgia. When this failed, Azerbaijanis set up an independent state. But Russian forces occupied the area in 1920. In 1922, the Communists set up a Transcaucasian Republic consisting of Armenia, Azerbaijan, and Georgia, and placed it under Russian control. In 1936, the areas became separate Soviet Socialist Republics within the Soviet Union.

POLITICS

Following the breakup of the Soviet Union in 1991, Azerbaijan became independent. In 1992, Abulfaz Elchibey became president in Azerbaijan's first contested election. In 1993 Elchibey fled and Heydar Aliev, former head of the Communist Party and the KGB in Azerbaijan, assumed the presidency. He was elected later that year and Azerbaijan joined the Commonwealth of Independent States (CIS).

Economic progress was slow, partly because of the conflict with Armenia over the enclave of Nagorno-Karabakh, a region in Azerbaijan where the majority

BAKU

Capital of Azerbaijan, a port on the west coast of the Caspian Sea. A trade and craft center in the Middle Ages, Baku prospered under the Shirvan shahs in the 15th century. Commercial petroleum production began in the 1870s. At the beginning of the 20th century, Baku lay at the center of the world's largest oil field. Industries: petroleum processing and equipment, shipbuilding, electrical machinery, and chemicals.

BAHRAIN

Red and white are traditional colors of the Gulf States. The white historically identified friendly Arab states. The five steps in the serration denote the five Pillars of Islam.

The Kingdom of Bahrain, a former Emirate and now a constitutional hereditary monarchy, is an archipelago consisting of more than 30 islands in the Persian Gulf. The largest of the islands, also called Bahrain, makes up seven-eighths of the country. Causeways link the island of Bahrain to the second largest island of Al Muharraq to the northeast and also to the Arabian peninsula.

Sandy, desert plains make up most of this small, low-lying island country. In the northern coastal areas of Bahrain, freshwater springs provide water for drinking and also for irrigation.

Area 268 sq mi [694 sq km]
Population 678,000
Capital (population) Manama (140,000)
Government Constitutional hereditary monarchy
Ethnic groups Bahraini 62%, others
Languages Arabic, English, Farsi, Urdu
Religions Muslim (Shia and Sunni) 81%, Christian 9%, other
Currency Bahraini dinar = 1000 fils
Website www.bahrain.gov.bh

CLIMATE

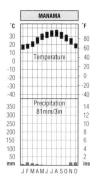

Bahrain has a humid climate. Winters are mild, with temperatures ranging from about 50°F [10°C] to 80°F [27°C]. Summers are hot and humid, with temperatures often soaring to more than 100°F [38°C]. The average annual rainfall is low. Northern Bahrain is the wettest area, with about 3 in [8 mm] a year. The rain occurs mainly in winter and rainfall is almost nonexistent in summer months.

HISTORY

Bahrain was part of a trading civilization called Dilmun, which prospered between about 2000 and 1800 BC. This civilization was linked to the Sumerian, Babylonian, and Assyrian civilizations to the north, and with the Indus Valley civilization in what is now Pakistan. Bahrain later came under Islamic Arab influence from the 7th century.

Portugal seized the archipelago from its Arab rulers in 1521, but the Persians conquered the islands in 1603, holding them against attacks by the Portuguese and Omanis. However, in 1782, the Al Khalifah Arabs from Saudi Arabia took over the islands and they have ruled ever since.

In the early 19th century, Britain helped Bahrain to prevent annexation by Saudi Arabian invaders. As a result, Bahrain agreed to let Britain take control of its foreign affairs. Bahrain effectively became a British protectorate, though it was not called one. In the 1920s and 1930s, the Bahrainis established welfare systems, which were later funded by revenue from petroleum, which was discovered in 1932.

Political reforms began in the 1950s and, in 1970, the Emir turned over some of his power to a Council of State, which became a Cabinet. Britain withdrew from the Persian Gulf region in 1971 and Bahrain became fully independent.

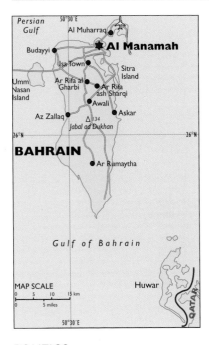

POLITICS

Bahrain adopted a new constitution in 1973. This created a National Assembly with 30 elected members. However, relations between the National Assembly and the ruling Al Khalifa family were difficult and the National Assembly was dissolved in 1975. The country was then ruled by the Emir and his cabinet, headed by the prime minister, the Emir's appointee.

In February 2002, a new constitution changed the country from an Emirate into a constitutional hereditary monarchy and the ruler Sheikh Hamad bin Isa Al-Khalifa became king. Elections for a new directly elected House of Deputies took place later that year, with women allowed to vote for the first time. The 40-member House of Deputies together with a second chamber, a Shura Council, consisting of experts appointed by the king, made up the National Assembly, Bahrain's first parliament since 1975.

Political problems in recent years have included tensions between the Sunni Muslims and the Shiite majority. During the First Gulf War, Bahrain supported Iraq against Iran and, in 1996, Bahrain accused Iran of supporting an underground Shiite organization. Relations with Iran improved in 2002–4.

The fact that the US Fifth Fleet uses Bahrain as its headquarters in the Persian Gulf has provoked terrorist incidents. Although the people have more freedom than others in the region, opposition groups continue to press for further progress, including greater powers for the elected House of Deputies. The opposition groups organized large public rallies in 2005.

ECONOMY

The people of Bahrain enjoy one of the highest standards of living in the Persian Gulf region. The average life expectancy at birth (2005 estimate) is 74 years and free medical services are available. The adult literacy rate is 89%.

Bahrain's prosperity is based on petroleum, although the country lacks major reserves. Petroleum and petroleum products accounted for 68% of the exports in 2002. Its aluminum smelting plant is the Persian Gulf's largest nonpetroleum industrial complex and aluminum, in all forms, accounted for 15% of the exports in 2002. Textiles and clothing accounted for another 8%.

Bahrain is a major banking and financial center, and it is home to numerous multinational companies that operate in the Persian Gulf region. It is a popular tourist destination.

MANAMA (AL-MANAMAH)

Capital of Bahrain, on the north coast of Bahrain Island, in the Persian Gulf. It was made a free port in 1958, and a deepwater harbor was built in 1962. It is the principal port and commercial center. Industries include petroleum refining, banking, and boatbuilding.

BANGLADESH

Bangladesh adopted this flag in 1971, following the country's break from Pakistan. The green is said to represent the fertility of the land. The red disk is the sun of independence. It commemorates the blood shed during the struggle for freedom.

The People's Republic of Bangladesh is one of the world's most densely populated countries. Apart from the hilly regions in the far northeast and southeast, most of the land is flat and covered by fertile alluvium spread over the land by the Ganges, Brahmaputra and Meghna rivers. These rivers overflow when they are swollen by the annual monsoon rains. Floods also occur along the coast, 357 mi [575 km] long, when tropical cyclones (the name for hurricanes in this region) drive seawater inland. These periodic storms cause great human suffering. The world's most devastating tropical cyclone ever recorded occurred in Bangladesh in 1970, when an estimated 1 million people were killed. Most of Bangladesh is cultivated, but forests cover about 16% of the land. They include bamboo forests in the northeast and mangrove forests in the swampy Sundarbans region in the southwest, which is a sanctuary for the Royal Bengal tiger.

Area 55,598 sq mi [143,998 sq km]
Population 141,340,000
Capital (population) Dhaka (3,839,000)
Government Multiparty republic
Ethnic groups Bengali 98%, tribal groups
Languages Bangali (official), English
Religions Islam 83%, Hinduism 16%
Currency Taka =100 paisas
Website www.bangladeshonline.com

relatively unscathed by the tsunami in the Indian Ocean.

HISTORY

For 300 years after the mid-8th century AD, Buddhist rulers governed eastern Bengal, the area that now makes up Bangladesh. In the 13th century, Muslims from the north extended their rule into Bengal and, in 1576, the area became part of the Muslim Mughal Empire which was ruled by the emperor Akbar. This empire, which also included India, Pakistan, and Afghanistan, began to break up in the early 18th century. Europeans, who had first made contact with the area in the 16th century, began to gain influence.

The East India Company, chartered by the English government in 1600 to develop trade in Asia, became the leading trade power in Bengal by the mid-18th century. In 1757, following the defeat of the nawab of Bengal in the Battle of Plessey, the East India Company effectively ruled Bengal. Discontent with the company led to the Sepoy Rebellion in 1857. In 1958, the British government took over the East India Company and its territory became known as British India.

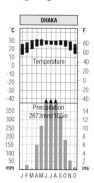

CLIMATE

Bangladesh has a tropical monsoon climate. Dry northerly winds blow during the winter, but, in summer, moist winds from the south bring monsoon rains. In 1998, around two-thirds of the entire country was submerged, causing extensive damage. In December 2004, Bangladesh emerged

POLITICS

In 1947, British India was partitioned between the mainly Hindu India and the Muslim Pakistan. Pakistan consisted of two parts, West and East Pakistan, which were separated by about 1,000 mi [1,600 km] of Indian territory. Differences developed between West and East Pakistan, since people in the east felt themselves victims of ethnic and economic discrimination by the Urdu and Punjabi-speaking peoples of the west.

In 1971, resentment turned to war when Bengali irregulars, aided by Indian troops, established the independent nation of "Free Bengal," with Sheikh Mujibur Rahman as head of state. The Sheikh's assassination in 1975—in one of the four military coups in the first 11 years of independence—led finally to a takeover by General Zia Rahman, who created an Islamic state before he, too, was assassinated in 1981. General Ershad took over in a coup in 1982. He resigned as army chief in 1986 to become a civilian president.

By 1990, protests from supporters of his two predecessors toppled Ershad from power and, after the first free parliamentary elections since independence, a coalition government was formed in 1991. Many problems arose in the 1990s, including the increasing strength of Muslim fundamentalism and the consequences of cyclone damage. In 1996, Sheikh Hasina Wajed of the Awami League became prime minister, but, in 1999, she was defeated by Khaleda Zia, leader of the Nationalist Party.

ECONOMY

Bangladesh is one of the world's poorest countries. Its economy depends mainly on agriculture, which employs more than half the workforce. Rice is the chief crop and Bangladesh is the world's fourth largest producer.

Other important crops include jute, sugar cane, tobacco, and wheat. Jute processing is the leading manufacturing industry and jute is the leading export. Other manufactures include leather, paper, and textiles.

DHAKA (DACCA)

Capital of Bangladesh, a port on the Ganges delta, eastern Bangladesh. Its influence grew as the 17th century Mogul capital of Bengal. In 1765 it came under British control. At independence (1947) it was made capital of the province of East Pakistan. Severely damaged during the war of independence from Pakistan, it became capital of Bangladesh (1971). Sights include the Dakeshwari Temple, Bara Katra Palace (1644), and mosques. It is in the center of the world's largest jute-producing area. Industries include engineering, textiles, printing, glass, chemicals.

BELARUS

In September 1991, Belarus adopted a red and white flag, replacing the flag used in the Soviet era. In June 1995, following a referendum in which Belarussians voted to improve relations with Russia, it was replaced with a design similar to the old flag, but without the hammer and sickle.

The Republic of Belarus is a land-locked country in Eastern Europe, formerly part of the Soviet Union. The land is low-lying and mostly flat. In the south, much of the land is marshy. This area contains Europe's largest marsh and peat bog, the Pripet Marshes.

A hilly region extends rom northeast to southwest and includes the highest point in Belarus, situated near the capital Minsk. This hill reaches a height of 1,122 ft [342 m] above sea level. Over 1,000 lakes, mostly small, dot the landscape. Forests cover large areas. Belarus and Poland jointly control a remnant of virgin forest, which contains a herd of rare wisent (European bison). This is the Belovezha Forest, which is known as the-Bialowieza Forest in Poland.

Area 80,154 sq mi [207,600 sq km]
Population 10,311,000
Capital (population) Minsk (1,677,000)
Government Multiparty republic
Ethnic groups Belarusian 81%, Russian 11%, Polish, Ukrainian, others
Languages Belarusian, Russian (both official)
Religions Eastern Orthodox 80%, others 20%
Currency Belarusian ruble = 100 kopecks
Website www.mfa.gov.by/eng

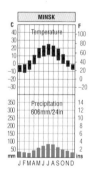

CLIMATE

The climate of Belarus is affected by both the moderating influence of the Baltic Sea and continental conditions to the east. The winters are cold and the summers warm.

HISTORY

Slavic people settled in what is now Belarus about 1,500 years ago. In the 9th century, the area became part of the first East Slavic state, Kievan Rus, which became a major European power in the 10th and 11th centuries. Mongol invaders captured the eastern part of Kievan Rus in the 13th century, while Germanic tribes threatened from the west. Belarus allied itself with Lithuania, which also became a powerful state. In 1386, the Lithuanian Grand Duke married the queen of Poland and Lithuanian-Polish kings ruled both countries until 1569, when Lithuania and Belarus merged with Poland. In the 18th century, Russia took most of eastern Poland, including Belarus. Yet the people of Belarus continued to maintain their individuality.

Following the Russian Revolution of 1917, a Communist government replaced tsarist rule in Russia, and, in March 1918, Belarus became an independent, non-Communist republic. Later that year, Russian Communists invaded Belarus, renaming it Byelorussia, a name derived from the Russian *Belaya Rus*, or White Russia. They established a Communist government there in 1919, and in 1922, the country became a founder republic of the Soviet Union. In 1939, Russia occupied what is now western Belarus, which had been part of Poland since 1919. Nazi troops occupied the area between 1941 and 1944, during which one in four citizens died. Byelorussia became a founding member of the United Nations in 1945.

POLITICS

In 1990, the Byelorussian parliament declared that its laws took precedence over those of the Soviet Union. On August 25 1991, many observers were very surprised that this most conservative and Communist-dominated of parliaments declared its independence. This quiet state of the Soviet Union played a supporting role in its deconstruction and the creation of the Commonwealth of Independent States (CIS). In September 1991, the republic changed its name back from the Russian form of Byelorussia to Belarus, its Belarusian form.

The Communists retained control in Belarus after independence. A new constitution introduced in 1994 led to presidential elections that brought Alexander Lukashenko to power. This enabled economic reform to get under way, though the country remained pro-Russian. Lukashenko favored a union with Russia and, in 1999, signed a union treaty committing the countries to setting up a confederal state. However, Russia insisted that a referendum would have to take place before any merger took place. In 2001, Lukashenko was reelected presi-

dent amid accusations of electoral fraud. A referendum in 2004 showed overwhelming support for the removal of the two-term limit on Lukashenko's rule Western observers alleged fraud saying that the vote was neither free nor fair.

In 2005 Belarus was listed by the US as Europe's last remaining outpost of tyranny.

ECONOMY

The World Bank classifies Belarus as an "upper-middle-income" economy. Like other former republics of the Soviet Union, it faces many problems in turning from Communism to a free-market economy.

Under Communist rule, many manufacturing industries were set up, making such things as chemicals, trucks and tractors, machine tools and textiles. Farming is important and major products include barley, eggs, flax, meat, potatoes and other vegetables, rye, and sugarbeet. Leading exports include machinery and transportation equipment, chemicals and food products.

MINSK

Capital of Belarus, on the River Svisloch. Founded c. 1060, it was under Lithuanian and Polish rule before becoming part of Russia in 1793. During World War II the city was almost completely destroyed and only a few historical buildings were left standing. The city was rebuilt as the showplace city of a modern republic. In 1974 Minsk was awarded the Soviet title of "Hero City" for its sufferings in World War II and speedy reconstruction. In 1991, Minsk became the administrative center of the newly formed CIS. Industries include textiles, machinery, motor vehicles, electronic goods.

BELGIUM

Belgium's national flag was adopted in 1830, when the country won its independence from the Netherlands. The colors came from the arms of the province of Brabant, in central Belgium, which rebelled against Austrian rule in 1787.

The Kingdom of Belgium is a densely populated country in western Europe. Behind the 39 mi [63 km] long coastline on the North Sea, lie its coastal plains. Some low-lying areas, called polders, are protected from the sea by dikes (sea walls).

Central Belgium consists of low plateaus and the only highland region is the Ardennes in the southeast. The Ardennes, reaching a height of 2,277 ft [694 m], consists largely of moorland, peat bogs and woodland. The country's chief rivers are the Schelde, which flows through Tournai, Gent (or Ghent) and Antwerp in the west, and the Sambre and the Meuse, which flow between the central plateau and the Ardennes.

Area 11,787 sq mi [30,528 sq km]
Population 10,348,000
Capital (population) Brussels (136,000)
Government Federal constitutional monarchy
Ethnic groups Belgian 89% (Fleming 58%, Walloon 31%), others
Languages Dutch, French, German (all official)
Religions Roman Catholic 75%, others 25%
Currency Euro = 100 cents
Website www.belgium.be

CLIMATE

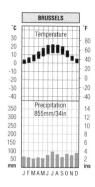

The moderating effects of the sea give much of Belgium a temperate climate, with mild winters and cool summers. Moist winds from the Atlantic Ocean bring significant amounts of rainfall throughout the year, especially in the Ardennes. During January and February, much snow falls in the Ardennes, where temperatures are more extreme. Brussels has mild winters and warm summers.

HISTORY

Due to its strategic position, Belgium has often been called the "cockpit of Europe." In the Middle Ages, the area was split into small states, but, with the Netherlands and Luxembourg, it was united and made prosperous by the dukes of Burgundy in the 14th and 15th centuries. Later, at various times, Belgium, came under Austrian, Spanish, and French rule.

From 1815, following the Napoleonic Wars, Belgium and the Netherlands were united as the "Low Countries" but, in 1830, a National Congress proclaimed independence from the Dutch. In 1831, Prince Leopold of Saxe-Coburg became Belgium's king.

The division between Belgium and the Netherlands rested on history rather than geography. Belgium was a mainly Roman Catholic country while the Netherlands was mainly Protestant. Both were neutral in foreign policy, but both were occupied by the Nazis from 1940 until September 1944.

After World War II, Belgium achieved rapid economic progress, first through collaboration with the Netherlands and-Luxembourg, which formed a customs union called Benelux, and later as a founder member of what is now the European Union. In 1960, Belgium

ber of Deputies has had 150 members, and the Senate 71. The regional assembly of Flanders had 118 deputies, while the assemblies of Brussels and Wallonia had 75 each.

ECONOMY

Belgium is a major trading nation, with a highly developed economy. Almost 75% of its trade is with other EU nations.

With few natural resources it must import a large percentage of the raw materials required for industry. Its main products include chemicals, processed food, and steel. The steelworks lie near to ports because they are powered by petroleum. In 2002, parliament voted to phase out the use of nuclear energy by 2025.

Agriculture employs less than 2% of the people, but Belgian farmers produce most of the food needed by the people. The chief crops are barley and wheat, but the most valuable activities are dairy farming and livestock rearing.

granted independence to the Belgian Congo (now the Democratic Republic of the Congo) and, in 1962, its supervision of Ruanda-Urundi (now Rwanda and Burundi) was ended.

POLITICS

Belgium has always been an uneasy marriage of two peoples: the majority Flemings, who speak a language closely related to Dutch, and the Walloons, who speak French. The dividing line between the two communities runs east–west, just south of Brussels, although the capital is officially bilingual.

Since the inception of the country, the Flemings have caught up and overtaken the Walloons in cultural influence as well as in numbers. In 1971, the constitution was revised and three economic regions were established: Flanders (Vlaanderen), Wallonia (Wallonie) and Brussels. However, tensions remained.

In 1993, Belgium adopted a federal system of government, with each of the three regions being granted its own regional assembly. Further changes in 2001 gave the regions greater tax-raising powers, plus responsibility for agriculture and the promotion of trade. Elections under this system were held in 1995 and 1999. Since 1995, the Cham-

BRUSSELS (BRUXELLES)

Capital of Belgium and of Brabant province, central Belgium. During the Middle Ages, it achieved prosperity through the wool trade and became capital of the Spanish Netherlands. In 1830 it became capital of newly independent Belgium. Places of interest include a 13th-century cathedral, the town hall, splendid art nouveau buildings, and academies of fine arts. The main commercial, financial, cultural, and administrative center of Belgium, it is also the headquarters of the European Union (EU) and of the North Atlantic Treaty Organization (NATO). Industries include textiles, chemicals, electronic equipment, electrical goods, brewing.

BELIZE

Above the shield is a mahogany tree, beside it stand two woodcutters denoting the two main ethnic groups of Belize. A ring of 50 laurel leaves marks the year 1950, the start of the liberation struggle. The country's motto is also shown—Sub umbra floreo (I flourish in the shade).

Belize is a small country in Central America, lying on the Caribbean Sea. It is a monarchy whose head of state is Britain's monarch. A governor-general represents the monarch, while an elected government, headed by a prime minister, actually rules the country day-to-day.

Behind the swampy coastal plain in the south, the land rises to the low Maya Mountains, which reach 3,675 ft [1,120 m] at Victoria Peak. Northern Belize is mostly low lying and swampy. The main river, the River Belize, flows across the center of the country. Rainforest covers large areas. A barrier reef stretches 185 mi (297 km) along the coast, the longest of its kind in the Western Hemisphere.

Area 22,966 sq km [8,867 sq miles]
Population 273,000
Capital (population) Belmopan (8,000)
Government Constitutional monarchy
Ethnic groups Mestizo 49%, Creole 25% ,Mayan Indian 11%, Garifuna 6%, others 9%
Languages English (official), Spanish, Creole
Religions Roman Catholic 50%, Protestant 27%, others
Currency Belizean dollar = 100 cents
Website www.belize.gov.bz

CLIMATE

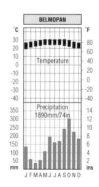

Belize has a humid tropical climate with high temperatures all year round. The average rainfall ranges from 52 in [1,300 mm] in the north to over 150 in [3,800 mm] in the south. Hurricanes sometimes occur. One in 2001 killed 22 people and left 12,000 homeless.

HISTORY

Between 300 BC and AD 1000 Belize was part of the Maya Empire, which was in decline long before Spanish explorers reached the coast in the 16th century. Spain claimed the area but did not

MAYA

Outstanding culture of classic American civilization, occupying south Mexico and north Central America. The civilization divides into three periods. The preclassic era was from 1000 BC to AD 300. The classic era, when it was at its height, was from the 3rd to the 9th centuries. The Maya built great temple cities with buildings surmounting stepped pyramids. They were skilful potters and weavers and productive farmers. They worshipped gods and ancestors and blood sacrifice was an important element of religion. The post-classic era extended from AD 100 to 1500 when the Maya civilization declined.

Much was destroyed after the Spanish conquest in the 16th century. The modern Maya, numbering about 4 million, live in the same area and speak a variety of languages related to those of their ancestors.

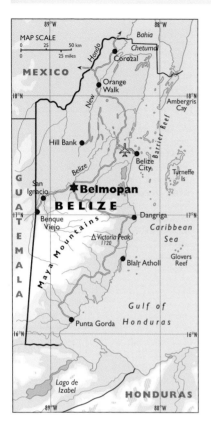

relations in the early 1990s led Guatemala to recognize Belize's independence and in 1992, Britain agreed to withdraw its troops from the country. Mayan land rights remain a contentious political issue.

High levels of unemployment are a major problem, as is a growing involvement in the South American drug trade which has brought with it increasing levels of violent crime

ECONOMY

The World Bank classifies Belize as a "lower-middle-income" developing country. Tourism has become the mainstay of the economy and in recent years cruise ships have called there, bringing extra income.

Agriculture is still important, and cane sugar is the chief commercial crop and export. Other crops include bananas, beans, citrus fruits, corn and rice. Forestry is of longstanding importance, the timber trade even features on the flag. Fishing is now the second biggest earner.

settle. In 1638 the first European settlement was founded by shipwrecked soldiers. Over the next 150 years Britain gradually took control of Belize and established sugar plantations using slave labor.

In 1862 Belize became the colony of British Honduras. Renamed Belize in 1973 it gained full independence in 1981. Guatemala, which had claimed the area since the early 19th century, opposed Belize's independence and British troops remained to prevent a possible invasion.

POLITICS

In 1983, Guatemala reduced its claim to the southern half of Belize. Improved

BELMOPAN

Capital of Belize, on the River Belize. It replaced Belize City, 50 mi [80 km] upstream, as capital in 1971, the latter having been largely destroyed by a hurricane in 1961. The building of Belmopan began in 1966 and was completed in 1971. As an inland city temperatures are high in the daytime but much cooler at night. Belmopan is the center of government. In its center are the National Assembly Building and the majority of central government offices. The residents of Belmopan are mostly government employees and their families. As the city is relatively new the inhabitants originate from all other areas of the country, with only a small indigenous population.

BENIN

The colors on this flag, used by Africa's oldest independent nation, Ethiopia, symbolize African unity. Benin adopted this flag after independence in 1960. A flag with a red (Communist) star replaced it between 1975 and 1990, after which Benin dropped its Communist policies.

The Republic of Benin, formerly called Dahomey, is one of Africa's smallest countries. It extends north-south for about 390 mi [620 km]. The coastline on the Bight of Benin, which is about 62 mi [100 km] long, is lined by lagoons. It lacks natural harbors and the harbor at Cotonou, the main port and commercial center, is artificial.

Behind the coastal lagoons is a flat plain. Beyond this plain is a marshy depression, but the land rises to a low plateau in central Benin. The highest land is in the northwest.

Savanna covers most of northern Benin. The north is home to many typical savanna animals, such as buffaloes, elephants, and lions. The north has two national parks, the Penjari and the "W," which Benin shares with Burkina Faso and Niger.

Area 43,483 sq mi [112,622 sq km]
Population 7,250,000
Capital (population) Porto-Novo (233,000)
Government Multiparty republic
Ethnic groups Fon, Adja, Bariba, Yoruba, Fulani
Languages French (official), Fon, Adja, Yoruba
Religions Traditional beliefs 50%, Christianity 30%, Islam 20%
Currency CFA franc = 100 centimes
Website www.gouv.bj/en/

CLIMATE

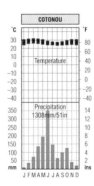

Benin has a hot, wet climate. The average annual temperature on the coast is 77°F [25°C], while the average annual rainfall is 52 in [1,330 mm]. The forested inland plains are wetter than the coast, but the rainfall decreases to the north, which has rainy summer season and a very dry winter.

HISTORY

The ancient kingdom of Dahomey, a prominent West African kingdom which developed in the 15th century, had its capital at Abomey in what is now south-western Benin. In the 17th century, the kings of Dahomey became involved in supplying slaves to European slave traders, including the Portuguese, who shipped many Dahomeans to the Americas. The shoreline of present-day Benin became part of what was called the Slave Coast. Many slaves were shipped to Brazil. Traces of the culture and religion of the slaves still survive in parts of the Americas, For example, the voodoo cult in Haiti originated in Dahomey.

After slavery was ended in the 19th century, France began to gain influence in the area. Around 1851, France signed a treaty with the kingdom of Dahomey and, in the 1890s, the area, which also included some other small African states, became a French colony. From 1904, they ruled Dahomey as part of a huge region called French West Africa, which also included what are now Burkina Faso, Guinea, Ivory Coast, Mauritania, Mali, Niger, and Senegal. Dahomey became an overseas territory of France in 1946 and a self-governing nation in the French Community in 1958. Full independence was achieved in 1960.

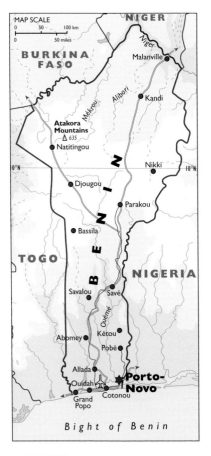

MAP SCALE
0   50   100 km
0   50 miles

NIGER

BURKINA FASO

Malanville

Kandi

Atakora Mountains
Δ 635

Natitingou

Nikki

Djougou

Parakou

Bassila

TOGO

NIGERIA

Savalou   Savé

Kétou

Abomey

Pobé

Allada

Porto-Novo

Ouidah   Cotonou

Grand Popo

Bight of Benin

in 1977, it became a one-party state. This regime, headed by President Kérékou, held power until 1989, when, following the lead of several East European countries in abandoning Communism, Kérékou announced that his country would also abandon Marxism-Leninism and, instead, follow liberal economic policies.

In 1990, a new democratic constitution with a presidential system was introduced. Presidential elections were held in 1991 and Nicéphore Soglo, a former World Bank executive and prime minister, defeated Kérékou. However, in 1996, Kérékou defeated Soglo and returned to power. Kérékou, who was reelected in 2001, worked to restore Benin's fragile economy. Many observers have praised Benin's transition from a Marxist-Leninist state into one of Africa's most stable democracies.

ECONOMY

Benin is a poor developing country. About half the population depends on agriculture, but farming is largely at subsistence level. The main food crops include beans, cassava, millet, rice, sorghum, and tropical yams. The chief cash crops are cotton, palm oil, and palm kernels. Forestry is also important.

Benin produces some petroleum, but manufacturing remains small-scale. It depends heavily upon Nigeria for trade.

POLITICS

Dahomey suffered from instability and unrest in the early years of independence. The first president, Hubert Maga, was removed in 1963 in a military coup led by General Christophe Soglo. A presidential council was set up in 1970 and Hubert Maga became one of three rotating presidents. But this regime was overthrown in 1973 by a coup led by Lt-Col Matthieu Kérékou. In 1975, Kérékou announced that the country would be renamed Benin after the powerful state known for its magnificent sculptures in southwestern Nigeria. Benin became a Marxist-Leninist People's Republic and,

PORTO-NOVO

Capital of Benin, a port on the Gulf of Guinea near the border with Nigeria. Settled by 16th-century Portuguese traders, it later became a shipping point for slaves to America.
It was made the country's capital at independence in 1960, but Cotonou is assuming increasing importance. Today it is a market for the surrounding agricultural region.

BHUTAN

The dragon dates to the 17th century. The jewels in the dragon's claws represent Bhutan's wealth. The white symbolizes purity. The gold represents the secular power of the Druk Gyalpo (Dragon King), and the orange, the spiritual power of Buddhism.

The Kingdom of Bhutan is a small, landlocked country in the eastern Himalayas, between India and the Tibetan plateau of China.

Southern Bhutan, along the border with India, is the lowest land region, ranging between about 160 to 2,950 ft [50 and 900 m] above sea level. North of the plains is a mountainous region between about 4,920 to 13,940 ft [1,500 and 4,250 m]. The northernmost region lies in the Great Himalayan range, reaching more than 23,950 ft [7,300 m]. Most people live in the fertile valleys of rivers which flow generally from north to south.

Area 47,000 sq km [18,147 sq mi]
Population 2,186,000
Capital (population) Thimphu (35,000)
Government Constitutional monarchy
Ethnic groups Bhutanese 50%, Nepalese 35%
Languages Dzongkha (official)
Religions Buddhism 75%, Hinduism 25%
Currency Ngultrum = 100 cetrum
Website www.bhutan.gov.bt

CLIMATE

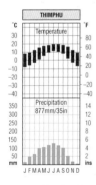

The altitude determines the climate. The southern plains have a subtropical, rainy climate, with an average annual rainfall of around 197 in [5,000 mm]. Dense vegetation covers much of the region, with savanna in the far south. Central Bhutan has a moderate climate, though winters are cold.

HISTORY

Tibetan invaders settled in the area around 1,200 years ago. In the early 17th century, Bhutan became a separate state when a Tibetan lama (Buddhist monk), who was both a spiritual and temporal ruler, took power. The country was divid-

ed into districts ruled by governors and fort commanders. The 19th century was plagued by civil wars in which rival governors battled for power. In 1907, Bhutan became a monarchy when Ugyen Wangchuk, the powerful governor of Tongsa district, made himself Maharajah (now King) and set up the country's first effective central government. The monarch was hereditary and the Maharajah's successors have ruled the country ever since.

In 1910, Britain took control of Bhutan's foreign affairs, but it did not interfere in internal affairs. This treaty was renewed with newly independent India in 1949. India also returned parts of Bhutan which had been annexed by Britain and agreed to help Bhutan develop its economy and, later, its defense.

POLITICS

Bhutan's remote but strategic position cut it off from the outside world for centuries and it only began to open up to outsiders in the 1970s. The roots of reform go back to 1952 when Jigme Dorji Wangchuk succeeded to the throne and a national assembly was established to advise the king. Slavery was abolished

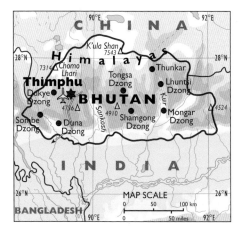

measures emphasizing Buddhist culture further antagonized the minority Nepalis. This led to violence in 1990, causing many Nepalis to flee. In 2005 the king announced that he would step down in 2008, when democratic elections would be held.

ECONOMY

Bhutan is a poor country. The rugged terrain makes the building of roads and other infrastructure difficult. Agriculture, mainly subsistence farming, cattle rearing, and forestry, accounts for 93% of the workforce. Barley, rice, and wheat are the chief food crops. Other products include citrus fruits, dairy products, and corn. Industry is small scale, some coal is mined in the south.

The country's economy is closely linked to that of India with nearly 90% of Bhutan's total exports going to India. Bhutan has considerable hydroelectric power potential and electricity is exported to India. However, economic development is hampered by Bhutan's desire to maintain its traditional culture. The controls placed by the government on outside groups have inevitably restricted foreign investment.

in 1958 and, in 1959, Bhutan admitted several thousand refugees after China had annexed Tibet. The first cabinet was set up in 1968.

In 1972, King Jigme Dorji Wangchuk died and was succeeded by his son, Jigme Singye Wangchuk. The new king continued Bhutan's policy of slow modernization. The first foreign tourists were admitted in 1974, although tourism was restricted to people on prepackaged or guided tours. Independent travel was discouraged as Bhutan sought to preserve its majority Buddhist culture. A television service was not introduced until 1999.

The king gave up some of the monarch's absolute powers in 1998, giving up his role as head of the government. Instead, he ruled in conjunction with the government, a National Assembly and a royal advisory council. In 2005, the government published a new draft constitution, which would make Bhutan a democracy with a parliament consisting of two elected houses. The parliament would have the right to impeach the king by a two-thirds vote.

Ethnic conflict has marred Bhutan's recent history. In 1986, a new law came into force making citizenship dependent on length of residence in Bhutan. Many ethnic Nepalis living in the south were made illegal immigrants, while other

THIMPHU

Bhutan's capital is located in the west of the country. The city is modern in age only (established in 1952) as all new buildings are built following traditional designs. Among its sights are the Memorial Chorten (dedicated to the king's late father Jigme Dorji Wangchuck) and the Tashicho Dzong a 350-year-old structure built by Shabdrung Ngawang Namgyal and refurbished in 1961 to house government departments and ministries.

BOLIVIA

This flag, which has been Bolivia's national and merchant flag since 1888, dates back to 1825 when Bolivia became independent. The red stands for Bolivia's animals and the courage of the army, the yellow for its mineral resources, and the green for its agricultural wealth.

The Republic of Bolivia is a land-locked country in South America. It can be divided into two regions. The west is dominated by two parallel ranges of the Andes Mountains. The western cordillera forms Bolivia's border with Chile. The eastern range runs through the heart of Bolivia. Between the two lies the Altiplano. The Altiplano is the most densely populated region of Bolivia and the site of its famous ruins, it includes the seat of government, La Paz, close to Lake Titicaca. Sucre, the legal capital, lies in the Andean foothills.

The east is a relatively unexplored region of lush, tropical rainforest, inhabited mainly by Native South Americans. In the southeast lies the Gran Chaco.

The windswept Altiplano is a grassland region. The semiarid Gran Chaco is a largely unpopulated vast lowland plain, drained by the River Madeira, a tributary of the Amazon. The region is famous for its quebracho trees which are a major source of tannin.

Area 424,162 sq mi [1,098,581 sq km]
Population 8,724,000
Capital (population) La Paz (seat of government, 940,000); Sucre (legal capital/seat of judiciary, 177,000)
Government Multiparty republic
Ethnic groups Mestizo 30%, Quechua 30%, Aymara 25%, White 15%
Languages Spanish, Aymara, Quechua (all official)
Religions Roman Catholic 95%
Currency Boliviano = 100 centavos
Website www.bolivia.com

HISTORY

American Indians have lived in Bolivia for at least 10,000 years. The main groups today are the Aymara and Quechua people.

When Spanish soldiers arrived in the early 16th century, Bolivia was part of the

CLIMATE

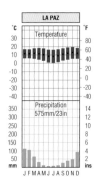

Bolivia's climate varies greatly according the to altitude with the highest Andean peaks perma-nently covered in snow. In con-trast, the eastern plains have a humid tropical climate. The main rainy sea-son takes place between December and February.

TITICACA

Lake in the Andes on the Peru-Bolivia border, draining south through the River Desaguader into Lake Poopó. At an altitude of 12,500 ft [3,810 m], it is the world's highest navigable body of water. The constant supply of water has enabled the region to grow crops since ancient times. The lake is home to giant edible frogs, and is famed for its totora reeds, from which the Uru make their float-ing island homes and fishing rafts. It covers an area of 3,200 sq mi [8,290 sq km] with a maximum depth of 920 ft [280 m].

another period of instability.

Elections were held in 1980, but the military again intervened until 1982, when civilian government was restored. Presidential elections were held in 1989, 1993, and 1997, when General Hugo Bánzer Suárez, who had ruled as a dictator in the 1970s, became president. In 2005, Evo Morales, a left-wing Aymaran Indian and peasant leader, was elected president.

ECONOMY

Bolivia is one of the poorest countries in South America. It has several natural resources, including tin, silver, and natural gas, but the chief activity is agriculture, which employs 47% of the people. Potatoes, wheat, and a grain called quinoa are important crops on the Altiplano, while bananas, cocoa, coffee, and corn are grown at the lower, warmer levels.

Manufacturing is small-scale and the main exports are mineral ores and fossil fuels. Coca, which is used to make cocaine, is exported illegally. In 2002–3, the production of coca plummeted, causing social unrest. In 2004, the people voted in favor of a government plan to export natural gas via a port in Peru.

Inca empire. Following the defeat of the Incas, Spain ruled from 1532 to 1825, when Antonio José de Sucre, one of revolutionary leader Simón Bolívar's generals, defeated the Spaniards.

Since independence, Bolivia has lost much territory to its neighbors. In 1932, Bolivia fought with Paraguay for control of the Gran Chaco region. Bolivia lost and most of this area passed to Paraguay in 1938.

POLITICS

Following the Chaco War, Bolivia entered a long period of instability. It had ten presidents, six of whom were members of the military, between 1936 and 1952, when the Revolutionary Movement replaced the military. The new government launched a series of reforms, which included the breakup of large estates and the granting of land to Amerindian farmers. Another military uprising occurred in 1964, heralding

LA PAZ

Administrative capital and largest city of Bolivia, in the west of the country. Founded by the Spanish in 1548 on the site of an Inca village, it was one of the centers of revolt in the War of Independence (1809–24). Located at 12,000 ft [3,600 m] in the Andes, it is the world's highest capital city. Industries include chemicals, tanning, flour milling, electrical equipment, textiles, brewing, and distilling.

Bosnia-Herzegovina adopted a new flag in 1998, because the previous flag was thought to be synonymous with the wartime Muslim regime. The blue background and white stars represent the country's links with the EU, and the triangle stands for the three ethnic groups in the country.

Bosnia-Herzegovina is one of the five republics to emerge from the former Federal People's Republic of Yugoslavia. Much of the country is mountainous or hilly, with an arid limestone plateau in the southwest. The River Sava, which forms most of the northern border with Croatia, is a tributary of the River Danube. Because of the country's odd shape, the coastline is limited to a short stretch of 13 mi [20 km] on the Adriatic coast.

Area 19,767 sq mi [51,197 sq km]
Population 4,008,000
Capital (population) Sarajevo (529,000)
Government Federal republic
Ethnic groups Bosnian 48%, Serb 37%, Croat 14%
Languages Bosnian, Serbian, Croatian
Religions Islam 40%, Serbian Orthodox 31%, Roman Catholic 15%, others 14%
Currency Convertible marka = 100 convertible pfenniga
Website www.fbihvlada.gov.ba/engleski/

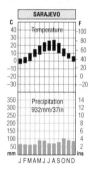

CLIMATE

The coast benefits from a Mediterranean climate. Summers are dry and sunny, while winters are moist and mild. Inland, the weather is more severe, with hot, dry summers and bitterly cold, snowy winters. The north experiences the most severe weather.

HISTORY

Slavs settled in the area that is now Bosnia-Herzegovina around 1,400 years ago. In the late 15th century, the area was taken by the Ottoman Turks. In 1878, the dual monarchy of Austria-Hungary gained temporary control over Bosnia-Herzegovina and it formally took over the area in 1908. The assassination of Archduke Franz Ferdinand of Austria-Hungary in Sarajevo, in June 1914, was the catalyst for the start of World War I. In 1918, Bosnia-Herzegovina became part of the Kingdom of the Serbs, Croats and Slovenes, renamed Yugoslavia in 1929. Germany occupied Yugoslavia during World War II, and Bosnia-Herzegovina came under a puppet regime in Croatia. A Communist government took over in Yugoslavia in 1945, and a new constitution in 1946 made the country a federal state, with Bosnia-Herzegovina as one of its six constituent republics.

Under Communism, Bosnia-Herzegovina was a potentially explosive area due to its mix of Bosnian Muslims, Orthodox Christian Serbs, and Roman Catholic Croats, as well as Albanian, gypsy, and Ukrainian minorities. The ethnic and religious differences started to exert themselves after the death of Yugoslavia's president Josip Broz Tito in 1980, and increasing indications that Communist economic policies were not working.

POLITICS

Free elections were held in 1990 and non-Communists won a majority, with a Muslim, Alija Izetbegovic, as president. In 1991, Croatia and Slovenia declared themselves independent republics and

In 1995, the warring parties agreed to a solution to the conflict, the Dayton Peace Accord—the dividing of the country into two self-governing provinces, one Bosnian Serb and the other Muslim-Croat, under a central, unified, multiethnic government. A NATO-led force helped stabilize the country, this was replaced in 2004 by a European force when problems were no longer political.

ECONOMY

The economy of Bosnia-Herzegovina was shattered by the war in the early 1990s. Manufactures include electrical equipment, machinery and transport equipment, and textiles. Farm products include fruits, corn, tobacco, vegetables, and wheat, but the country has to import food.

seceded from Yugoslavia. Bosnia-Herzegovina held a referendum on independence in 1992. Most Bosnian Serbs boycotted the vote, but the Muslims and Croats voted in favor and Bosnia-Herzegovina proclaimed its independence. War then broke out.

At first, the Muslim-dominated government allied itself uneasily with the Croat minority, but it was at once under attack from local Serbs, supported by their conationals from beyond Bosnia-Herzegovina's borders. In their "ethnic cleansing" campaign, heavily equipped Serb militias drove poorly armed Muslims from towns they had long inhabited. By early 1993, the Muslims controlled less than a third of the former federal republic, and even the capital, Sarajevo, became disputed territory, with constant shelling.

The Muslim-Croat alliance rapidly disintegrated and refugees approached the million mark. Tougher economic sanctions on Serbia in April 1993 had little effect on the war in Bosnia. A small UN force attempted to deliver relief supplies to civilians and maintain "safe" Muslim areas to no avail.

SARAJEVO

Capital of Bosnia-Herzegovina, on the River Miljacka. It fell to the Turks in 1429, and flourished as a commercial center in the Ottoman Empire. Passing to the Austro-Hungarian Empire in 1878, the city was a center of Serb and Bosnian resistance to Austrian rule. In June 1914 a Serb nationalist assassinated Austrian Archduke Franz Ferdinand in the city, an act that precipitated World War I. In 1991 Sarajevo became the focal point of the civil war between Bosnian-Serb troops and Bosnian government forces. The city lay under prolonged siege, often without water, electricity, or basic medical supplies. After the 1995 Dayton Peace Accord, it in effect became a Bosnian city , though with a reduced population as many Serbs fled.

BOTSWANA

The black-and-white zebra stripe in the center of Botswana's flag symbolizes racial harmony. The blue represents rainwater, because water supply is the most vital need in this dry country. This flag was adopted in 1966, when Botswana became independent from Britain.

The Republic of Botswana is a land-locked country which lies in the heart of southern Africa. The majority of the land is flat or gently rolling, with an average height of about 3,280 ft [1,000 m]. More hilly country lies in the east. The Kalahari, a semi-desert area covers much of Botswana.

Most of the south has no permanent streams. But large depressions occur in the north. In one, the Okavango River, which flows from Angola, forms a large delta, an area of swampland. Another depression contains the Makgadikgadi Salt Pans. During floods, the Botletle River drains from the Okavango Swamps into the Makgadikgadi Salt Pans.

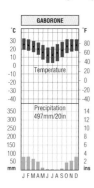

CLIMATE

Temperatures are high during the summer, which runs from October to April, but the winter months are much cooler. Night-time temperatures in winter sometimes drop below freezing. The average rainfall ranges from over 16 in [400 mm] in the east to less than 8 in [200 mm] in the southwest.

Gaborone, the capital of Botswana, lies in the wetter eastern part of the country, where the majority of the population lives. The rainy season occurs during summer, between the months of November and March. Frosts sometimes occur in parts of the east when the temperature drops below freezing.

Area 224,606 sq mi [581,730 sq km]
Population 1,562,000
Capital (population) Gaberone (186,000)
Government Multiparty republic
Ethnic groups Tswana (or Setswana) 79%, Kalanga 11%, Basarwa 3%, others
Languages English (official), Setswana
Religions Traditional beliefs 85%, Christianity 15%
Currency Pula = 100 thebe
Website www.gov.bw

HISTORY

The earliest inhabitants of the region were the San, who are also called Bush-men. They had a nomadic way of life, hunting wild animals and collecting plant foods.

The Tswana, who speak a Bantu lan-guage, now form the majority of the population. They are cattle owners, who settled in eastern Botswana more than 1,000 years ago. Their arrival led the San to move into the Kalahari region. Today,

KALAHARI

A desert region in southern Africa covering parts of Botswana, Namibia and South Africa between the Orange and Zambezi rivers. Thorn scrub and forest grow in some parts of the desert and it is possible to graze animals in the rainy season. The Kalahari is inhabited by the San as well as by Africans and Europeans primarily engaged in rearing cattle. It has an area of approximately 100,000 sq mi [260,000 sq km]).

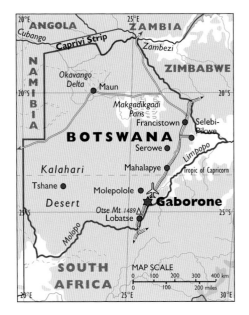

the virus. The average life expectancy fell from 60 to 40 years. Botswana does however have one of Africa's most progressive programs in place to deal with the disease.

ECONOMY

In 1966, Botswana was one of Africa's poorest countries, depending on meat and live cattle for its exports. But the discovery of minerals, including coal, cobalt, copper, and nickel, has helped to diversify the economy. The mining of diamonds at Orapa started in 1971 and was the chief factor in the transformation of the economy. By 1997, Botswana had become the world's leading producer, overtaking Australia and the Democratic Republic of Congo. Diamonds accounted for about 74% of Botswana's exports, followed by copper-nickel matte, textiles, and meat products. Another major source of income comes from tourists, the majority of whom come from South Africa, which continues to have a great influence on Botswana.

The development of mining and tourism has reduced the relative importance of farming, though agriculture still employs about a fifth of the population. The most important type of farming is livestock raising, particularly cattle, which are mostly reared in the wetter east. Crops include beans, corn, millet, sorghum, and vegetables.

the San form a tiny minority, most of whom live in permanent settlements and work on cattle ranches.

POLITICS

Britain ruled the area as the Bechuanaland Protectorate between 1885 and 1966. When the country became independent, it adopted the name of Botswana. Since then, unlike many African countries, Botswana has been a stable multiparty democracy.

The economy has undergone a steady process of diversification under successive presidents.. Botswana's first president was Sir Seretse Khama, who died in 1980, and his successor was Sir Ketumile Masire, who served from 1980 until 1998, when he retired in favor of Festus Mogae. Despite a severe drought, the economy expanded and the government introduced major social programs. Tourism also grew as huge national parks and reserves were established. However, by the early 2000s, Botswana had the world's highest rate of HIV infection— around one in five of the population had

GABORONE

Capital of Botswana, close to the border with South Africa. First settled in the 1890s, it served as the administrative headquarters of the former Bechuanaland Protectorate. In 1966 it became the capital of an independent Botswana.

BRAZIL

The green on the flag symbolizes Brazil's rainforests and the yellow diamond its mineral wealth. The blue sphere bears the motto "Order and Progress." The 27 stars, arranged in the pattern of the night sky over Rio de Janeiro, represent the states and the federal district.

The Federative Republic of Brazil is the world's fifth largest country. Structurally, it has two main regions. In the north is the vast Amazon basin, once an inland sea and now drained by a river system that carries one-fifth of the world's running water. The largest area of river plain is in the upper part of the basin, along the frontiers with Bolivia and Peru. Downstream, the flood plain is relatively narrow.

The Brazilian Highlands make up the country's second main region and consist largely of hard crystalline rock dissected into rolling uplands. They include the heartland (Mato Grosso) and the whole western flank of the country from the bulge to the border with Uruguay. The undulating plateau of the northern highlands carries poor soils.

The typical vegetation is thorny scrub which, in the south, merges into wooded savanna. Conditions are better in the south, where rainfall is more reliable. More than 60% of the population lives in the four southern and southeastern states, the most developed part of Brazil, though accounting only for 17% of Brazil's total area.

CLIMATE

Manaus has high temperatures all through the year. The rainfall is heavy, though the period from June to September is drier than the rest of the year. The capital, Brasília, and the city Rio de Janeiro also have tropical climates, with much more marked dry seasons than Manaus. The far south has a temperate climate. The northeastern interior is the

driest region, with an average annual rainfall of only 10 in [250 mm] in places. The rainfall is also unreliable and severe droughts are common in this region.

The Amazon basin contains the world's largest rainforests, which the

BRAZIL

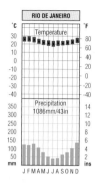

RIO DE JANEIRO

Temperature

Precipitation
1086mm/43in

J F M A M J J A S O N D

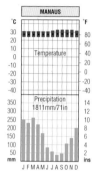

MANAUS

Temperature

Precipitation
1811mm/71in

J F M A M J J A S O N D

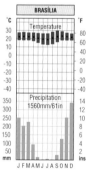

BRASÍLIA

Temperature

Precipitation
1560mm/61in

J F M A M J J A S O N D

Brazilians call the *selvas*. The forests contain an enormous variety of plant and animal species. But many species are threatened by loggers and those who wish to exploit the forests. The destruction of the forest is also ruining the lives of the last surviving groups of Amazonian Indians.

Forests grow on the northeastern coasts, but the dry interior has large areas of thorny scrub. The southeast contains fertile farmland and large ranches.

HISTORY

The Portuguese explorer Pedro Alvarez Cabral claimed Brazil for Portugal in 1500. While Spain was occupied in western South America, the first Portuguese colonists settled in the northeast in the 1530s. They were followed by other settlers, missionaries, explorers, and prospectors who gradually penetrated the country during the 17th and 18th centuries. They found many groups of Amerindians, some of whom lived seminomadic lives, hunting, fishing, and gathering fruits, while others lived in farming villages, growing cassava and other crops.

Area 3,287,338 sq mi [8,514,215 sq km]
Population 184,101,000
Capital (population) Brasilia (2,016,000)
Government Federal republic
Ethnic groups White 55%, Mulatto 38%, Black 6%, others 1%
Languages Portuguese (official)
Religions Roman Catholic 80%
Currency Real = 100 centavos
Website www.turismo.gov.br

The Portuguese enslaved many Amerindians who were used for plantation work, while others were driven into the interior. The Portuguese also introduced about 4 million African slaves, notably in the sugar-cane-growing areas in the northeast. For many decades following the early settlements, Brazil was mainly a sugar-producing colony, with most plantations centered on the rich coastal plains of the northeast. These areas later produced cotton, cocoa, rice, and other crops. In the south, colonists penetrated the interior in search of slaves and minerals, especially gold and diamonds. The city of Ouro Preto in Minas Gerais was built and Rio de Janeiro grew as a port for the area.

Initially little more than a group of rival provinces, Brazil began to unite in 1808, when the Portuguese royal court, transferred from Lisbon to Rio de Janeiro. The eldest son of King Joas VI of Portugal was chosen as the "Perpetual Defender" of Brazil by a national congress. In 1822, he proclaimed the independence of the country and was chosen as the constitutional emperor with the title of Pedro I. He became increasingly unpopular and was forced to abdicate in 1831. He was succeeded by his five-year-old son, Pedro II, who officially took office in 1841. Pedro's liberal policies included the gradual abolition of slavery.

During the 19th century, São Paulo state became the center of a huge coffee-growing industry. While the fortunes made in mining helped to develop Rio de Janeiro, profits from coffee were invested in the city of São Paulo. Immigrants from Italy and Germany settled in the south, introducing farming in the fertile valleys, in coexistence with the cattle ranchers and gauchos of the plains. The second half of the 19th century saw the development of the wild rubber industry in the Amazon basin, where the city of Manaus, with its world-famous opera house, served as a center and market. Although Manaus lies 1,000 mi [1,600 km] from the mouth of the Amazon, rubber from the hinterland could be shipped out directly to world markets in oceangoing steamers. Brazil enjoyed a virtual monopoly of the rubber trade until the early 20th century, when Malaya began to compete, later with massive success.

A federal system was adopted for the United States of Brazil in the 1881 constitution and Brazil became a republic in 1889. Until 1930, the country experienced very strong economic expansion and prospered, but social unrest in 1930 resulted in a major revolt. From then on the country was under the control of President Getulio Vargas, who established a strong corporate state similar to that of fascist Italy, although Brazil entered World War II on the side of the Allies. Democracy, often corrupt, prevailed from 1956, 1964 and 1985. In between there were five military presidents of illiberal regimes.

POLITICS

A new constitution came into force in October 1988—the eighth since Brazil became independent from Portugal in 1822. The constitution transferred powers from the president to the congress and paved the way for a return to democracy. In 1989, Fernando Collor de Mello was elected to cut inflation and combat corruption. But he made little progress and in 1992, with inflation soaring, his vice-president, Itamar Franco, took over as president. He served until 1994 when the Social Democrat Fernando Henrique Cardoso, a former finance minister, was elected president.

In elections in 2002, Luiz Inácio Lula da Silva, leader of the left-wing Workers' Party, was elected president. Popularly known as "Lula," he had promised many social reforms. In office, he proved to be a pragmatist, following moderate economic policies. In 2005, his government was damaged by corruption charges.

ECONOMY

Brazil's total volume of production is one of the largest in the world, but many people, including poor farmers and residents of the *favelas* (city slums), do not share in the country's fast economic growth. Widespread poverty, together with high inflation and unemployment, cause political problems.

Industry is the most valuable activity, employing about 20% of the workforce. Brazil is among the world's top producers of bauxite, chrome, diamonds, gold, iron ore, manganese, and tin. Its manufactures include airplanes, automobiles, chemicals, processed food, raw sugar, iron and steel, paper and textiles.

Agriculture employs 28% of workers. Coffee is a major export. Other leading products include bananas, citrus fruits, cocoa, corn, rice, soybeans and sugar cane. Brazil is the top producer of eggs, meat, and milk in South America.

Forestry is a major industry, though the exploitation of the rainforests, with 1.5% to 4% of Brazil's forest being destroyed every year, is a disaster for the entire world.

BRASÍLIA

Capital city, located in west central Brazil. Although the city was originally planned in 1891, building did not start until 1956. The city was laid out in the shape of an airplane, and Oscar Niemeyer designed the modernist public buildings. It was inaugurated as the capital in 1960, in order to develop Brazil's interior.

BULGARIA

This flag, first adopted in 1878, uses the colors associated with the Slav people. The national emblem, incorporating a lion—a symbol of Bulgaria since the 14th century—was first added to the flag in 1947. It is now added only for official government occasions.

The Republic of Bulgaria is a country in the Balkan Peninsula, facing the Black Sea in the east. There are two main lowland regions. The Danubian lowlands in the north consist of a plateau that descends to the Danube, which forms much of the boundary with Romania. The other lowland region is the warmer valley of the River Maritsa, where cotton, fruits, grains, rice, tobacco and vines are grown.

Separating the two lowland areas are the Balkan Mountains (Stara Planina), rising to heights of over 6,500 ft [2,000 m]. North of the capital Sofia (Sofiya), the Balkan Mountains contain rich mineral veins of iron and non-ferrous metals.

In south-facing valleys overlooking the Maritsa Plain, plums, tobacco, and vines are grown. A feature of this area is Kazanluk, from which attar of roses is exported worldwide to the cosmetics industry. South and west of the Maritsa Valley are the Rhodope (or Rhodopi) Mountains, which contain lead, zinc, and copper ores.

Area 42,823 sq mi [110,912 sq km]
Population 7,518,000
Capital (population) Sofia (1,139,000)
Government Multiparty republic
Ethnic groups Bulgarian 84%, Turkish 9%, Gypsy 5%, Macedonian, Armenian, others
Languages Bulgarian (official), Turkish
Religions Bulgarian Orthodox 83%, Islam 12%, Roman Catholic 2%, others
Currency Lev = 100 stotinki
Website www.government.bg/English

HISTORY

Most of the Bulgarian people are descendants of Slavs and nomadic Bulgar tribes who arrived from the east in the 6th and 7th centuries. A powerful Bulgar kingdom was set up in 681, but the country became part of the Byzantine Empire in the 11th century.

Ottoman Turks ruled Bulgaria from 1396 and ethnic Turks still form a sizable minority in the country. In 1879, Bulgaria became a monarchy, and in 1908

CLIMATE

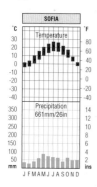

The average temperature in Sofia is 60-70°F (15°-21°C) in the summer and between 30° and 40°F (-1° to 5°C) in winter. Other regions experience more extreme ranges of temperature but winters are rarely severe. Rainfall is moderate all through the year.

BALKAN MOUNTAINS

Major mountain range of the Balkan Peninsula, extending from eastern Serbia through central Bulgaria to the Black Sea. The range is a continuation of the Carpathian Mountains. It is rich in minerals and forms a climatic barrier for the interior. The highest pass is Shipka Pass, c. 4,166 ft [1,270 m], and the highest peak is Botev, 7,793 ft [2,375 m].

when the monarchy was abolished, became prime minister. He left office when his party lost the elections in 2005.

ECONOMY

According to the World Bank, Bulgaria in the 1990s was a "lower-middle-income" developing country. Bulgaria has some deposits of minerals, including lignite, manganese, and iron ore. Manufacturing is the leading economic activity, though problems arose in the early 1990s, because much industrial technology was outdated. The main products are chemicals, processed foods, metal products, machinery and textiles. Manufactures are the leading exports. Bulgaria trades mainly with countries in Eastern Europe.

Wheat and corn are the chief crops of Bulgaria. Fruit, oilseeds, tobacco, and vegetables are also important. Livestock farming, particularly the rearing of dairy and beef cattle, sheep, and pigs, is an important source of revenue.

became fully independent. Bulgaria was an ally of Germany in World War I (1914–18) and again in World War II (1939–45). In 1944, Soviet troops invaded Bulgaria. After the war, the monarchy was abolished and the country became a Communist ally of the Soviet Union.

POLITICS

In the period after World War II, and especially under President Zhikov from 1954, Bulgaria became all too dependent on the Soviet Union. In 1990, the Communist Party held on to power under increasing pressure by ousting Zhikov, renouncing its leading role in the nation's affairs and changing its name to the Socialist Party, before winning the first free elections since the war, albeit unconvincingly and against confused opposition. With improved organization, the Union of Democratic Forces defeated the old guard the following year and began the unenviable task of making the transition to a free-market economy. Subsequent governments faced numerous problems, including inflation, food shortages, rising unemployment, strikes, a large foreign debt, a declining manufacturing industry, increased prices for raw materials, and a potential drop in the expanding tourist industry. In 2001, the former king, Siméon Saxe-Coburg-Gotha, who had left Bulgaria in 1948

SOFIA (SOFIJA)

Capital of Bulgaria and Sofia province, in west central Bulgaria, at the foot of the Vitosha Mountains. Known for its hot mineral springs, Sofia was founded by the Romans in the 2nd century AD. From 1018 to 1185, it was ruled by the Byzantine Empire (as Triaditsa). Sofia passed to the second Bulgarian Empire (1186–1382), and then to the Ottoman Empire (1382–1878). In 1877, Sofia was captured by Russia and chosen as the capital of Bulgaria by the Congress of Berlin. Industries include steel, machinery, textiles, rubber, chemicals, metallurgy, leather goods, and food processing.

BURKINA FASO

This flag was adopted in 1984, when Upper Volta was renamed Burkino Faso. The red, green, and yellow colors used on this flag symbolize the desire for African unity. This is because they are used on the flag of Ethiopia, Africa's oldest independent country.

The Democratic People's Republic of Burkina Faso is a landlocked country, a little larger than the United Kingdom, in West Africa. But Burkina Faso has only one-sixth of the population of the UK.

Burkina Faso consists of a plateau, between about 650 and 2,300 ft [300 m –700 m] above sea level. The plateau is cut by several rivers. Most of the rivers flow south into Ghana or east into the River Niger. During droughts, some of the rivers stop flowing, becoming marshes.

The northern part of the country is covered by savanna, consisting of grassland with stunted trees and shrubs. It is part of a region called the Sahel, where the land merges into the Sahara Desert. Overgrazing of the land and deforestation are common problems in the Sahel, causing desertification in many areas of the country.

Woodlands border the rivers and parts of the southeast region are swampy. The southeast contains the "W" National Park, which Burkina Faso shares with Benin and Niger, and the Arly Park. A third wildlife area is the Po Park situated south of Ouagadougou.

Area 105,791 sq mi [274,000 sq km]
Population 13,575,000
Capital (population) Ouagadougou (637,000)
Government Multiparty republic
Ethnic groups Mossi 40%, Gurunsi, Senufo, Lobi, Bobo, Mande, Fulani
Languages French (official), Mossi, Fulani
Religions Islam 50%, traditional beliefs 40%, Christianity 10%
Currency CFA franc = 100 centimes
Website www.burkinaembassy-usa.org

while it is hot and humid from May to September.

HISTORY

The people of Burkina Faso are divided into two main groups. The Voltaic group includes the Mossi, who form the largest single group, and the Bobo. The other main group is the Mande family. Burkina Faso also contains some Fulani herders and Hausa traders, who are related to the people of northern Nigeria. In early times, the ethnic groups in Burkina Faso were divided into kingdoms and chiefdoms. The leading kingdom, which was ruled by an absolute

CLIMATE

Burkina Faso has three main seasons. From October to February, it is relatively cool and dry. From March to April, it is hot and dry,

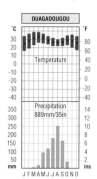

OUAGADOUGOU

FULANI

A west African people numbering 6 million. Their language belongs to the Niger-Congo group. Originally a pastoral people, they helped to spread Islam throughout West Africa from the 16th century, establishing an empire that lasted until British colonialism in the 19th century.

Captain Blaise Campaore seized power. Campaore became president in unopposed elections in 1991. Elections in 1992 were the first multiparty ballots since 1978. In the 1998 elections, Campaore gained a landslide victory. More than 7% of the population have HIV, the second highest rate of infection in Africa (after Uganda).

ECONOMY
Burkina Faso is one of the world's 20 poorest countries and has become extremely dependent on foreign aid. Approximately 90% of the people earn their living by

monarch called the Moro Naba, was that of the Mossi. It has existed since the 13th century. The semiautonomous states fiercely resisted domination by the larger Mali and Songhai Empires.

The French conquered the Mossi capital of Ouagadougou in 1897 and they made the area a protectorate. In 1919, the area became a French colony called Upper Volta. In 1947, Upper Volta gained semiautonomy within the French Union, and in 1958 became an autonomous republic within the French Community.

POLITICS
Upper Volta achieved independence in 1960 and adopted a strong presidential form of government. Persistent drought and austerity measures led to a military coup in 1966. Civilian rule partially returned in 1970 but the military, led by Sangoule Lamizana, regained power in 1974. Lamizana became president after elections in 1978, but was overthrown in 1980. Parliament and the constitution were suspended and a series of military regimes ensued. In 1983 Thomas Sankara gained power in a bloody coup.

In 1984, as a symbolic break from the country's colonial past, Sankara changed Upper Volta's name to Burkina Faso "land of the incorruptible." In 1987, Sankara was assassinated and

farming or by raising livestock. Grazing land covers around 37% of the land and farmland covers around 10%.

Most of Burkina Faso is dry with thin soils. The country's main food crops are beans, corn, millet, rice and sorghum. Cotton, peanuts and shea nuts, whose seeds produce a fat used to make cooking oil and soap, are grown for sale abroad. Livestock is also important.

The country has few resources and manufacturing is on a small scale. There are deposits of manganese, zinc, lead, and nickel in the north of the country, but exploitation awaits improvements to the transportation system. Many young men work abroad in Ghana and Ivory Coast. The money they send to their families is important to the country's economy.

OUAGADOUGOU

Capital city lying in the center of Burkina Faso. Founded in the late 11th century as capital of the Mossi empire, it remained the center of Mossi power until captured by the French in 1896. Industries include handcrafts, textiles, food processing, peanuts, and vegetable oil.

BURMA (MYANMAR)

The colors on Burma's flag were adopted in 1948 when the country became independent from Britain. The socialist symbol, added in 1974, includes a ring of 14 stars representing the country's 14 states. The gearwheel represents industry and the rice plant symbolizes agriculture.

The Union of Burma is now officially known as the Union of Myanmar; its name was changed in 1989. Mountains border the country in the east and west, with the highest mountains in the north. Burma's highest mountain is Hkakabo Razi, which is 19,294 ft [5,881 m] high. Between these ranges is central Burma, which contains the fertile valleys of the Irrawaddy and Sittang rivers. The Irrawaddy delta on the Bay of Bengal is one of the world's leading rice-growing areas. Burma also includes the long Tenasserim coast in the southeast.

Area 261,227 sq mi [676,578 sq km]
Population 42,720,000
Capital (population) Rangoon (2,513,000)
Government Military regime
Ethnic groups Burman 68%, Shan 9%, Karen 7%, Rakhine 4%, Chinese, Indian, Mon
Languages Burmese (official), minority ethnic groups have their own languages
Religions Buddhism 89%, Christianity, Islam
Currency Kyat = 100 pyas
Website www.burmaproject.org

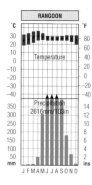

CLIMATE

Burma has a tropical monsoon climate. There are three seasons. The rainy season runs from late May to mid-October. A cool, dry season follows, between late October and the middle part of February. The hot season lasts from late February to mid-May, though temperatures remain high during the humid rainy season.

HISTORY

Conflict between the Burmans and Mons dominated Burma's early history. In 1044 the Burman King Anawratha unified the Irrawaddy delta region. In 1287 Kublai Khan conquered the Burman capital, Pagan. Burma was divided: the Shan controlled north Burma, while the resurgent Mons held the south. In the

16th century, the Burmans subjugated the Shan. In 1758 Alaungpaya reunified Burma, defeating the Mons kingdom and establishing the Konbaung dynasty.

Wars with British India marked much of the 19th century. The first war in 1824 resulted in the British gaining the coastal regions of Tenasserim and Arakan. The second war in 1852 saw the British gain control of the Irrawaddy delta. British India annexed Burma in the third war in 1885. In 1937 Burma gained limited self-government. Helped by the Burmese Independent Army, led by Aung San, Japan conquered the country in 1942. The installation of a puppet regime led Aung San to form a resistance movement. In 1947 Aung San was murdered. Burma achieved independence in 1948.

POLITICS

The socialist AFPFL government, led by U Nu, faced secessionist revolts by communists and Karen tribesmen. In 1958 U Nu invited General Ne Win to reestablish order. Civilian rule returned in 1960, but in 1962 Ne Win mounted a successful coup. His military dictatorship faced mass insurgency. In 1974 Ne

and placed Aung San Suu Kyi under house arrest. In 1997 SLORC became the State Peace and Development Council (SPDC). In 1998, NLD calls for the reconvening of Parliament led to mass detention of political opponents by the SPDC. In 2002 the SPDC released Aung San Suu Kyi from house arrest. She was arrested again in 2003.

In 2004 a United Nations report criticized the regime for holding more than 1,800 political detainees and for its failure to release opposition leader Aung San Suu Kyi from house arrest.

In November 2005 Burma announced that it was moving the seat of government to Pyinmana 250 mi [400 km] north of Rangoon, with immediate effect. Officials would not commit to whether this new site would become the capital with officials saying that everything would be made public at the appropriate time. The reason given for the move was that Pyinmana is in the center of the country

ECONOMY

Agriculture is the main activity, employing 66% of the workforce. The chief crop is rice. Peanuts, corn, plantains, pulses, seed cotton, sesame seeds and sugar cane are also produced. Forestry is important and teak is a major product. Fish and shellfish are another industry. The varied natural resources are mostly underdeveloped, but Burma is famous for its precious stones, especially rubies. It is almost self-sufficient in petroleum and natural gas.

RANGOON (YANGON)

Capital of Burma (Myanmar), a seaport on the Rangoon River. The site of a Buddhist shrine, it became capital in 1886, when the British annexed the country. It was the scene of heavy fighting between British and Japanese forces in World War II. It is the country's chief trade center.

Win became president. Mass demonstrations forced Ne Win to resign in 1988, but the military retained power under the guise of the State Law and Order Restoration Council (SLORC), led by General Saw Muang. In 1989 the country's name changed to Myanmar. The National League for Democracy (NLD), led by Aung San Suu Kyi, won elections in 1990, but SLORC annulled the result

BURUNDI

This flag was adopted in 1967 when Burundi became a republic. It contains three red stars rimmed in green, symbolizing the nation's motto of "Unité, Travail, Progrès." The green represents hope for the future, the red the struggle for independence, and the white the desire for peace.

The Republic of Burundi is a small country in east-central Africa. A section of the Great Rift Valley, lies in the west. It contains part of Lake Tanganyika, whose shoreline is 2,533 ft [772 m] above sea level. East of the Rift Valley is a mountain zone, rising to 8,760 ft [2,670 m]. The land descends to the east in a series of steppe-like plateaus. Burundi forms part of the Nile-Congo watershed and contains the headwaters of the River Kagera, the most remote source of the Nile.

Grassland covers much of Burundi, because much of the original forest has been cleared by farming and overgrazing. New forests are now being planted to halt the loss of soil fertility caused by erosion.

Area 10,747 sq mi [27,834 sq km]
Population 6,231,000
Capital (population) Bujumbura (235,000)
Government Republic
Ethnic groups Hutu 85%, Tutsi 14%, Twa (Pygmy)
Languages French, Kirundi (both official)
Religions Roman Catholic 62%, traditional beliefs 23%, Islam 10%, Protestant 5%
Currency Burundi franc = 100 centimes
Website www.burundi-embassy-berlin.com

CLIMATE

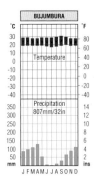

Bujumbura has an average annual temperature of 73°F [23°C]. June to August and December-January are dry, but the rest of the year is rainy. The mountains and central plateaus are distinctly cooler and wetter than the Rift Valley floor, but the rainfall decreases to the east.

HISTORY

The first known inhabitants of the area were the Twa, a pygmy group of hunting and gathering people, who now make up just 1% of the population. Around 1,000 years ago, a Bantu-speaking, iron-using farming people from the west, the Hutu, began to settle, pushing the Twa into remote areas. A third group, the cattle-owning Tutsi from the northeast, arrived around 600 years ago. They gradually took control of the area and, although in the minority, formed the ruling class. The Tutsi created a feudal state, making the Hutu serfs. The explorers Richard Burton and John Hanning Speke visited the area in 1858 in their quest to find the source of the Nile.

A powerful Tutsi kingdom under Mwami (king) Rugamba, that developed in the late 18th century, had broken up by the 1880s. Germany conquered what are now Burundi and Rwanda, in the late 1890s. The area, called Ruanda-Urundi, became part of German East Africa. But after Germany's defeat in World War I, Belgium took control.

In 1961, the people of Urundi voted to become a monarchy under Mwami Mwambutsa IV, who had ruled since 1915, while the people of Ruanda voted to become a republic.

POLITICS

The two territories finally became fully independent as Burundi and Rwanda on

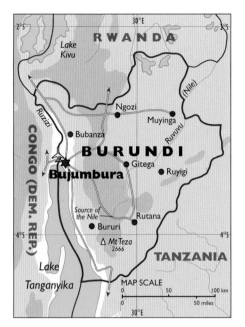

In 1992, a new constitution gave the country a multiparty system and, in 1993, Melchior Ndadaye, a Hutu, beat Buyoya in presidential elections. But supporters of Bagaza assassinated Ndadaye. In 1994, the new president, Cyprien Ntaryamira, a Hutu, was killed in an airplane crash, together with the Rwandan president, causing more ethnic violence. In 1996, Buyoya staged another coup and suspended the constitution. In 1999, peace talks began which led, in 2001, to the setting up of a transitional, power-sharing government. However, some Hutu rebel groups refused to sign the cease-fire. In 2003, Domitien Ndayizeye succeeded Buyoya as president, under the power-sharing agreement. In 2004, the disarming of rebels and soldiers began. In 2005, the people voted in favor of the new power-sharing constitution and hopes were high of an end to the conflict.

July 1, 1962. Since then, Burundi has suffered great conflict caused by ethnic rivalry between the Hutu majority and the Tutsi. Around 300,000 people have perished with many thousands displaced or as refugees. In 1965, Mwambutsa refused to appoint a Hutu prime minister, although the Hutu were in the majority. An attempted coup was brutally put down. In 1966, Mwambutsa was deposed by his son who became Mwami Ntare V, but Tutsi prime minister Michel Micombero deposed Ntare and declared Burundi to be a republic, with himself as president.

Between 1966 and 1972, most Hutu and some Tutsi were removed from high office. This culminated in a rebellion, when between 100,000 and 200,000 mostly Hutu were killed. In 1976, Jean-Baptiste Bagaza, a Tutsi, deposed Micombero. In 1981, Burundi became a one-party state, but Bagaza was deposed in 1987 by a coup led by Pierre Buyoya. Another uprising in 1988 led to the slaughter of thousands of Hutus.

ECONOMY

Burundi is one of the world's poorest countries. About 94% of the people depend on farming, mainly at subsistence level. The main food crops are bananas, beans, cassava, corn, and sweet potatoes. Cattle, goats and sheep are raised and fishing is important.

The economy depends on coffee and tea, which account for 90% of foreign exchange earnings, and cotton.

BUJUMBURA

Capital and chief port of Burundi, east-central Africa, at the northeast end of Lake Tanganyika. Founded in 1899 as part of German East Africa, it was the capital of Ruanda-Urundi after World War I and remained capital of Burundi upon independence in 1962.

CAMBODIA

Red is the traditional color of Cambodia. The blue symbolizes the water resources that are so important to the people, three-quarters of whom depend on farming for a living. The silhouette is the historic temple at Angkor Wat.

The Kingdom of Cambodia is a country in Southeast Asia. Low mountains border the country except in the southeast. Most of Cambodia consists of plains drained by the River Mekong, which enters Cambodia from Laos in the north and exits through Vietnam in the southeast. The northwest contains Tonlé Sap (or Great Lake). In the dry season, this lake drains into the River Mekong. But in the wet season, the level of the Mekong rises and water flows in the opposite direction from the river into Tonlé Sap—the lake then becomes the largest freshwater lake in Asia.

Area 69,898 sq mi [181,035 sq km]
Population 13,363,000
Capital (population) Phnom Penh (1,000,000)
Government Constitutional monarchy
Ethnic groups Khmer 90%, Vietnamese 5%, Chinese 1%, others
Languages Khmer (official), French, English
Religions Buddhism 95%, others 5%
Currency Riel = 100 sen
Website www.cambodia.gov.kh

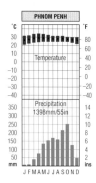

CLIMATE

Cambodia has a tropical monsoon climate, with high temperatures all through the year. The dry season, when winds blow from the north or northeast, runs from November to April. During the rainy season, from May to October, moist winds blow from the south or southeast. The high humidity and heat often make conditions unpleasant. The rainfall is heaviest near the coast, and rather lower inland.

HISTORY

From 802 to 1431,.the Hindu-Buddhist Khmer people ruled a great empire. Its zenith came in the reign of Suryavarman II (1113–50), who built the great funerary temple of Angkor Wat. Together with Angkor Thom, the Angkor site contains the world's largest group of religious buildings. The wealth of the kingdom rested on fish from the lake and rice from the flooded lowlands, for which an extensive system of irrigation channels and strong reservoirs was developed. Thai forces captured Angkor in 1431 and forests covered the site. Following its rediscovery in 1860, it has been gradually restored and is now a major tourist attraction.

France ruled the country from 1863 as part of Indo-China until it achieved independence in 1954. In a short period of stability during the late 1950s and 1960s, the country developed its small-scale agricultural resources and rubber plantations. It remained predominantly rural, but achieved self-sufficiency in food, with some exports.

POLITICS

In 1969, US airplanes bombed North Vietnamese targets in Cambodia. In 1970, King Norodom Sihanouk was overthrown and Cambodia became a republic. Under assault from South Vietnamese troops, the Communist Vietnamese withdrew deep into Cambodia.

Rouge continued hostilities and were banned in 1994. In 1997, Hu Sen, the second prime minister, engineered a coup against Prince Norodom Ranariddh (Sihanouk's son) and the first prime minister Ranariddh went into exile but returned in 1998. Elections in 1998 resulted in victory for Hu Sen, but Ranariddh alleged electoral fraud. A coalition government was formed in December 1998, with Hu Sen as prime minister. In 2001, the government set up a court to try leaders of the Khmer Rouge. In 2004, Sihanouk abdicated due to ill health and was succeeded by his son Prince Norodom Sihamoni.

ECONOMY

Cambodia is a poor country whose economy has been wrecked by war. By 1986, it was only able to supply 80% of its needs. Recovery has been slow. Farming is the main activity and rice, rubber, and corn important. Tourism is increasing—the impressive Angkor temples are a major attraction.

US raids ended in 1973, but fighting continued as Cambodia's Communists in the Khmer Rouge fought against the government. The Khmer Rouge, led by Pol Pot, were victorious in 1975. They began a reign of terror, murdering government officials and educated people. Up to 2 million people were estimated to have been killed. After the overthrow of Pol Pot by Vietnamese forces in 1979, civil war raged between the puppet government of the People's Republic of Kampuchea (Cambodia) and the US-backed government of Democratic Kampuchea, a coalition of Prince Sihanouk, the Khmer Liberation Front, and the Khmer Rouge, who, from 1982, claimed to have abandoned their Communist ideology.

Devastated by war and denied almost any aid, Cambodia continued to decline. It was only the withdrawal of Vietnamese troops in 1989, sparking fear of a Khmer Rouge revival, that forced a settlement. In October 1991, a UN-brokered peace plan for elections in 1993 was accepted by all parties. A new constitution was adopted in September 1993, restoring democracy and the monarchy. Sihanouk again became king. However, the Khmer

PHNOM PENH (PHNUM PÉNH)

Capital of Cambodia, in the south of the country, a port at the confluence of the rivers Mekong and Tonlé Sap. Founded in the 14th century, the city was the capital of the Khmers after 1434. In 1865, it became the capital of Cambodia. Occupied by the Japanese during World War II, it was extensively damaged during the Cambodian civil war. After the Khmer Rouge took power in 1975, the population was drastically reduced when many of its inhabitants were forcibly removed to work in the countryside. Industries include rice milling, brewing, and distilling.

CAMEROON

Cameroon uses the colors that appear on the flag of Ethiopia, Africa's oldest independent nation. These colors symbolize African unity. The flag is based on the tricolor adopted in 1957. The design with the yellow liberty star dates from 1975.

The Republic of Cameroon in West Africa got its name from the Portuguese word *camarões*, or prawns. This name was used by Portuguese explorers who fished for prawns along the coast.

Behind the narrow coastal plains on the Gulf of Guinea, the land rises to a series of plateaus. In the north, the land slopes down toward the Lake Chad (Tchad) basin. The mountain region in the southwest of the country includes Mount Cameroon, a volcano which erupts from time to time. The vegetation varies greatly from north to south. The deserts in the north merge into dry and moist savanna in central Cameroon, with dense tropical rainforests in the humid south.

Area 183,568 sq mi [475,442 sq km]
Population 16,064,000
Capital (population) Yaoundé (649,000)
Government Multiparty republic
Ethnic groups Cameroon Highlanders 31%, Bantu 27%, Kirdi 11%, Fulani 10%, others
Languages French and English (both official), many others
Religions Christianity 40%, traditional beliefs 40%, Islam 20%
Currency CFA franc = 100 centimes
Website www.camnet.cm

Portuguese explorers, who were seeking a sea route to Asia around Africa, reached the Cameroon coast. From the 17th century, southern Cameroon was a center of the slave trade, but slavery was ended in the early 19th century. In 1884, the area became a German protectorate. Germany lost Cameroon during World War I

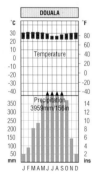

CLIMATE

The rainfall is heavy, especially in the highlands. The rainiest months near the coast are from June to September. The rainfall decreases to the north and the far north has a hot, dry climate. Temperatures are high on the coast, whereas the inland plateaus are cooler.

HISTORY

Among the early inhabitants of Cameroon were groups of Bantu-speaking people. (There are now more than 160 ethnic groups, each with their own language.) In the late 15th century,

YAOUNDÉ

Capital of Cameroon, West Africa. Located in beautiful hills on the edge of dense jungle, German traders founded it in 1888. During World War I, it was occupied by Belgian troops, and later acted as capital (1921–60) of French Cameroon. Since independence, it has grown rapidly as a financial and administrative center with strong Western influences. It is the site of the University of Cameroon (1962). The city also serves as a market for the surrounding region, notably in coffee, cacao, and sugar.

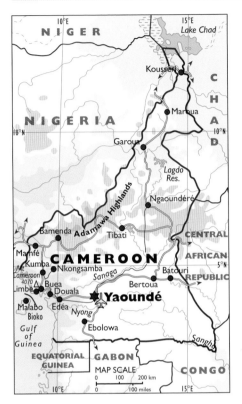

became the 52nd member of the Commonwealth. In 2002, the International Court of Justice gave Cameroon sovreignty over the disputed petroleum-rich Bakassi peninsula. But Nigeria failed to reach the deadline for the handover of the area in 2004.

Presidential elections in 2004 saw Paul Biya win a new seven-year term with more than 70% of the vote. The result was accepted by Commonwealth observers, but opposition parties alleged widespread fraud.

ECONOMY

Like most countries in tropical Africa, Cameroon's economy is based on agriculture, which employs 73% of the people. The chief food crops include cassava, corn, millet, sweet potatoes, and tropical yams.

Cameroon is fortunate in having some petroleum, the country's chief export, and bauxite. Although Cameroon has few manufacturing and processing industries, its mineral exports and its self-sufficiency in food production make it one of the wealthier countries in tropical Africa. Another important industry is forestry, ranking second among the exports, after petroleum. Other exports are cocoa, coffee, aluminum and cotton.

(1914–18). The country was then divided into two parts, one ruled by Britain and the other by France.

POLITICS

In 1960, French Cameroon became the independent Cameroon Republic. In 1961, after a vote in British Cameroon, part of the territory joined the Cameroon Republic to become the Federal Republic of Cameroon. The other part joined Nigeria. In 1972, Cameroon became a unitary state called the United Republic of Cameroon. It adopted the name Republic of Cameroon in 1984, but the country had two official languages. Opposition parties were legalized in 1992, and Paul Biya was elected president in 1993 and 1997. In 1995, partly to placate English-speakers, Cameroon

---

DOUALA

Chief port of Cameroon, on the Bight of Biafra, West Africa. As Kamerunstadt, it was capital of the German Kamerun Protectorate (1885–1901), became Douala (1907), and was capital of French Cameroon (1940–46). Industries include ship repairing, textiles, and palm oil. It has a population of 1,239,100.

73

CANADA

Canada's flag, with its simple 11-pointed maple leaf emblem, was adopted in 1965 after many attempts to find an acceptable design. The old flag, used from 1892, was the British Red Ensign, but this flag became unpopular with Canada's French community.

A vast confederation of ten provinces and three territories, Canada is the world's second largest country after Russia, with an even longer coastline—about 155,000 mi [250,000 km]. It is sparsely populated because it contains vast areas of virtually unoccupied mountains, cold forests, tundra, and polar desert in the north and west. About 80% of the population of Canada lives within about 186 mi [300 km] of the southern border.

Forests of cedars, hemlocks, and other trees grow on the western mountains, with firs and spruces at the higher levels. The mountain forests provide habitats for bears, deer, and mountain lions, while the surefooted Rocky Mountain goats and bighorn sheep roam above the tree line (the upper limit of tree growth).

The interior plains were once grassy prairies. While the drier areas are still used for grazing cattle, the wetter areas are used largely for growing wheat and other cereals. North of the prairies are the boreal forests which, in turn merge into the treeless tundra and Arctic wastelands in the far north. The lowlands in southeastern Canada contain forests of deciduous trees, such as beech, hickory, oak and walnut.

CLIMATE

Canada has a cold climate. In winter, temperatures fall below freezing point throughout most of the country. But the southwestern coast has a relatively mild climate. Along the Arctic Circle, the temperatures are, on average, below freezing for seven months a year. By contrast, hot winds from the Gulf of Mexico warm southern Ontario and the St. Lawrence

Area 3,849,653 sq mi [9,970,610 sq km]
Population 32,508,000
Capital (population) Ottawa (774,000)
Government Federal multiparty constitutional monarchy
Ethnic groups British origin 28%, French origin 23%, other European 15%, Amerindian/Inuit 2%, others
Languages English and French (both official)
Religions Roman Catholic 46%, Protestant 36%, Judaism, Islam, Hinduism
Currency Canadian dollar = 100 cents
Website http://canada.gc.ca

River lowlands in summer. As a result, southern Ontario has a frost-free season of nearly six months.

The coasts of British Columbia are wet, with an average annual rainfall of more than 98 in [2,500 mm] in places. The prairies however are arid or semi-arid, with an average annual rainfall of 10 to 20 in [250–500 mm]. The rainfall in southeastern Canada ranges from around 31 in [800 mm] in southern Ontario to about 59 in [1,500 mm] on the coasts of Newfoundland and Nova

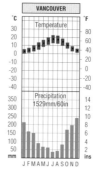

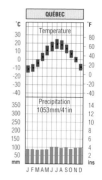

Scotia. Heavy snow falls in eastern Canada in winter.

HISTORY

Canada's first people, ancestors of the Native Americans, arrived in North America from Asia around 40,000 years ago. Later arrivals were the Inuit (Eskimos), who also came from Asia. Norse voyagers and fishermen were probably the first to visit Canada, but John Cabot's later discovery of North America in 1497 led to the race to annex lands and wealth, with France and Britain the main contenders.

The creation of the British Commonwealth in 1931 made Canada a sovereign nation under the crown. Canada is now a constitutional monarchy. Under the Constitution Act of 1982, Queen Elizabeth II is head of state and a symbol of the close ties between Canada and Britain. The British monarch is represented by an appointed governor-general, but the country is ruled by a prime minister, and an elected, two-chamber parliament.

POLITICS

Canada combines the cabinet system with a federal form of government, with each province having its own government. The federal government can reject any law passed by a provincial legislature, though this seldom happens in practice. The territories are self-governing, but the federal government plays a large part in their administration.

Canada and the United States of America have the largest bilateral trade flow in the world. Economic cooperation was further enhanced in 1993 when Canada, the United States, and Mexico set up NAFTA (North American Free Trade Agreement).

A constant problem facing those who want to maintain the unity of Canada is the persistence of French culture in Québec, which has fuelled a separatist movement seeking to turn the province into an independent French-speaking republic. More than two-thirds of the population of Québec are French speakers. In 1994, the people of Québec voted the separatist Parti Québécois into provincial office. The incoming prime minister announced that independence for Québec would be the subject of a referendum in 1995. In that referendum, 49.4% voted "Yes" (for separation) while 50.5% voted "No."

Provincial elections in 1998 resulted in another victory for the Parti Québécois. But while the separatist party won 75 out of the 125 seats in the provincial assembly, it won only 43% of the popular vote, compared with 44% for the antisecessionist Liberal Party and 12% for the floating Action Démocratique de Québec. Also significant was a ruling by Canada's highest court that, under Canadian law, Québec does not have the right to secede unilaterally. The court ruled that, should a clear majority of the people in the province vote by "a clear majority" to a "clear question" in favor of independence, the federal government and the other provinces would have to negotiate Québec's secession.

Other problems involve the rights of the aboriginal Native Americans and the Inuit, who together numbered about 470,000 in 1991. In 1999, a new Inuit territory was created. Called Nunavut, it

OTTAWA

Capital of Canada, in southeast Ontario, on the Ottawa River and the Rideau Canal. Founded in 1826 as Bytown, it acquired its present name in 1854. Queen Victoria chose it as capital of the United Provinces in 1858, and in 1867 it became the national capital of the Dominion of Canada. Industries include glass-making, printing, publishing, sawmilling, pulp-making, clocks, and watches.

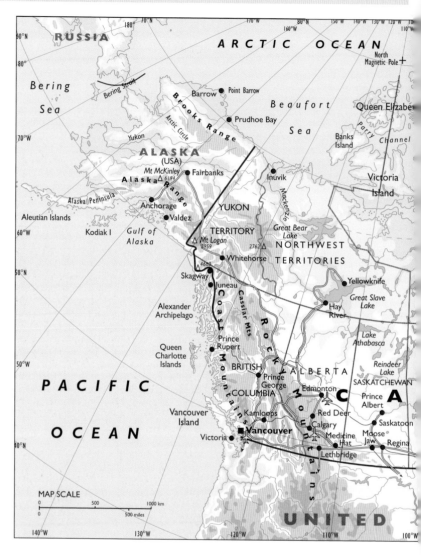

is made up of 64% of the former North-west Territories, and covers 649,965 sq mi [2,201,400 sq km]. The population in 1991 was about 25,000, 85% of whom were Inuit. Nunavut, whose capital is Iqaluit (formerly Frobisher Bay), will depend on future aid, but its mineral reserves and the prospects of an eco-tourist industry hold out promise for the future.

ECONOMY

Canada is a highly developed and pros-perous country. Although farmland cov-ers only 8% of the country, Canadian farms are highly productive, Canada is

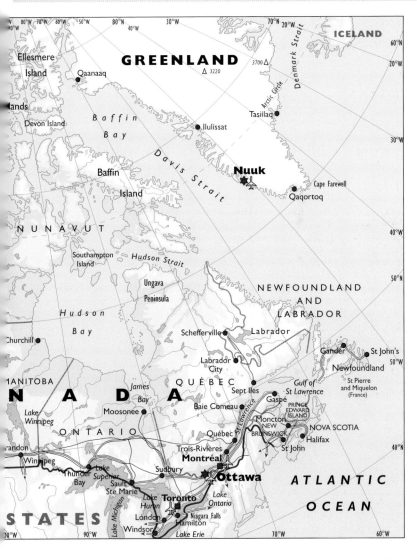

one of the world's leading producers of barley, wheat, meat, and milk. Forestry and fishing are other important industries. It is rich in natural resources, especially petroleum and natural gas. Canada exports minerals, including copper, gold, iron ore, uranium, and zinc. Manufacturing is important, mainly in the cities where 79% of the population lives. Canada processes farm and mineral products. It also produces cars, chemicals, electronic goods, machinery, paper, and timber products.

Tourism is an important source of income with both winter and summer popular tourist seasons.

CARIBBEAN ISLANDS

ANGUILLA

Area 37 sq mi [96 sq km]
Population 13,000
Capital The Valley
Government Overseas territory of the UK
Ethnic groups Black 90%, Mixed 5%, White 3%
Languages English
Religions Anglican 29%, Methodist 24%,
Protestant 30%, Roman Catholic 6%
Currency East Caribbean dollar = 100 cents
Website www.anguilla-vacation.com

Most northerly of the Leeward Islands. Settled in the 17th century by English colonists, it became part of the St. Kitts-Nevis-Anguilla group. Declared independent in 1967, it readopted British colonial status in 1980, and is now a self-governing dependency. The economy of the flat, coral island is based on fishing and tourism.

ANTIGUA & BARBUDA

Area 171 sq mi [442 sq km]
Population 68,000
Capital St. John's
Government Constitutional monarchy
Ethnic groups Black, British, others
Languages English, local dialects
Religions Christian
Currency East Caribbean dollar = 100 cents
Website www.antigua-barbuda.org

Part of the Lesser Antilles in the Leeward Islands. Antigua is atypical of the Leeward Islands as it has no rivers or forests. Barbuda, by contrast, is a wooded, low coral atoll. Only 1,400 people live on the game reserve island of Barbuda, where lobster fishing is the main occupation, and none on the rocky island of Redondo. Antigua and Barbuda gained internal self-government in 1967, and independence in 1981. The islands are dependent on tourism.

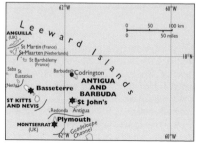

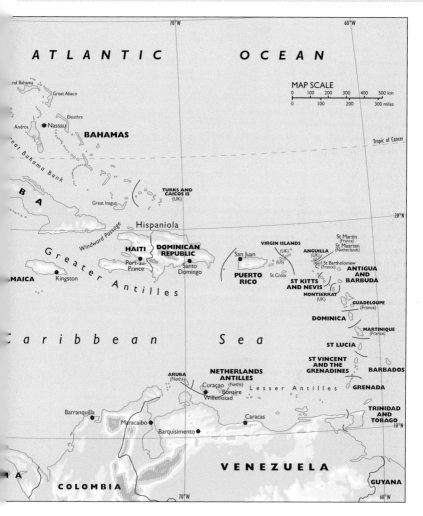

ARUBA

Dutch island in the Caribbean, off the coast of northwest Venezuela. It was part of the Netherlands Antilles until 1986. Independence was revoked in 1990, at Aruba's request, and it is now an autonomous part of the Netherlands. Petroleum refining, phosphates, and tourism are the key sources of revenue.

Area 75 sq mi [193 sq km]
Population 71,000
Capital Oranjestad
Government Parliamentary democracy
Ethnic groups Mixed White/Caribbean Amerindian 80%
Languages Dutch (official), Papiamento, English, Spanish
Religions Roman Catholic 82%, Protestant 8%
Currency Aruban guilder/florin = 100 cents
Website www.aruba.com

BAHAMAS

Area 5,358 sq mi [13,878 sq km]
Population 297,000
Capital Nassau
Government Constitutional parliamentary democracy
Ethnic groups Black 85%, White 12%, others
Languages English (official), Creole
Religions Christian
Currency Bahamian dollar = 100 cents
Website www.bahamas.gov.bs

Small independent state in the West Indies including over 700 islands, the largest of which is Grand Bahama. Mainly limestone and coral, the rocky terrain provides little chance for agricultural development. Most of the islands are low, flat, and riverless with mangrove swamps. The climate is subtropical with average temperatures of 70–90°F [21–32°C]. In 1973 the Bahamas gained independence. Revenue comes from tourism, fishing, salt, and rum.

CAYMAN ISLANDS

Area 102 sq mi [264 sq km]
Population 42,000
Capital Georgetown
Government Overseas territory of the UK
Ethnic groups Mixed 40%, White 20%, Black 20%
Languages English
Religions Protestant, Roman Catholic
Currency Caymanian dollar = 100 cents
Website www.gov.ky

British dependency in the West Indies, comprising Grand Cayman, Little Cayman, and Cayman Brac, 200 mi (325 km) northwest of Jamaica, in the Caribbean Sea. The islands were discovered by Columbus in 1503, and ceded to Britain in the 17th century. The islanders voted against independence in 1962. Tourism, international finance, and turtle and shark fishing, are sources of revenue.

BARBADOS

Area 166 sq mi] [430 sq km
Population 277,000
Capital Bridgetown
Government Parliamentary democracy
Ethnic groups Black 90%, Asian, White
Languages English
Religions Protestant 67%, Roman Catholic 4%
Currency Barbadian dollar = 100 cents
Website www.barbados.org

Island state in the Windward Islands, West Indies. Barbados' warm climate encouraged the growth of its two largest industries: sugarcane and tourism. It was settled by the British in 1627, and dominated by British plantation owners (using African slave labor until the abolition of slavery) for the next 300 years. It gained independence in 1966.

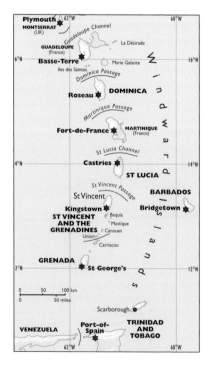

DOMINICA

Area 290 sq mi [751 sq km]
Population 70,000
Capital Roseau
Government Parliamentary democracy
Ethnic groups Black, Mixed black and European, European, Syrian, Carib Amerindian
Languages English (official), French patois
Religions Roman Catholic 77%, Protestant 15%
Currency East Caribbean dollar = 100 cents
Website www.avirtualdominica.com

An independent island nation in the east Caribbean Sea, it is the largest of the Windward Islands. The present population are mainly the descendants of African slaves. Dominica is mountainous and heavily forested, and the climate is tropical. It achieved complete independence as a republic within the Commonwealth in 1978. Dominica is one of the poorest Caribbean countries. Agriculture dominates the economy.

GRENADA

Area 133 sq mi [344 sq km]
Population 89,000
Capital St. George's
Government Constitutional monarchy
Ethnic groups Black 82%, others
Languages English (official), French patois
Religions Roman Catholic 53%, Protestant 47%
Currency East Caribbean dollar = 100 cents
Website www.gov.gd

Independent island nation in the southeast Caribbean Sea, the most southerly of the Windward Islands, 160 km [100 mi] north of Venezuela. It consists of Grenada and the smaller islands of the Southern Grenadines dependency. It is volcanic in origin, with a ridge of mountains running north–south. The climate is tropical with occasional hurricanes. The agrarian economy is based on cocoa, bananas, sugar, spices, and citrus fruits. It depends greatly on tourism.

GUADELOUPE

Area 658 sq mi [1,705 sq km]
Population 440,000
Capital Basse-Terre
Government Overseas department of France
Ethnic groups Black or mulatto 90%, White 5%
Languages English (official), Spanish, Creole
Religions Roman Catholic 95%, Hindu and pagan African 4%
Currency Euro = 100 cents
Website www.guadeloupe.pref.gouv.fr

French overseas department (since 1946), consisting of the islands of Basse-Terre, Grande-Terre, and several smaller islands in the Leeward Islands. Discovered in 1493 by Columbus, and settled by France in 1635. Briefly held by Britain and Sweden, it reverted to French rule in 1816. The chief crops are sugarcane and bananas. Industries include distilling and tourism.

MARTINIQUE

Area 425 sq mi [1,102 sq km]
Population 426,000
Capital Fort de France
Government Overseas department of France
Ethnic groups African and African-white-Indian mixture 90%
Languages French, Creole patois
Religions Roman Catholic 85%, Protestant 10%
Currency Euro = 100 cents
Website www.martinique.org

Island in the Windward group of the Lesser Antilles. Martinique was inhabited by Carib Indians until they were displaced by French settlers after 1635. The island became a permanent French possession after the Napoleonic Wars. Of volcanic origin, it is the largest of the Lesser Antilles. In 1902, a volcanic eruption completely destroyed the original capital, St. Pierre.

MONTSERRAT

Area 40 sq mi [102 sq km]
Population 9,000
Capital Plymouth
Government Overseas territory of the UK
Ethnic groups Black, White
Languages English
Religions Christian
Currency East Caribbean dollar = 100 cents
Website www.gov.ms

Montserrat is one of the Leeward Islands in the Lesser Antilles. It is dominated by an active volcano in the Soufrière Hills. The British colonized in 1632. It formed part of the Leeward Island colony from 1871 until 1956, when it became a separate, dependent territory of the UK. In 1997, the British government offered aid to the remaining islanders after increased volcanic activity. Sources of income include tourism, offshore finance, and cotton.

NETHERLANDS ANTILLES

Area 309 sq mi [800 sq km]
Population 216,000
Capital Willemstadt (Curacao)
Government Autonomous country within the Kingdom of the Netherlands
Ethnic groups Mixed Black 85%, others
Languages Papiamento 65%, English 16%, Dutch 7% (official)
Religions Roman Catholic 72%, others
Currency Netherlands Antillean guilder = 100 cents
Website www.gov.an

Group of five main islands (and part of a sixth) in the West Indies. The islands were settled by the Spanish in 1527 and captured by the Dutch in 1634. They were granted internal self-government in 1954. The group includes Aruba, Bonaire, Curaçao, Saba, Saint Eustatius, and the southern half of Saint Maarten. Petroleum refining, petrochemicals, and tourism provide revenue.

ST. KITTS & NEVIS

Area 261 sq km [101 sq mi]
Population 39,000
Capital Basseterre (St. Kitts)
Government Constitutional monarchy
Ethnic groups Black
Languages English
Religions Anglican, Roman Catholic
Currency East Caribbean dollar = 100 cents
Website www.stkittsnevis.net

Self-governing state in the Leeward Islands, West Indies. It comprises the islands St. Kitts and Nevis. The English settled in 1623 and the French in 1624. The Treaty of Paris (1783) settled Anglo-French disputes over possession, and the islands gained self-government in 1967. Nevis held a referendum on independence in May 1998. Industries include tourism, sugar and cotton.

ST. LUCIA

Area 539 sq km [208 sq mi]
Population 162,000
Capital Castries
Government Parliamentary democracy
Ethnic groups Black 90%
Languages English (official), French patois
Religions Roman Catholic 68%, Protestant
Currency East Caribbean dollar = 100 cents
Website www.stlucia.gov.lc

Volcanic island in the Windward group, West Indies. The island changed hands 14 times between France and Britain before being ceded to Britain in 1814. It finally achieved full self-government in 1979. Mountainous (the twin peton peaks are a scenic highlight), lush and forested, its tourist income is growing rapidly, especially from cruise ships. The principal export is bananas.

ST. VINCENT & GRENADINES

Area 150 sq mi [388 sq km]
Population 117,000
Capital Kingstown
Government Parliamentary democracy
Ethnic groups Black 66%, Mixed 19%
Languages English, French patois
Religions Anglican 47%, Methodist 28%, Roman Catholic 13%
Currency East Caribbean dollar = 100 cents
Website www.svgtourism.com

Island state of the Windward Islands, between St. Lucia and Grenada comprising the volcanic island of St. Vincent and five islands of the Grenadine group, including Mustique. In 1783 the British deported most of the native Carib population, who were replaced by African slave labor. St. Vincent was part of the British Windward Islands colony (1880-1958) and of the West Indies Federation (1958–62). It gained self-government in 1969 and full independence in the Commonwealth in 1979.

US VIRGIN ISLANDS

Area 134 sq mi [347 sq km]
Population 125,000
Capital Charlotte Amalie (St. Thomas)
Government Unincorporated territory of the US
Ethnic groups Black 76%, White 13%
Languages English 75%, Spanish or Spanish Creole 17%, French or French Creole 7%
Religions Baptist 42%, Roman Catholic 34%
Currency US dollar = 100 cents
Website www.usvi.net

Group of 68 islands in the Lesser Antilles. Chief islands are St. Croix and St. Thomas. Spanish from 1553, the islands were Danish until 1917, when the USA bought them for US$25 million, to protect the northern approaches to the newly completed Panama Canal. Tourism is the biggest earner.

TURKS & CAICOS

Area 166 sq mi [430 sq km]
Population 19,000
Capital Cockburn Town (Grand Turk Island)
Government Overseas territory of the UK
Ethnic groups Black 90%
Languages English (official)
Religions Baptist 40%, Methodist 16%, Anglican 18%, Church of God 12%
Currency US dollar = 100 cents
Website www.turksandcaicosislands.gov.tc

Turks and Caicos are two island groups of the British West Indies. They include more than 40 islands, eight of which are inhabited. Discovered in 1512 by Ponce de León, the islands were British from 1766. They were then administered via Jamaica from 1873–1959, and became a separate Crown Colony from 1973. Most food products are imported. Exports include salt, sponges, and shellfish. The islands' main sources of income are tourism and offshore banking.

BRITISH VIRGIN ISLANDS

Area 58 sq mi [151 sq km]
Population 22,000
Capital Road Town (Tortola)
Government Overseas territory of the UK
Ethnic groups Black 84%
Languages English (official)
Religions Protestant 86%, Roman Catholic 10%
Currency US dollar = 100 cents
Website www.bvitourism.com

British colony in the West Indies. It is a group of 36 islands, which form part of the Antilles group between the Caribbean Sea and the Atlantic Ocean; Tortola is the main island. First settled in the 17th century, the islands formed part of the Leeward Islands colony until 1956. The chief economic activity is tourism.

CENTRAL AFRICAN REPUBLIC

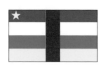

The red, yellow, and green colors on this flag were originally used by Ethiopia, Africa's oldest independent nation. They symbolize African unity. The blue, white and red recall the flag of France, the country's colonial ruler. This flag was adopted in 1958.

The Central African Republic is a remote landlocked country in central Africa. It lies on a plateau, mostly 1,970–2,620 ft [600–800 m] above sea level, forming a watershed between the headwaters of two river systems. In the south, the rivers flow into the navigable River Ubangi (a tributary of the Congo). The Ubangi and the Bomu form much of its southern border. In the north, most rivers are headwaters of the River Chari, which flows north into Lake Chad.

Wooded savanna covers much of the country, with open grasslands in the north and rainforests in the southwest. The country has many forest and savanna animals, such as buffalo, leopards, lions, and elephants, and many bird species. About 6% of the land is protected in national parks and reserves, but tourism is on a small scale because of the republic's remoteness.

Area 240,534 sq mi [622,984 sq km]
Population 3,742,000
Capital (population) Bangui (553,000)
Government Multiparty republic
Ethnic groups Baya 33%, Banda 27%, Mandjia 13%, Sara 10%, Mboum 7% Mbaka 4%, others
Languages French (official), Sangho
Religions Traditional beliefs 35%, Protestant 25%, Roman Catholic 25%, Islam 15%
Currency CFA franc = 100 centimes
Website www.banguinet.net

reduced by slavery, and the country is still thinly populated. France first occupied the area in 1887, and in 1894 established the colony of Ubangi-Shari at Bangui. In 1906 the colony was united with Chad, and in 1910 was subsumed into French Equatorial Africa (which included Chad, Congo, and Gabon). Forced-labor rebellions occurred in

CLIMATE

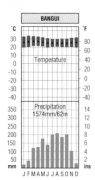

The climate is warm throughout the year, with an average annual rainfall in Bangui totalling 62 in [1,574 mm]. The north is drier, with an average annual rainfall total of about 31 in [800 mm].

HISTORY

Little is known of the country's early history. Between the 16th and 19th centuries, the population was greatly

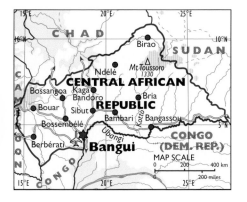

1928, 1935, and 1946. During World War II Ubangi-Shari supported the Free French. Post-1945 the colony received representation in the French parliament. In 1958 the colony voted to become a self-governing republic within the French community, and became the Central African Republic.

POLITICS
In 1960, Central African Republic declared independence, but the next six years saw a deterioration in the economy, and increasing government corruption and inefficiency under President David Dacko. It became a one-party state in 1962. In 1966 Colonel Jean Bédel Bokassa assumed power in a bloodless coup. He abrogated the constitution and dissolved the National Assembly.

In 1976 Bokassa transformed the republic into an empire, and proclaimed himself Emperor Bokassa I. The country was renamed the Central African Empire. His rule became increasingly brutal, and in 1979 he was deposed in a French-backed coup led by Dacko. Dacko, faced with continuing unrest, was replaced by André Kolingba in 1981. The army quickly banned all political parties.

The country adopted a new, multiparty constitution in 1991. Elections were held in 1993. An army rebellion in 1996 was finally put down in 1997 with the assistance of French troops. An attempt-ed coup in 2001 was put down, with Libyan help, by President Ange-Félix Patassé, who had served as president since 1993. But a coup in 2003 brought General François Bozize to power and Patassé went into exile in Togo. A new constitution was introduced in 2004, followed by elections in 2005 which were won by General Bozize.

ECONOMY
The World Bank classifies the Central African Republic as a "low-income" developing country. Approximately 10% of the land is cultivated and over 80% of the workforce are engaged in subsistence agriculture. The main food crops are bananas, corn, cassava, millet and tropical yams. Coffee, cotton, timber, and tobacco are the main cash crops

Diamonds, the only major mineral resource, are the most valuable single export. Manufacturing is on a very small scale. Products include beer, cotton fabrics, footwear, leather, soap, and sawn timber. The Central African Republic's development has been greatly impeded by its remote position, its poor transportation system and its untrained workforce. The country is heavily dependent on aid, especially from France.

BANGUI

Capital of the Central African Republic, on the River Ubangi, near the border with the Democratic Republic of Congo. Founded in 1889 by the French, it is the nation's chief port for international trade. Places of interest include the triumphal arch dedicated to Jean-Bédel Bokassa, the Boganda Museum, and the central market. Industries include textiles, shoes, food processing, beer, and soap.

CHAD

Chad's flag was adopted in 1959 as the country prepared for independence in 1960. The blue represents the sky, the streams in southern Chad, and hope. The yellow symbolizes the sun and the Sahara in the north. The red represents national sacrifice.

Chad is Africa's fifth largest country. It is more than twice as big as France (the former colonial power). Southern Chad is crossed by rivers that flow into Lake Chad, on the western border with Nigeria. The capital, Ndjamena, lies on the banks of the River Chari. Beyond a large depression (northeast of Lake Chad) are the Tibesti Mountains, which rise steeply from the sands of the Sahara Desert. The mountains contain Chad's highest peak, Emi Koussi, at 11,204 ft [3,415 m]. The far south contains forests, while central Chad is a region of savanna, merging into the dry grasslands of the Sahel. Plants are rare in the northern desert. Droughts are common in north central Chad. Long droughts, overgrazing, and felling for firewood have exposed the Sahel's soil and wind erosion is increasing desertification.

Area 495,752 sq mi [1,284,000 sq km]
Population 9,539,000
Capital (population) Ndjamena (530,000)
Government Multiparty republic
Ethnic groups 200 distinct groups: mostly Muslim in the north and center; mostly Christian or animist in the south
Languages French and Arabic (both official), many others
Religions Islam 51%, Christianity 35%, animist 7%
Currency CFA franc = 100 centimes
Website www.chadembassy.org

CLIMATE

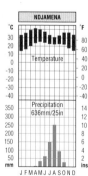

Central Chad has a hot tropical climate. There is a marked dry season from November to April. The south is wetter, with an average annual rainfall of about 39 in [1000 mm]. Conversely, the hot northern desert has an average annual rainfall of less than 5 in [130 mm].

HISTORY

Chad straddles two often conflicting worlds: the north, populated by nomadic or seminomadic Muslim peoples, such as Arabs and Tuaregs; and the dominant south, where a sedentary population practice Christianity or traditional religions, such as animism. Lake Chad was an important watering point for the trans-Saharan caravans. Around AD 700 North African nomads founded the Kanem Empire. In the 14th century, the kingdom of Bornu expanded to incorporate Kanem. In the late 19th century the region fell to Sudan.

The first major European explorations were by the French in 1890. The French

LAKE CHAD (TCHAD)

Lake in north-central Africa, mainly in the Republic of Chad and partly in Nigeria, Cameroon, and Niger. The chief tributary is the River Chari. The lake has no outlet. Depending on the season, the area of the surface varies from 3,850–10,000 sq mi [10,000 to 26,000 sq km] with a maximum depth: of 25 ft [7.6 m].

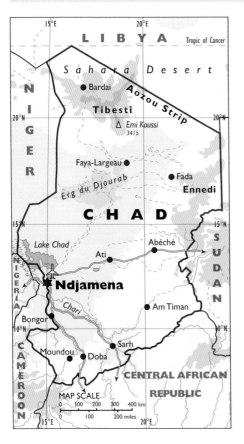

defeated the Sudanese in 1900, and in 1908 Chad became the largest province of French Equatorial Africa. In 1920 it became a separate colony.

POLITICS

In 1958 Chad gained autonomous status within the French Community, and in 1960 achieved full independence. Divisions between north and south rapidly surfaced. In 1965, President François Tombalbaye declared a one-party state and the northern Muslims, led by the Chad National Liberation Front (Frolinat), rebelled. By 1973 the government, helped by the French, quashed the revolt. In 1980 Libya occupied northern Chad. In 1982, two leaders of Frolinat, Hissène Habré and Goukouni Oueddi, formed rival regimes. Splits soon emerged and Libya's bombing of Chad in 1983 led to the deployment of 3,000 French troops. Libyan troops retreated, retaining only the uranium-rich Aozou Strip. A ceasefire took effect in 1987. In 1990, Habré was removed in a coup led by Idriss Déby. In 1994, the Aozou Strip was awarded to Chad. In 1996, a new democratic constitution was adopted and multiparty elections confirmed Déby as president. He was reelected in 2001. In 2002 a peace treaty, signed by the government and the Movement for Democracy and Justice, ended three years of civil war. In 2004-5, Chad forces clashed with pro-Sudanese militia as the conflict in Sudan's Darfur province spilled over the border.

ECONOMY

Chad is one of the world's poorest countries. Agriculture dominates the economy; more than 80% of the workforce are engaged in farming, mainly at subsistence level. Peanuts, millet, rice, and sorghum are major crops in the wetter south. The most valuable crop is cotton.

NDJAMENA

Capital of Chad, a port on the River Chari. Founded by the French in 1900, it was known as Fort Lamy until 1973. Ndjamena grew rapidly after independence in 1960. An important market for the surrounding region, which produces livestock, dates, and cereals. The main industry is meat processing.

CHILE

Chile's flag was adopted in 1817. It was designed in that year by an American serving in the Chilean army who was inspired by the US Stars and Stripes. The white represents the snow-capped Andes, the blue the sky, and the red the blood of the nation's patriots.

The Republic of Chile stretches 2,650 mi [4,260 km] from north to south, while the maximum east-west distance is only 270 mi [430 km]. The Andes mountains form Chile's eastern borders with Argentina and Bolivia. Ojos del Salado, at 22,516 ft [6,863 m], is the second-highest peak in South America. Easter Island lies 2,200 mi [3,500 km] off Chile's west coast.

Western Chile contains three main land regions. In the north is the sparsely populated Atacama Desert, stretching 1,000 mi [1,600 km] south from the Peruvian border. The Central Valley, which contains the capital, Santiago, Valparaíso and Concepción, is by far the most densely populated region. In the south, the land has been heavily glaciated, the coastal uplands have been worn into islands, while the inland valleys are arms of the sea.

In the far south, the Strait of Magellan separates the Chilean mainland from Tierra del Fuego. Punta Arenas is the world's southernmost city.

Area 292,133 sq mi [756,626 sq km]
Population 15,824,000
Capital (population) Santiago (4,789,000)
Government Multiparty republic
Ethnic groups Mestizo 95%, Amerindian 3%
Languages Spanish (official)
Religions Roman Catholic 89%, Protestant 11%
Currency Chilean peso = 100 centavos
Website www.chileangovernment.cl

CLIMATE

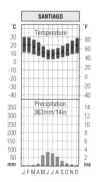

Chile is divided into three main climate zones. The Atacama Desert in the north has an arid climate, but temperatures are moderated by the cold Peru Current. Central Chile has a Mediterranean climate with hot, dry summers and mild, moist winters. The south has a cool and stormy climate prone to alpine conditions.

HISTORY

Amerindian people reached the southern tip of South America at least 8,000 years ago. In 1520, the Portuguese navigator Ferdinand Magellan became the first European to sight Chile. The country became a Spanish colony in the 1540s. Under Spain, the economy in the north was based on mining, while huge ranches, or *haciendas*, were set up in central Chile. After Chile became independent in 1818, mining continued to flourish in the north, while Valparaiso developed as a port, exporting produce from central Chile to California and Australia. During a war (1879–83), it gained mineral-rich areas from Peru and Bolivia. Industrial growth, fueled by revenue from nitrate exports, began in the early 20th century.

POLITICS

After World War II, Chile faced economic problems, partly caused by falls in world copper prices. A Christian Democrat was elected president in 1964, but was replaced by Salvador Allende Gossens in 1970. Allende's administration, the world's first democratically elected

a dictator, banning all political activity in a repressive regime. A new constitution took effect from 1981, allowing for an eventual return to democracy. Elections took place in 1989. President Patrico Aylwin took office in 1990, but Pinochet secured continued office as commander-in-chief of the armed forces. Eduardo Frei was elected president in 1993 and he was succeeded by a socialist, Ricardo Lagos, who narrowly defeated a conservative candidate in January 2000. In 1999, General Pinochet, who was visiting Britain for medical treatment, was faced with extradition to Spain to answer charges that he had presided over acts of torture when he was Chile's dictator. In 2000, he was allowed to return to Chile where, in 2001, he was found to be too ill to stand trial. New charges were brought against him in 2004 and, in 2005 he was placed under house arrest.

ECONOMY

The World Bank classifies Chile as a "lower-middle-income" developing country. Mining is important. Minerals dominate Chile's exports. The most valuable activity is manufacturing. Products include processed foods, metals, iron and steel, wood products, and textiles.

Agriculture employs 18% of the workforce. The chief crop is wheat. Beans, fruits, corn and livestock products are also important. Chile's fishing industry is one of the world's largest.

SANTIAGO

Capital of Chile, in central Chile on the River Mapocho, 55 mi [90 km] from the Atlantic coast. Founded in 1541, it was destroyed by an earthquake in 1647. Most of Santiago's architecture dates from after 1850. It is the nation's administrative, commercial, and cultural center, accounting for nearly a third of the population.

Marxist government, was overthrown in a CIA-backed coup in 1973. General Augusto Pinochet Ugarte took power as

CHINA

China's flag was adopted in 1949, when the country became a Communist People's Republic. Red is the traditional color of both China and Communism. The large star represents the Communist Party program. The smaller stars symbolize the four main social classes.

The People's Republic of China is the world's third largest country. Most people live on the eastern coastal plains, in the highlands or the fertile valleys of the rivers Huang He and Yangtze, Asia's longest river, at 3,960 mi [6,380 km].

Western China includes the bleak Tibetan plateau, bounded by the Himalayas. Everest, the world's highest peak, lies on the Nepal-Tibet border. Other ranges include the Tian Shan and Kunlun Shan. China also has deserts, such as the Gobi.

Large areas in the west are covered by sparse grasses or desert. The most luxuriant forests are in the southeast, such as the bamboo forest habitat of the rare giant panda.

CLIMATE

The capital, Beijing, in northeast China, has cold winters and warm summers, with moderate rainfall. Shanghai, in the east-central region, has milder winters and more rain. The southeast region has a wet, subtropical climate. In the west, the climate is more cold and severe.

Area 3,705,387 sq mi [9,596,961 sq km]
Population 1,298,848,000
Capital (population) Beijing (7,362,000)
Government Single-party Communist republic
Ethnic groups Han Chinese 92%, many others
Languages Mandarin Chinese (official)
Religions Atheist (official)
Currency Renminbi yuan= 10 jiao = 100 fen
Website www.china.org.cn/english

HISTORY

The first documented dynasty was the Shang (1523–1030 BC), when bronze casting was perfected. The Zhou dynasty (1030–221 BC) was the age of Chinese classical literature, in particular Confucius and Lao Tzu. China was unified by Qin Shihuangdi, whose tomb near Xian contains the famous terracotta army. The Qin dynasty (221–206 BC) also built the majority of the Great Wall. The Han dynasty (202 BC–AD 220) developed the Empire, a bureaucracy based on Confucianism, and also introduced Buddhism. China then split into three kingdoms (Wei, Shu, and Wu) and the influence of Buddhism and Taoism grew. The T'ang dynasty from 618–907 was a golden era of artistic achievement, especially in poetry and the fine arts. Genghis Khan conquered most of China in the 1210s and established the Mongul Empire. Kublai Khan founded the Yüan

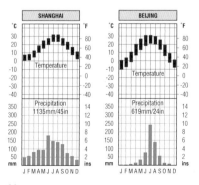

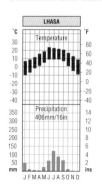

BEIJING (PEKING)

Capital of the People's Republic of China, between the Pei and Hun rivers, northeast China. A settlement since 1000 BC, Beijing served as China's capital from 1421 to 1911. After the establishment of the Chinese Republic (1911–12), Beijing remained the political center of China. The seat of government was transferred to Nanking in 1928. Beijing ("northern capital") became known as Pei-p'ing ("northern peace"). Occupied by the Japanese in 1937, it was restored to China in 1945 and came under Communist control in 1949. Its name was restored as capital of the People's Republic. The city comprises two walled sections: the Inner (Tatar) City, including the Forbidden City (imperial palace complex), and the Outer (Chinese) city. Beijing is the political, cultural, educational, financial, and transportation center of China. Heavy industry expanded after the end of the Civil War, and products now include textiles, iron, and steel.

The Communist Party of China initially allied with the Kuomintang. In 1926, the Kuomintang, led by Chiang Kai-shek, emerged victorious and turned on their Communist allies. In 1930 a rival communist government was established, but was uprooted by Kuomintang troops and began the Long March (1934). Japan, taking advantage of the turmoil, established the puppet state of Manchukuo (1932). Chiang was forced to ally with the Communists. Japan launched a full-

HONG KONG
(XIANGGANG SPECIAL ADMINISTRATIVE REGION)

Former British Crown Colony off the coast of southeast China; the capital is Victoria on Hong Kong Island. Hong Kong comprises: Hong Kong Island, ceded to Britain by China in 1842; the mainland peninsula of Kowloon, acquired in 1860; the New Territories on the mainland, leased for 99 years in 1898; and some 230 islets in the South China Sea. The climate is subtropical, with hot, dry summers. In 1984, the UK and China signed a Joint Declaration in which it was agreed that China would resume sovereignty over Hong Kong in 1997. It also provided that Hong Kong would become a special administrative region, with its existing social and economic structure unchanged for 50 years. It would remain a free port. The last British governor, Chris Patten (1992–97), introduced a legislative council. The handover to China was completed on July 1, 1997. Chief Executive Tung Chee-hwa was sworn in and a provisional legislative council appointed. Hong Kong is a vital international financial center with a strong manufacturing base.

dynasty (1271–1368), an era of dialog with Europe. The Ming dynasty (1368–1644) reestablished Chinese rule and is famed for its fine porcelain. The Manchu Qing dynasty (1644–1912) began by vastly extending the empire, but the 19th century was marked by foreign interventions, such as the Opium War (1839–42), when Britain occupied Hong Kong. Popular disaffection culminated in the Boxer Rebellion (1900). The last Emperor (Henry Pu Yi) was overthrown in a revolution led by Sun Yat-sen and a republic established (1912).

China rapidly fragmented between a Beijing government supported by warlords, and Sun Yat-sen's nationalist Kuomintang government in Guangzhou.

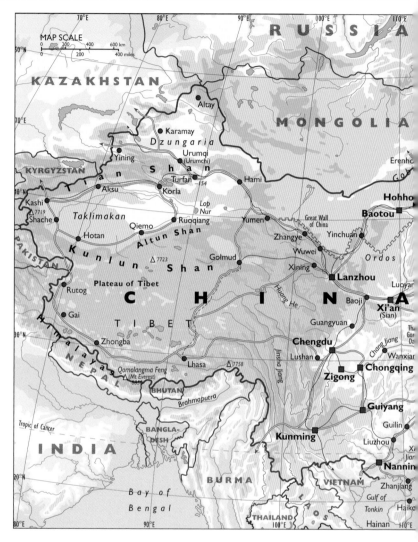

scale invasion in 1937, and conquered much of north and east China. From 1941 Chinese forces, with Allied support, began to regain territory. At the end of World War II, civil war resumed: nationalists supported by the USA and Communists by Russia. The Communists, with greater popular support, triumphed and the Kuomintang fled to Taiwan.

POLITICS

Mao Zedong established the People's Republic of China on October 1, 1949. In 1950 China seized Tibet. Domestically, Mao began to collectivize agriculture and nationalize industry. In 1958 the Great Leap Forward planned to revolutionize industrial production. The Cultural Revolution (1966–76) mobilized

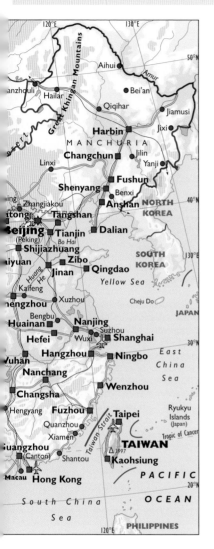

links with the West. In 1989 a prodemocracy demonstration was crushed in Tiananmen Square. In 1997 Jiang Zemin succeeded Deng as paramount leader and in 2002 was himself succeeded by Hu Jintao. Providing that the country's leaders continue to follow their pragmatic path, many experts predict a major economic blossoming for China in the 21st century.

ECONOMY

By 2005, China had the world's sixth largest economy. However China remains a poor country. Agriculture still employs nearly half of the people, although only 10% of the land is farmed.

The government announced in 2004 that it planned to slow down the country's rapid economic growth to help the rural poor, who had become relatively worse off as China's industries expanded.

MACAU (MACAO)

Former Portuguese overseas province in southeast China, 40 mi [64 km] west of Hong Kong, on the River Pearl estuary; it consists of the 2 sq mi [6 sq km] Macau Peninsula and the islands of Taipa and Colôane. Santa Nome de Deus de Macao connects via a narrow isthmus to the Chinese province of Guangzhou. Vasco da Gama discovered Macau in 1497, and the Portuguese colonized the island in 1557. In 1849, Portugal declared it a free port. In 1887, the Chinese government recognized Portugal's right of "perpetual occupation." Competition from Hong Kong and the silting of Macau's harbor led to the port's decline at the end of the 19th century. In 1974, Macau became a Chinese province under Portuguese administration. It returned to China in 1999. Gambling and tourism are dominant in the economy.

Chinese youth against bourgeois culture. By 1971 China had a seat on the UN security council and its own nuclear capability. Following Mao's death (1976), a power struggle developed between the Gang of Four and moderates led by Deng Xiaoping; the latter emerged victorious. Deng began a process of modernization, forging closer

COLOMBIA

The yellow on Colombia's flag depicts the land, which is separated from the tyranny of Spain by the blue, symbolizing the Atlantic Ocean. The red symbolizes the blood of the people who fought to make the country independent. The flag has been used since 1806.

Colombia is the only South American country to have coastlines on both the Pacific Ocean and the Caribbean Sea. Cartagena is the main Caribbean port. Colombia is dominated by three ranges of the Andean Mountains. On the edge of the western Cordillera lies the city of Cali. The Central Cordillera is a chain of volcanoes that divide the valleys of the rivers Magdalena and Cauca. It includes the city of Medellín. The eastern Cordillera contains the capital, Bogotá, at 9,200 ft [2,800 m]. East of the Andes lie plains drained by headwaters of the Amazon and Orinoco rivers.

Vegetation varies from dense rainforest in the southeast to tundra in the snow-capped Andes. Coffee plantations line the western slopes of the eastern Cordillera. The ancient forests of the Caribbean lowlands have mostly been cleared. Savanna (*llanos*) covers the northeastern plains.

Area 439,735 sq mi [1,138,914 sq km]
Population 42,311,000
Capital (population) Bogotá (6,545,000)
Government Multiparty republic
Ethnic groups Mestizo 58%, White 20%, Mulatto 14%, Black 4%
Languages Spanish (official)
Religions Roman Catholic 90%
Currency Colombian peso = 100 centavos
Website www.gobiernoenlinea.gov.co/ingles

CLIMATE

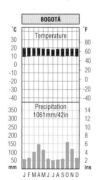

Altitude greatly affects the climate. The Pacific lowlands have a tropical, rainy climate, but Bogotá has mild annual temperatures. The lowlands of the Caribbean and the Magdalena valley both have dry seasons.

HISTORY

The pre-Colombian Chibcha civilization lived undisturbed in the eastern cordillera for thousands of years. In 1525 the Spanish established the first European settlement at Santa Marta. By 1538 conquistador Gonzalo Jiménez de Quesada conquered the Chibcha and established the city of Bogotá. Colombia became part of the New Kingdom of Granada, whose territory also included Ecuador, Panama, and Venezuela.

In 1819 Simón Bolívar defeated the Spanish at Boyacá, and established Greater Colombia. Bolívar became president. In 1830, Ecuador and Venezuela

MEDELLÍN

City in northwest central Colombia; capital of Antioquia department and the second-largest city in Colombia. Founded in the early 17th century by Spainish refugees, it became the center of Colombia's illegal cocaine trade. The surrounding region has gold and silver mines. Industries include food processing, coffee, chemicals, and steel. It has a population of 2,026,789.

(PSC) leader Andrés Pastrana Arango won the 1998 presidential elections and, in an effort to end the 30-year guerrilla war, negotiated with the Revolutionary Armed Forces of Colombia (FARC) and the National Liberation Army (ELN). Pastrana granted FARC a safe haven in southeast Colombia.

In 1999, the worst earthquake in Colombia's history killed more than 1,000 people and left thousands homeless. In 2002 Pastrana declared war on FARC, sending the army into FARC's "safe haven." Alvaro Uribe defeated Pastrana in 2002 presidential elections. Uribe promised even tougher action against terrorism.

ECONOMY

Colombia is a lower-middle income developing country. It is the world's second-largest coffee producer. Other crops include bananas, cocoa, and corn. Colombia also exports coal, petroleum, emeralds, and gold. In 1997 a collapse in the world coffee and banana markets led to a budget deficit. In 1998 Colombia devalued the peso, triggering the longest strike (20 days) in Colombia's history.

gained independence. In 1885, the Republic of Colombia was formed. Differences between republican and federalist factions proved irreconcilable and the first civil war from 1899–1902 killed nearly 100,000 people. In 1903, aided by the United States, Panama gained independence. The second civil war, La Violencia, from 1949–57 was even more bloody.

In 1957 Liberal and Conservative parties formed a National Front Coalition, which held power until 1974. Throughout the 1970s, Colombia's illegal trade in cocaine grew steadily, creating wealthy drug barons. In the 1980s, armed cartels (such as the Cali) destabilized Colombia with frequent assassinations of political and media figures.

A new constitution in 1991 protected human rights. Social Conservative Party

BOGOTÁ

Capital of Colombia, in the center of the country on a fertile plateau. Bogotá was founded in 1538 by the Spanish on the site of a Chibcha Indian settlement. In 1819 it became the capital of Greater Colombia, part of which later formed Colombia. Today, it is a center for culture, education, and finance. Industries include tobacco, sugar, flour, textiles, engineering, and chemicals.

CONGO

Congo's red flag, with the national emblem of a crossed hoe and mattock (a kind of pickax), was dropped in 1990, when the country officially abandoned the Communist policies it had followed since 1970. This new flag was adopted in its place.

The Republic of Congo lies on the River Congo in west-central Africa. The Equator runs through the center of the country. Congo has a narrow coastal plain on which stands its main port, Pointe Noire, which itself lies on the Gulf of Guinea. Behind the plain are forested highlands through which the River Niari has carved a fertile valley. To the east lies Malebo (formerly Stanley) Pool, a large lake where the River Congo widens.

Central Congo consists of luxuriant savanna. Tree species include the valuable okoumé and mahogany. The north contains large swamps in the tributary valleys of the Congo and Ubangi rivers..

Area 132,046 sq mi [342,000 sq km]
Population 2,998,000
Capital (population) Brazaville (938,000)
Government Military regime
Ethnic groups Kongo 48%, Sangha 20%, Teke 17%, M'bochi 12%
Languages French (official), many others
Religions Christianity 50%, animist 48%, Islam 2%
Currency CFA franc = 100 centimes
Website www.congo-site.com

CLIMATE

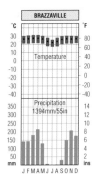

Most of the country has a humid, equatorial climate, with rain throughout the year. Brazzaville has a dry season between June and September. The narrow treeless coastal plain is drier and cooler than the rest of the country, because the cold Benguela current flows northward along the coast.

HISTORY

The Loango and Bakongo kingdoms dominated the Congo when the first European arrived in 1482. Between the 15th and 18th centuries, part of Congo belonged to the huge Kongo kingdom, whose center lay to the south. Portuguese explorers reached the coast of Congo in the 15th century and the area soon became a trading region, the main commodities being slaves and ivory. The slave trade continued until the 19th century.

European exploration of the interior did not occur until the late 19th century. In 1880 Pierre Savorgnan de Brazza

RIVER CONGO

The River Congo (formerly the River Zaïre) in central and west Africa; the second-longest in the continent. It rises in southern Democratic Republic of Congo and flows in a massive curve to the Atlantic Ocean for 2,900 mi [4,670 km]. Its rate of flow and size of drainage basin make it Africa's largest untapped source of hydroelectric power. The chief ocean port is Matadi, and the major river ports are Kinshasa and Kisangani. The main headstream is the Lualaba, and the Kasai and Ubangi are among its many large tributaries.

ever, in 1997, Sassou-Nguesso, assisted by his personal militia and also by troops from Angola, launched an uprising which overthrew Lissouba, who fled the country, taking refuge in Burkina Faso. But forces loyal to Lissouba fought back, starting a civil war. Cease-fires were agreed in 1999 and, in 2002, Sassou-Nguesso was elected president, winning 89% of the vote. A peace accord was signed in 2003.

ECONOMY

The World Bank classifies Congo as a "lower-middle-income" developing country. Agriculture is the most important activity, employing about 60% of the workforce. But many farmers function merely at a subsistence level. The chief food crops include bananas, cassava, corn, plantains, rice and tropical yams, while the leading cash crops include cocoa, coffee, and sugarcane.

Congo's main exports are petroleum (which makes up 90% of the total) and timber. Manufacturing is relatively unimportant at the moment, hampered as it is by poor transportation links. Inland, rivers form the main lines of communication, and Brazzaville is linked to the port of Pointe-Noire by the Congo-Ocean Railroad.

explored the area and it became a French protectorate. It became known as Middle Congo, a country within French Equatorial Africa, which also included Chad, Gabon, and Ubangi-Shari (now called Central African Republic). In 1910 Brazzaville became the capital of French Equatorial Africa. In 1960 the Republic of Congo gained independence.

POLITICS

In 1964 Congo adopted Marxism-Leninism as the state ideology. The military, led by Marien Ngouabi, seized power in 1968. Ngouabi created the Congolese Workers Party (PCT) and was assassinated in 1977. The PCT retained power under Colonel Denis Sassou-Nguesso. In 1990 it renounced Marxism and Sassou-Nguesso was deposed. The Pan-African Union for Social Democracy (UPADS), led by Pascal Lissouba, won multiparty elections in 1992. How-

BRAZZAVILLE

Capital and largest city of the Congo, West Africa, on the River Congo, below Stanley Pool. Founded in 1880, it was capital of French Equatorial Africa (1910–58) and a base for Free French forces in World War II. It has a university (1972) and a cathedral. It is a major port, connected by rail to the main Atlantic seaport of Pointe-Noire.

CONGO (DEMOCRATIC REPUBLIC OF)

The Democratic Republic of Congo adopted a new flag in 2006. The blue symbolizes peace, the red line is to recall the blood of the millions whose deaths were caused by the war and the yellow is for the mining of deposits of Democratic Republic of Congo.

Democratic Republic of Congo is Africa.".s third-largest country. It is dominated by the River Congo. North-central Congo consists of a high plateau. In the east, the plateau rises to 16,762 ft [5,109 m] in the Ruwenzori mountains. Lakes Albert and Edward form much of Congo's border with Uganda. Lake Kivu lies along its border with Rwanda. Lake Tanganyika forms the border with Tanzania. All the lakes lie in an arm of the Great Rift Valley. Dense equatorial rainforests grow in the north, with savanna and swamps in the south.

Area 905,350 sq mi [2,344,858 sq km]
Population 58,318,000
Capital (population) Kinshasa (4,665,000)
Government Single-party republic
Ethnic groups Over 200; the largest are Mongo, Luba, Kongo, Mangbetu-Azande
Languages French (official), tribal languages
Religions Roman Catholic 50%, Protestant 20%, Islam 10%, others
Currency Congolese franc = 100 centimes
Website www.monuc.org

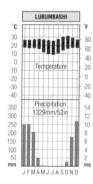

CLIMATE

Much of the country has an equatorial climate with high temperatures and heavy rainfall throughout the year. The south has a more subtropical climate.

HISTORY

From the 14th century, large Bantu kingdoms emerged. In 1482, a Portuguese navigator became the first European to reach the mouth of the River Congo. In the 19th century, ivory and slave traders formed powerful states. Henry Morton Stanley's explorations from 1874–77 established the route of the Congo. In 1878, King Leopold II of Belgium employed Stanley to found colonies along the Congo. In 1885, Leopold established the Congo Free State. His empire grew, and concessionaires gained control of the lucrative rubber trade. In 1908, Belgium established direct control as the colony of Belgian Congo. European companies exploited African labor to develop copper and diamond mines.

POLITICS

In 1960, the Republic of the Congo gained independence and Patrice Lumumba became prime minister. Joseph Mobutu, commander in chief of the Congolese National Army, seized power later that year. Lumumba was imprisoned and later murdered. In 1964, Belgian Congo plunged into civil war and Belgian troops intervened. In 1965, Mobutu proclaimed himself president and began a campaign of "Africanization:" Leopoldville became Kinshasa in 1966; the country and river were renamed Zaïre in 1971; Katanga became Shaba in 1972; and Mobutu adopted the name Mobutu Sese Seko. Zaïre became a one-party state. Mobutu was reelected unopposed in 1974 and 1977, finally accepting opposition parties in 1990, though elections were repeatedly deferred. In 1995, millions of Hutus fled from Rwanda into east Zaïre, to escape

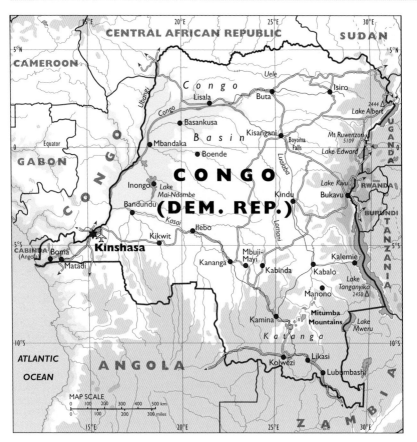

possible Tutsi reprisals. In 1996 rebels, led by Laurent Kabila, overthrew Mobuto. Zaïre became the Democratic Republic of Congo. In 1998, Congo descended into civil war between government forces and Tutsi-dominated Congolese Rally for Democracy (RCD). The Lusaka Peace Agreement (1999) brought a cease-fire and 5,500 UN peacekeeping troops, but fighting continued. By 2001 the civil war had claimed more than 2.5 million lives. In 2001, Kabila was assassinated. He was succeeded by his son Joseph Kabila who, under a peace agreement, was installed as interim president of a transitional government in 2003. Unrest continued into 2005.

ECONOMY

DR Congo is a low-income developing country. It is the world's leading producer of cobalt and the second-largest producer of diamonds. Agriculture employs 71% of the workforce, mainly at subsistence level. Palm oil is the most vital cash crop.

KINSHASA

Capital of Democratic Republic of Congo, a port on the River Congo. Founded in 1881, it replaced Boma as the capital of the Belgian Congo in 1923. An important music center.

COSTA RICA

Costa Rica's flag is based on the blue-white-blue pattern used by the Central American Federation (1823–39). This Federation consisted of Costa Rica, El Salvador, Guatemala, Honduras, and Nicaragua. The red stripe, which was adopted in 1848, reflects the colors of France.

The Republic of Costa Rica in Central America is bordered by Nicaragua to the north and Panama to the south. It has coastlines on both the Pacific Ocean and on the Caribbean Sea. Central Costa Rica consists of mountain ranges and plateaus with many volcanoes. The Meseta Central, where the capital, San José is situated, and the Valle del General in the southeast, have rich, volcanic soils and are the most densely populated parts of Costa Rica.

The highlands descend to the Caribbean lowlands and the Pacific Coast region, with its low mountain ranges. San José stands at about 3,840 ft [1,170 m] above sea level.

Evergreen forests cover around 50% of Costa Rica. Oaks grow in the highlands, palm trees along the Caribbean coast, and mangrove swamps are common on the Pacific coast.

Area 19,730 sq mi [51,100 sq km]
Population 3,957,000
Capital (population) San José (337,000)
Government Multiparty republic
Ethnic groups White (including Mestizo) 94%, Black 3%, Amerindian 1%, Chinese 1%, others
Languages Spanish (official), English
Religions Roman Catholic 76%, Evangelical 14%
Currency Costa Rican colón = 100 céntimos
Website www.visitcostarica.com

HISTORY

Christopher Columbus reached the Caribbean coast in 1502 and named the land Costa Rica, Spanish for "rich coast." Rumors of treasure attracted many Spaniards to settle in the country from 1561.

Spain ruled the country until 1821, when Spain's Central American colonies broke away to join the Mexican empire in 1822. In 1823, the Central American states broke with Mexico and set up the Central American Federation, the members being Costa Rica, Guatemala, Honduras, Nicaragua, and El Salvador. Later, this large union broke up and Costa Rica

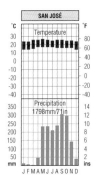

CLIMATE

The Meseta Central benefits from a pleasant climate and an average annual temperature of 68°F [20°C], compared with more than 81°F [27°C] on the coast. The coolest months are December and January. The northeast trade winds bring heavy rain to the Caribbean coast. There is only half as much rainfall in the highlands and on the Pacific coastlands as occurs on the Caribbean coast.

SAN JOSÉ

Capital and largest city of Costa Rica, in central Costa Rica, capital of San José province. Founded around 1736, it succeeded Cartago as capital in 1823, and soon became the center of a prosperous coffee trade. Products include coffee, sugarcane, cacao, vegetables, fruit, and tobacco.

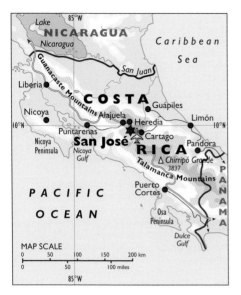

perous countries in Central America. There are high educational standards, and a high average life expectancy of 78 years.

Agriculture employs 24% of the workforce. Major crops include coffee, pineapples, bananas, and sugar. Other crops include beans, citrus fruits, cocoa, potatoes, rice, and corn. Cattle ranching is important.

The country's resources include its forests, but it lacks minerals apart from some bauxite and manganese. Manufacturing is increasing with electronics to the fore. It also has hydroelectric power.

Tourism is a fast-growing industry with ecotourism gaining in importance. The United States is Costa Rica's chief trading partner.

became fully independent in 1838. From the late 19th century, Costa Rica experienced a number of revolutions, with periods of both dictatorship and democracy. In 1917–19 General Tinoco formed a dictatorship, then in 1948, following a revolt, the armed forces were abolished.

POLITICS

Jose Figueres Ferrer served as president from 1953 to 1958 and again from 1970 to 1974. In 1987 President Oscar Arias Sanchez was awarded the Nobel Peace Prize for his efforts to end the civil wars in Central America.

In 2002 Abel Pancho won the presidential elections. Costa Rica's image was tarnished in 2004 when three former presidents were imprisoned on charges of corruption. Despite this, Costa Rica is seen as an example of political stability in the region and continues to maintain this without having to resort to armed forces.

ECONOMY

Costa Rica is classified by the World Bank as a "lower-middle-income" developing country. It is one of the most pros-

COSTA RICAN BUTTERFLIES

In Costa Rica the butterfly is omnipresent, whether on the plains or at a volcanic summit. Of the 20,000 butterfly species that exist worldwide about 1,000 or 5% can be found in Costa Rica. The butterfly is a day-flying insect of the order Lepidoptera. The adult has two pairs of scale-covered wings that are often brightly colored. The female lays eggs on a selected food source and the caterpillar larvae emerge within days or hours. The larvae have chewing mouthparts and often do great damage to crops until they reach the "resting phase" of the life cycle, the pupa (chrysalis). Within the pupa, the adult (imago) is formed with wings, wing muscles, antennae, a slender body, and sucking mouth parts. The adults mate soon after emerging from the chrysalis, and the four-stage life cycle begins again.

CROATIA

Croatia adopted a red, white, and blue flag in 1848. Under Communist rule, a red star appeared at the center. In 1990, the red star was replaced by the present coat of arms, which symbolizes the various parts of the country.

The Republic of Croatia was part of Yugoslavia until becoming independent in 1991. The region bordering the Adriatic Sea is called Dalmatia. It includes the coastal ranges, which contain large areas of bare limestone, reaching 6,276 ft [1,913 m] at Mount Troglav. Other highlands lie in the northeast. Most of the rest of the country consists of the fertile Pannonian Plains, which are drained by Croatia's two main rivers, the Drava and the Sava.

Area 21,829 sq mi [56,538 sq km]
Population 4,497,000
Capital (population) Zagreb (779,000)
Government Multiparty republic
Ethnic groups Croat 90%, Serb 5%, others
Languages Croatian 96%
Religions Roman Catholic 88%, Orthodox 4%, Islam 1%, others
Currency Kuna = 100 lipas
Website www.vlada.hr/default.asp?ru=2

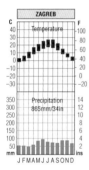

CLIMATE

The coastal area has a climate akin to that of the Mediterranean, with hot, dry summers and mild, moist winters. Inland, the climate becomes more continental. Winters are cold, while temperatures often soar to 100°F [38°C] in the summer months.

HISTORY

Slav people settled in the area around 1,400 years ago. In 803, Croatia became part of the Holy Roman Empire. In 1102, the king of Hungary also became king of Croatia, creating a union that lasted 800 years. In 1526, much of Croatia and Hungary came under the Ottoman Turks. At about the same time, the Austrian Habsburgs gained control of the rest of Croatia. In 1699, the Habsburgs drove out the Turks and Croatia again came under Hungarian rule. In 1809,

Croatia became part of the Illyrian provinces of Napoleon I of France, but the Habsburgs took over in 1815.

In 1867, Croatia became part of the dual monarchy of Austria-Hungary and in 1868 Croatia signed an agreement with Hungary guaranteeing Croatia some of its historic rights. During World War I, Austria-Hungary fought on the side of the defeated Axis powers, and, in 1918, the empire was broken up. Croatia declared its independence and joined with neighboring states to form the Kingdom of the Serbs, Croats and Slovenes. Serbian domination provoked Croatian opposition.

In 1929, the king changed the country's name to Yugoslavia and began to rule as a dictator. He was assassinated in 1934 by a Bulgarian employed by a Croatian terrorist group, provoking more hostility between Croats and Serbs.

Germany occupied Croatia in World War II. After the war, Communists took power in Yugoslavia, with Josip Broz Tito as leader. After Tito's death in 1980, economic and ethnic rivalries threatened stability. In the early 1990s, Yugoslavia split into five nations. One of them, Croatia, declared itself independent in 1991.

announced that it would prosecute suspected war criminals and cooperate with the war crimes tribunal in The Hague.

ECONOMY

The wars of the 1990s disrupted Croatia's economy. Tourism on the Dalmatian coast had been a major industry and is making a gradual return. The manufacturing industries are the chief exports. Manufactures include cement, chemicals, refined petroleum and petroleum products, ships, steel, and wood products.

Agriculture is important and major farm products include fruits, livestock, corn, soybeans, sugarbeet, and wheat.

POLITICS

After Serbia supplied arms to Serbs living in Croatia, war broke out between the two republics, causing great damage, large-scale movements of refugees and disruption of the economy, including the vital tourist industry.

In 1992, the United Nations sent a peacekeeping force to Croatia, effectively ending the war with Serbia. However, in 1992, war broke out in Bosnia-Herzegovina and Bosnian Croats occupied parts of the country. In 1994, Croatia helped to end the Croat-Muslim conflict in Bosnia-Herzegovina and, in 1995, after retaking some areas occupied by Serbs, it contributed to the drawing up of the Dayton Peace Accord, which ended the civil war.

Croatia's arch-nationalist president, Franco Tudjman, died in December 1999. In January 2000, Tudjman's Croatian Democratic Union was defeated in a general election by a more liberal, westward-leaning alliance of Social Democrats and Social Liberals. Stipe Mesic, the last head of state of the former Yugoslavia before it disintegrated in 1991, was elected president. In 2000, the government

ZAGREB

Capital of Croatia, on the River Sava. Founded in the 11th century, it became capital of the Hungarian province of Croatia and Slavonia during the 14th century. The city was an important center of the 19th-century Croatian nationalist movement. In 1918 it was the meeting place of the Croatian Diet (parliament), which severed all ties with Austria-Hungary. In World War II, Zagreb was the capital of the Axis-controlled, puppet Croatian state. It was wrested from Axis control in 1945, and became capital of the Croatian Republic of Yugoslavia. Following the breakup of Yugoslavia in 1992, Zagreb remained capital of the newly independent state of Croatia. The city has many places of historical interest, including a Gothic cathedral and a Baroque archiepiscopal palace. Zagreb has a university (founded 1669) and an Academy of Arts and Sciences (1861).

CUBA

Cuba's flag, the "Lone Star" banner, was designed in 1849, but it was not adopted as the national flag until 1901, after Spain had withdrawn from the country. The red triangle represents the Cuban people's bloody struggle for independence.

The Republic of Cuba is the largest island country in the Caribbean Sea. It consists of one large island, Cuba, the Isle of Youth (Isla de la Juventud) and about 1,600 small islets.

Mountains and hills cover about a quarter of Cuba. The highest mountain range, the Sierra Maestra in the southeast, reaches 6,562 ft [2,000 m] above sea level at the Pico Real del Turquino. The rest of the land consists of gently rolling country or coastal plains, crossed by fertile valleys that have been carved by the short, mostly shallow and narrow rivers.

Farmland covers about half of Cuba and 66% of this is given over to sugarcane. Pine forests still grow, especially in the southeast. Mangrove swamps line some coastal areas.

Area 42,803 sq mi [110,861 sq km]
Population 11,309,000
Capital (population) Havana (2,192,000)
Government Socialist republic
Ethnic groups Mulatto 51%, White 37%, Black 11%
Languages Spanish (official)
Religions Christianity
Currency Cuban peso = 100 centavos
Website www.cubagob.gov.cu/ingles/

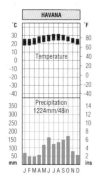

CLIMATE

Cuba lies in the tropics. But sea breezes moderate the temperature, warming the land in winter and cooling it in summer.

HISTORY

In 1492 Christopher Columbus discovered the island and Spaniards began to settle there from 1511. Spanish rule ended in 1898, when the United States defeated Spain in the Spanish-American War. The United States ruled Cuba from 1898 until 1902, when the people elected Tomás Estrada Palma as president of the independent Republic of Cuba, though American influence remained strong. In 1933, an army sergeant named Fulgencio Batista seized power and ruled as dictator. However, under a new constitution, he was elected president in 1940, serving until 1944. He again seized power in 1952 and became dictator once more, but, on January 1, 1959, he fled Cuba following the overthrow of his regime by a revolutionary force led by a young lawyer, Fidel Castro. Many Cubans who were opposed to Castro left the country, settling in the United States.

POLITICS

The United States opposed Castro's policies, so he turned to the Soviet Union for assistance. In 1962, the US learned that nuclear missile bases armed by the Soviet Union had been established in Cuba. The US ordered the Soviet Union to remove the missiles and bases. After a few days, during which many people feared that a world war might break out, the Soviet Union agreed to American demands.

Cuba's relations with the Soviet Union remained strong until 1991, when the Soviet Union was broken up. The loss

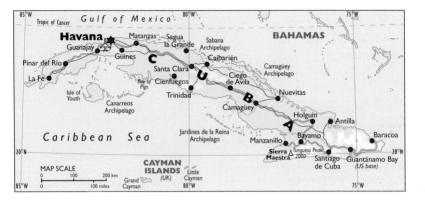

of Soviet aid greatly damaged Cuba's economy and the new situation undermined Castro's considerable social achievements. However, in February 1993, elections showed a high level of support for his left-wing policies. In 1998, hopes of a thaw in relations with the United States were raised when the US government announced that it was lifting the ban on flights to Cuba. The Pope, making his first visit to Cuba, criticized the "unjust and ethically unacceptable" US blockade on Cuba. In 2000, the United States lifted its food embargo on Cuba. The last Russian base in Cuba closed in 2002. In 2004, following a United States crackdown on currency and travel, Cuba declared that US dollars would no longer be accepted as payments for goods and services.

ECONOMY

The World Bank classifies Cuba as a "lower-middle-income" country. Sugar cane remains Cuba's outstandingly important cash crop, accounting for more than 60% of the country's exports. It is grown on more than half of the island's cultivated land and Cuba is one of the world's top ten producers of the product. Before 1959, the sugarcane was grown on large estates, many of them owned by US companies. Following the Revolution, they were nationalized and the Soviet Union and Eastern European countries replaced the United States as the main market. The other main crop is tobacco, which is grown in the northwest. Cattle raising, milk production, and rice cultivation have also been encouraged to help diversify the economy, and the Castro regime has devoted considerable efforts to improving the quality of rural life, making standards of living more homogeneous throughout the island.

Minerals and concentrates rank second to sugar among Cuba's exports, followed by fish products, tobacco and tobacco products, including the famous cigars, and citrus fruits. In the 1990s, Cuba sought to increase its trade with Latin America and China. Tourism is a major source of income, but the industry was badly hit following the terrorist attacks on the United States in 2001.

HAVANA (LA HABANA)

Capital of Cuba, on the northwest coast. It is the largest city and port in the West Indies. Havana was founded by the Spanish explorer Diego Velázquez in 1515, and moved to its present site in 1519. It became Cuba's capital at the end of the 16th century. Industries include petroleum refining, textiles, sugar, and cigars.

CYPRUS

This flag became the official flag when the country became independent from Britain in 1960. It shows an outline map of the island, with two olive branches. Since Cyprus was divided, the separate communities have flown the Greek and Turkish flags.

The Republic of Cyprus is an island nation which lies in the northeastern Mediterranean Sea. Geographers regard it as part of Asia, but it resembles southern Europe in many ways. Cyprus has scenic mountain ranges, including the Kyrenia Range in the north and the Troodos Mountains in the south, which rise to 6,401 ft [1,951 m] at Mount Olympus.

The island also contains fertile lowlands used extensively for agriculture, including the broad Mesaoria Plain. Pine forests grow on the mountain slopes.

Area 3,572 sq mi [9,251 sq km]
Population 776,000
Capital (population) Nicosia (198,000)
Government Multiparty republic
Ethnic groups Greek Cypriot 77%, Turkish Cypriot 18%, others
Languages Greek and Turkish (both official), English
Religions Greek Orthodox 78%, Islam 8%
Currency Cypriot pound = 100 cents
Website www.cyprus.gov.cy

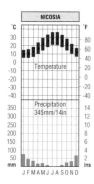

CLIMATE

Cyprus experiences hot, dry summers and mild, wet winters. Summers are hotter than those further west in the Mediterranean as Cyprus lies close to the hot mainland of southwestern Asia.

HISTORY

The history of Cyprus dates back to 7000 BC. Greeks settled on Cyprus around 3,200 years ago. By 1050 BC Cyprus was fully established as a Greek island having embraced the language and culture of Greece. In 333 BC it became part of the empire of Alexander the Great and in 58 BC part of the Roman Empire. From AD 330, the island was under the Byzantine empire.

Cyprus was defeated in 1191 by Richard the Lionheart and the island was sold to the Knights Templar. Catholicism became the official religion. The island was under Venetian control from 1489 and fortifications were added to the towns of Nicosia and Famagusta. In the 1570s, it became part of the Turkish Ottoman Empire and Islam was introduced.

FAMAGUSTA

A fortified port on the east coast of Northern Cyprus. Famagusta was founded by the King Ptolemy II in the 3rd century BC. Christian refugees fleeing the Holy Lands in 1291 transformed it into a rich city . In 1571 it fell to the Ottoman Empire. The old walled town is now only partially inhabited. Under the British from 1878-1960 it developed as a tourist resort. In 1974 it came under siege by Greek Cypriots. 11,000 Turkish Cypriots defended the city until the arrival of Turkish troops. Nearby, stands the ancient city of Salamis, a spectacular archaeological site.

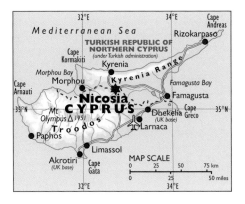

Turkish rule continued until 1878 when Cyprus was leased to Britain although it was still part of the Ottoman Empire. When the Ottomans entered World War I in 1914, on the side of Germany, the island was annexed by Britain. It was proclaimed a Crown Colony in 1925.

In the 1950s, Greek Cypriots, who made up four-fifths of the population, led by Greek Orthodox Archbishop Makarios, began a campaign for enosis (union) with Greece. A secret guerrilla force called EOKA attacked the British who exiled Makarios.

POLITICS

Cyprus became an independent country in 1960 with Makarios as president. Britain retained two military bases. The constitution of Cyprus provided for power-sharing between the Greek and Turkish Cypriots. But the constitution proved unworkable and fighting broke out. In 1964 the UN sent in a peace-keeping force.

In 1974, Cypriot forces led by Greek officers overthrew Makarios. This led Turkey to invade northern Cyprus, a territory occupying about 40% of the island. Many Greek Cypriots fled from the north, which, in 1983, was proclaimed an independent state called the Turkish Republic of Northern Cyprus. However, the only country to recognize

its status was Turkey. The UN regards Cyprus as a single nation under the Greek-Cypriot government in the south. It is estimated that more than 30,000 Turkish troops are deployed in northern Cyprus. Despite UN-brokered peace negotiations, there are still frequent border clashes between the two communities.

In 2002, the European Union invited Cyprus to become a member. In April 2004, the people voted on a UN plan to reunify the island. The Turkish Cypriots voted in favor of the plan, but the Greek Cypriots voted against. As a result of this, only the south was admitted to membership of the EU on May 1, 2004.

ECONOMY

Cyprus got its name from the Greek word kypros, meaning copper, but little copper remains. The chief minerals are asbestos and chromium. The most valuable activity in Cyprus is tourism.

Industry employs 37% of the workforce and manufactures include cement, clothes, footwear, tiles, and wine. In the early 1990s, the United Nations reclassified Cyprus as a developed rather than developing country, though the economy of the Turkish-Cypriot north lags behind that of the more prosperous Greek-Cypriot south.

NICOSIA (LEVKOSÍA)

Capital of Cyprus, in the center of the island. Known to the ancients as Ledra, the city was later held by Byzantines, French crusaders, and Venetians. The Ottoman Turks occupied the city from 1571 to 1878, when it passed to Britain. It is now divided into Greek and Turkish sectors. Industries: cigarettes, textiles, and footwear.

CZECH REPUBLIC

After independence, on January 1, 1993, the Czech Republic adopted the former flag of Czechoslovakia. It features the red and white of Bohemia in the west, together with the blue of Moravia and Slovakia. Red, white and blue are the colors of Pan-Slavic liberation.

The Czech Republic is the western three-fifths of the former country of Czechoslovakia. It contains two regions: Bohemia in the west and Moravia in the east. Mountains border much of the country in the west. The Bohemian basin in the north-center is a fertile lowland region, with Prague, the capital city, as its main center. Highlands cover much of the center of the country, with lowlands in the southeast. Some rivers, such as the Elbe (Labe) and Oder (Odra) flow north into Germany and Poland. In the south, rivers flow into the Danube Basin.

Area 78,866 sq mi [30,450 sq km]
Population 10,246,000
Capital (population) Prague (1,193,000)
Government Multiparty republic
Ethnic groups Czech 81%, Moravian 13%, Slovak 3%, Polish, German, Silesian, Gypsy, Hungarian, Ukrainian
Languages Czech (official)
Religions Atheist 40%, Roman Catholic 39%, Protestant 4%, Orthodox 3%, others
Currency Czech koruna = 100 haler
Website www.czech.cz

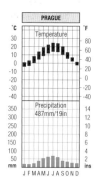

CLIMATE

The climate of the Czech Republic is influenced by its landlocked position in east-central Europe. The country experiences a humid continental climate, with warm summers and cold winters. The average rainfall is moderate, with 20 in to 30 in [500 mm to 750 mm] annually in lowland areas.

HISTORY

The ancestors of the Czech people began to settle in what is now the Czech Republic around 1,500 years ago. Bohemia, in the west, became important in the 10th century as a kingdom within the Holy Roman Empire. By the 14th century, Prague was one of Europe's major cultural cities. Religious wars in

the first half of the 15th century led many Czech people to become Protestant. From 1526, the Roman Catholic Habsburgs from Austria began to rule the area, but, in 1618, a Czech Protestant rebellion started the Thirty Years' War. From 1620, most Czechs were made to convert to Catholicism and adopt German as their language.

Czech nationalism grew throughout the 19th century. During World War I, Czech nationalists advocated the creation of an independent nation. At the end of the war, when Austria-Hungary collapsed, the new republic of Czechoslovakia was founded. The 1920s and 1930s were generally a period of stability and economic progress, but problems arose concerning the country's minority groups. Many Slovaks wanted a greater degree of self-government, while Germans living in the Sudetenland, in western Czechoslovakia, were unhappy under Czech rule.

In 1938, Sudetenland was turned over to Germany and, in March 1939, Germany occupied the rest of the country. By 1945, following the Nazi defeat, a coali-

tion government, including Czech Communists, was formed to rule the country. In 1948, Communist leaders seized control and made the country an ally of the Soviet Union in the Cold War. In 1968, the Communist government introduced reforms, which were known as the "Prague spring." However, Russian and other East European troops invaded and suppressed the reform group.

POLITICS

When democratic reforms were introduced in the Soviet Union in the 1980s, the Czechs also demanded change. In 1989, the Federal Assembly elected Václav Havel, a noted playwright and dissident, as the country's president and, in 1990, free elections were held. The smooth transition from Communism to democracy was called the "Velvet Revolution." The road to a free-market economy was not easy, with resulting inflation, falling production, strikes, and unemployment, though tourism has partly made up for some of the economic decline. Political problems also arose when Slovaks began to demand independence. Finally, on January 1, 1993, the more statist Slovakia broke away from the free-market Czech Republic. However, the split was generally amicable and border adjustments were negligible. The Czechs and Slovaks maintained a customs union and other economic ties. Meanwhile the Czech government continued to develop ties with Western Europe when it became a member of NATO in 1992. On May 1, 2004 the Czech Republic became a member of the European Union.

ECONOMY

Under Communist rule the Czech Republic became one of the most industrialized parts of Eastern Europe. The country has deposits of coal, uranium, iron ore, magnesite, tin, and zinc. Manufactures include such products as chemicals, iron and steel, and machinery, but the country also has light industries making such things as glassware and textiles for export. Manufacturing employs about 40% of the Czech Republic's entire workforce.

Farming is important. The main crops include barley, fruit, hops for beer-making, corn, potatoes, sugarbeet, vegetables, and wheat. Cattle and other livestock are raised. The country was admitted into the Organization for Economic Cooperation and Development (OECD) in 1995.

PRAGUE (PRAHA)

Capital of the Czech Republic, on the River Vltava. Founded in the 9th century, it grew rapidly after Wenceslaus I established a German settlement in 1232. In the 14th century it was the capital of Bohemia. It was the capital of the Czechoslovak Republic (1918–93). Occupied in World War II by the Germans it was liberated by Soviet troops in 1945. Prague was the center of Czech resistance to the Soviet invasion of 1968. Sights include Hradcany Castle and Charles Bridge. An important commercial center, industries include engineering, iron, and steel.

DENMARK

Denmark's flag is called the Dannebrog, or "the spirit of Denmark." It may be the oldest national flag in continuous use. It represents a vision thought to have been seen by the Danish King Waldemar II before the Battle of Lyndanisse, which took place in Estonia in 1219.

The Kingdom of Denmark is the smallest country in Scandinavia. It consists of a peninsula called Jutland (Jylland), which is joined to Germany, and more than 400 islands, 89 of which are inhabited. The land is flat and mostly covered by rocks dropped there by huge ice sheets during the last Ice Age. The highest point in Denmark is on Jutland and is only 568 ft [173 m].

Area 16,639 sq mi [43,094 sq km]
Population 5,413,000
Capital (population) Copenhagen (488,000)
Government Parliamentary monarchy
Ethnic groups Scandinavian, Inuit, Faeroese, German
Languages Danish (official), English, Faerose
Religions Evangelical Lutheran 95%
Currency Danish krone = 100 øre
Website http://denmark.dk

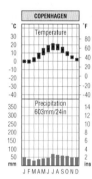

CLIMATE

Denmark has a cool but pleasant climate. During cold spells in the winter The Sound between Sjælland and Sweden may freeze over. Summers are warm. Rainfall occurs throughout the year.

HISTORY

Danish Vikings terrorized much of Western Europe for about 300 years after AD 800. Danish kings ruled England in the 11th century. Control of the entrances to the Baltic Sea contributed to the power of Denmark in the Middle Ages, when the kingdom dominated its neighbors and expanded its territories to include Norway, Iceland, Greenland, and the Faroe Islands. The link with Norway was broken in 1814, and with Iceland in 1944. But Greenland and the Faroes retained connections with Denmark. The granite island of Bornholm, off the southern tip of Sweden, also remains a Danish possession. This island was occupied by

Germany in World War II, but it was liberated by the Soviet Union and returned to Denmark in 1946. Denmark was also occupied by Germany in 1940, but it was liberated in 1945. The Danes then set about rebuilding their industries and restoring their economy.

POLITICS

Denmark is a generally comfortable mixture of striking political opposites. The Lutheran tradition and the cradle of Hans Christian Andersen's fairy tales coexist with open attitudes to pornography and one of the highest illegitimacy rates in the West.

The country is one of the "greenest" of the developed nations, with a pioneering Ministry of Pollution. In 1991, it became the first government anywhere to fine industries for emissions of carbon dioxide, the primary "greenhouse" gas.

It joined the North Atlantic Treaty Organization (NATO) in 1949, and in 1973 it joined the European Community (now the European Union). However, it remains one of the European Union's least enthusiastic members and was one of the four countries that did not adopt the euro, the single EU currency, on

ECONOMY

Denmark has few mineral resources, though there is now some petroleum and natural gas from the North Sea. It is one of Europe's wealthiest industrial nations. Farming employs only 4% of workers, but it is highly scientific and productive with dairy farming and pig and poultry breeding chief areas.

From a firm agricultural base, Denmark has developed a wide range of industries. Some, including brewing, meat canning, fish processing, pottery, textiles, and furniture making, use Danish products, but others, such as shipbuilding, petroleum refining, engineering, and metal-working, depend on imported raw materials. Copenhagen is the chief industrial center and draws more than a million tourists each year. At the other end of the scale is Legoland, the famous miniature town of plastic bricks, built at Billand, northwest of Vejle in eastern Jutland. It was here that Lego was created before it became the world's bestselling construction toy and a prominent Danish export.

January 1, 2002. In 1972, in order to join the EC, Denmark had become the first Scandinavian country to break away from the other major economic grouping in Europe, the European Free Trade Association (EFTA), but it continued to cooperate with its five Scandinavian partners through the consultative Nordic Council which was set up in 1953.

The Danes enjoy some of the world's highest living standards, although the cost of welfare provisions was high. The election of a Liberal-Conservative coalition in 2001 led to cutbacks. Under Prime Minister Anders Fogh Rasmussen, who won a second term in 2005, the government also tightened immigration controls, causing criticism by the UN High Commissioner for Refugees.

Denmark granted home rule to the Faeroe Islands in 1948, although in 1998, the government of the Faeroes announced plans for independence. In 1979, home rule was also granted to Greenland, which demonstrated its newfound independence by withdrawing from the European Community in 1985. Denmark is a constitutional monarchy, with a hereditary monarch, and its constitution was amended in 1953 to allow female succession to the throne.

COPENHAGEN (KØBENHAVN)

Capital and chief port of Denmark on east Sjaelland and north Amager Island, in the Øresund. A trading and fishing center by the early 12th century, it became Denmark's capital in 1443. It has a 17th-century stock exchange, the Amalienborg palace (home of the royal family) and the Christiansborg Palace. Other sights include the Tivoli Amusement Park and the Little Mermaid sculpture. The commercial and cultural center of the nation, it has shipbuilding, chemical and brewing industries.

DENMARK

FAEROE ISLANDS

The Faeroe (or Faroe) Islands are an autonomous region of Denmark. Situated in the North Atlantic Ocean between Iceland and the Shetland Islands, the region consists of 18 islands, 17 of which are inhabited. The main islands are Streymoy, on which the capital Tórshavn stands, Vágar, Suduroy and Sandoy. The islands have rugged coasts with abundant birdlife. Most people live on the small coastal lowlands.

Area 545.3 sq mi [1,399 sq km]
Population 46,962
Capital (population) Torshavn (17.939)
Government Self-governing overseas administrative division of Denmark
Ethnic groups Scandinavian
Languages Faroese, Danish
Religions Evangelical Lutheran
Currency Danish krone = 100 ore
Website www.tourist.fo

CLIMATE
Winters are mild under the influence of the North Atlantic Drift, the extension of the warm Gulf Stream. Summers are cool and the weather is often overcast. Fogs are common and winds are often strong.

HISTORY
Irish monks settled on the islands in the 6th century AD, though most of the islanders today are descendants of Vikings. The Faeroe Islands were part of the Kingdom of Norway from the 11th century until 1380. The islands came under Danish control when Norway joined the Kingdom of Denmark. British troops occupied the islands during World War II (1939–45), but they reverted to Danish control after the war.

POLITICS
The Danes granted the islands self-government in 1948, making them a self-governing overseas administrative division of Denmark, with Denmark responsible for defense and foreign relations. The islands have their own parliament, consisting of 32 members elected on a proportional basis from the seven constituencies. The parliament elects the leader of the majority group as prime minister. Two members are elected to the Danish parliament.

In 2001, a planned referendum on independence was cancelled when the Danish government stated that it would cut off all financial aid to the islands within four years of independence.

ECONOMY
The people enjoy a high standard of living. Fishing is the main activity and the country has benefited, since the 1990s, from increased production, together with high export prices. Faeroese fishermen have traditionally hunted pilot whales, but animal rights activists have called for the abandonment of the cull.

The government hopes to diversify the economy and explore the prospects of offshore oil and natural gas fields. The islands benefit from a Danish subsidy which accounts for 15% of the GDP.

TÓRSHAVN

Capital city in the center of the Faeroe Islands. In 800, Norwegian settlers replaced an Irish settlement on the islands. Tórshavn became the central meeting place. Christianity was introduced in about 1000. The isles came under Norwegian rule in 1035 and under Danish rule in 1816.

Tórshavn became the seat of government in 1856. The Faeroese seafishing trade started in Tórshavn. Since 1974 the town has joined with Kaldbak, Hoyvík, Argir and Kollafjørur.

GREENLAND

Situated in the northwest Atlantic Ocean and lying mostly within the Arctic Circle, Greenland is regarded by geographers as the world's largest island. It is almost three times larger than the second largest island, New Guinea.

An ice sheet, the world's second largest after Antarctica, covers more than 85% of its area with an average depth of 5,000 ft [1,500 m]. Settlement is confined to the rocky southwest coast.

Area 836,330 sq mi [2,166,086 sq km]
Population 56,375
Capital (population) Nuuk (13,400)
Government Self-governing overseas administrative division of Denmark
Ethnic groups Greenlander 88% (Inuit and Greenland-born whites), Danish and others 12%
Languages Greenlandic (East Inuit), Danish, English
Religions Evangelical Lutheran
Currency Danish krone = 100 ore
Website http://dk.nanoq.gl

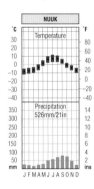

CLIMATE

The southwest coast, where the capital Nuuk (Godthåb) is situated, is warmed by Atlantic currents, even so, it has more than seven months with average temperatures below freezing.

HISTORY

European discovery of Greenland is credited to Erik the Red, who settled in 982, founding a colony that lasted more than 500 years. Greenland became a Danish possession in 1380 and an integral part of the Danish kingdom in 1953.

POLITICS

It was taken into the European Community (EC) in 1973, despite a majority of Greenlanders voting against this. In 1979, after another referendum, home rule was introduced, with full internal self-government in 1981. In 1985, Greenland withdrew from the EC, halving the Community's land area.

ECONOMY

Greenland still relies heavily on Danish aid and Denmark is its main trading partner. The chief rural occupations are sheep rearing and fishing, with shrimp, prawns and mollusks being exported. The only major manufacturing industry is fish canning, which has drawn many Inuit to the towns. Few Inuit now follow the traditional life of nomadic hunting.

Most Greenlanders live precariously between the primitive and the modern. Yet a nationalist mood prevails, buoyed by rich fish stocks, lead and zinc from Uummannaq in the northwest, untapped uranium in the south and, possibly, petroleum in the east. In addition, an adventure-oriented tourist industry is expanding. In 1997, the nationalist resurgence was evident when Greenland made Inuit name forms official.

NUUK
(Danish, GODTHÅB)

Capital and largest town of Greenland. The mouth of a group of fjords on the southwest coast. Founded in 1721, it is the oldest Danish settlement in Greenland. Places of interest include the Greelandic National Museum and Archives, the Katuaq Cultural Centre (with Greenland's only cinema) and Niels Lynges' House.

DJIBOUTI

Based on the banner of the African People's League for Independence. Blue is the color of the Issas people and also the sky and the sea. Green is the color of the Afars people and symbolizes fertile earth. The white triangle signifies peace and equality, the red star is a symbol of unity.

The Republic of Djibouti is a small country on the northeast coast of Africa, the capital is also Djibouti. Djibouti occupies a strategic position around the Gulf of Tadjoura, where the Red Sea meets the Gulf of Aden. Behind the coastal plain lie the Mabla Mountains, rising to Moussa Ali at 6,654 ft [2,028 m]. Djibouti contains the lowest point on the African continent, Lake Assal, at 509 ft [155 m] below sea level.

Nearly 90% of the land is semidesert, and shortage of pasture and water make farming difficult.

Area 8,958 sq mi [23,200 sq km]
Population 467,000
Capital (population) Djibouti (317,000)
Government Multiparty republic
Ethnic groups Somali 60%, Afar 35%
Languages Arabic and French (both official)
Religions Islam 94%, Christianity 6%
Currency Djiboutian franc = 100 centimes
Website www.presidence.dj

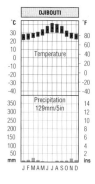

CLIMATE
Djibouti has one of the hottest and driest climates with summer temperatures regularly exceeding 100°F [42°C]. Average annual rainfall is only 5 in [130 mm]. In the wooded Mabla Mountains, the average annual rainfall reaches 20 in [500 mm].

HISTORY
Islam arrived in the 9th century. The subsequent conversion of the Afars led to conflict with Christian Ethiopians who lived in the interior. By the 19th century, Somalian Issas moved north and occupied much of the Afars' traditional grazing land.

France gained influence in 1862, with its interest centered around Djibouti, the French commercial rival to the port of Aden. French Somaliland was established in 1888.

A referendum in 1967 saw 60% of the electorate vote to retain links with France, though most Issas favored independence. The country was renamed the French Territory of the Afars and Issas.

POLITICS
In 1977 the Republic of Djibouti gained full independence, and Hassan Gouled Aptidon of the Popular Rally for Progress (RPP) was elected president. He declared a one-party state in 1981.

RED SEA
Narrow arm of the Indian Ocean between northeast Africa and the Arabian Peninsula, connected to the Mediterranean by the Gulf of Suez and the Suez Canal. With the building of vessels too large for the canal and the construction of pipelines, the Red Sea's importance as a trade route has diminished. Its widest point is 200 mi [320 km]. It covers an area of 169,000 sq mi [438,000 sq km].

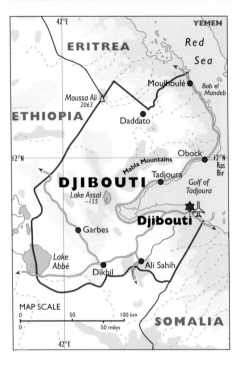

boycotted 1993 elections, and Aptidon was reelected for a fourth six-year term. FRUD rebels continued an armed campaign for political representation. In 1996, government and FRUD forces signed a peace agreement, recognizing FRUD as a political party.

In 1999, Ismael Omar Gelleh succeeded Aptidon as president in the country's first multiparty presidential elections. He pursues a policy of closer links with France, which still has a strong military presence in Djibouti. In addition it is forging closer ties with the United States, with the only US military base in sub-Saharan Africa stationed there.

ECONOMY

Djibouti is a poor nation, heavily reliant on food imports and revenue from the capital city. A free trade zone, it has no major resources and manufacturing is on a very small scale. The only important activity is livestock raising, and 50% of the population are pastoral nomads.

Its location at the mouth of the Red Sea is of great economic importance as it serves as a vital transshipment point.

Protests against the Issas-dominated regime forced the adoption of a multiparty constitution in 1992. The Front for the Restoration of Unity and Democracy (FRUD), supported primarily by Afars,

GULF OF ADEN

A body of water that makes up the western arm of the Arabian Sea, meeting the Red Sea at the Babu l-Mandeb strait. The gulf runs in a west–east direction, between Yemen and Somalia, meeting Djibouti at the western end. It is about 560 mi [900 km] long, and 310 mi [500 km] wide at the eastern end, between Ra's Asir of Somalia and the city of al-Mukalla in Yemen. The Gulf of Aden is an important route for commercial shipping.

DJIBOUTI (JIBUTI)

Capital of Djibouti, on the western shore of an isthmus in the Gulf of Tadjoura, northeast Africa. Founded in 1888, it became capital in 1892, and a free port in 1949. Ethiopian emperor Menelik II built a railway from Addis Ababa, and Djibouti became the chief port for handling Ethiopian trade. While Eritrea was federated with Ethiopia (1952–93), it lost this status to the Red Sea port of Assab. Now home to two-thirds of the country's population.

DOMINICAN REPUBLIC

Blue represents liberty, red stands for blood shed during the struggle for liberation, and the white cross is a symbol of sacrifice. The coat of arms features a Bible open at the Gospel of St. John, symbolizing the Trinitarian movement that led the movement for independence.

The Dominican Republic is the second largest of the Caribbean nations in both area and population, it shares the island of Hispaniola with Haiti, the Dominican Republic occupying the eastern two-thirds.

Of the steep-sided mountains that dominate the island, the country includes the northern Cordillera Septentrional, the huge Cordillera Central, which rises to Pico Duarte, at 10,414 ft [3,175 m] the highest peak in the Caribbean, and the southern Sierra de Baoruco. Between them and to the east lie fertile valleys and lowlands, including the Vega Real and the coastal plains where the main sugar plantations are found.

Area 18,730 sq mi [48,511 sq km]
Population 8,834,000
Capital (population) Santo Domingo (2,061,000)
Government Multiparty republic
Ethnic groups Mulatto 73%, White 16%, Black 11%
Languages Spanish (official)
Religions Roman Catholic 95%
Currency Dominican peso = 100 centavos
Website www.dominicanrepublic.com

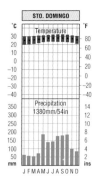

CLIMATE

Typical of the Caribbean region, the climate is humid and hot throughout the year close to sea level, while cooler conditions prevail in the mountains. Rainfall is heavy, especially in the northeast.

HISTORY

Christopher Columbus "discovered" the island on December 5, 1492. Its Amerindian population was soon to be decimated. The city of Santo Domingo, now the capital and chief port, was founded by Columbus' brother Bartholomew four years later and is the oldest in the Americas. For a long time a Spanish colony, Hispaniola was initially the centerpiece of their empire, but was later to become a poor relation.

In 1795, it became French, then Spanish again in 1809. But in 1821, when it was called Santo Domingo, it won its independence. Haiti held the territory from 1822 until 1844 when, on restoring sovereignty, it became the Dominican Republic.

Growing American influence culminated in occupation between 1916 and 1924. This was followed by a period of corrupt dictatorship. From 1930 until his assassination in 1961, the country was ruled by Rafael Trujillo, one of Latin America's best-known dictators, who imprisoned or killed many of his opponents. A power struggle developed between the military, the upper class, those who wanted the country to become a democracy and others who favored making it a Communist regime.

POLITICS

In 1962 Juan Bosch became president, but was ousted in 1963. Bosch supporters tried to seize power in 1965, but were met with strong military opposition. This led to US military intervention in 1965. In 1966, a new constitution was adopted

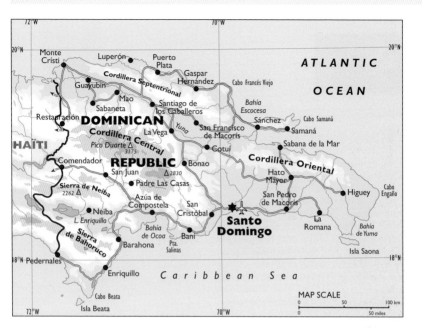

and Joaquín Balaguer was elected president (1966–78, 1986–96). Elections have been known to be violent and the United States has kept a watchful eye.

Leonel Fernandez was elected president for a second time in 2004. He had campaigned on a ticket that promised to tackle inflation and once in office he introduced austerity measures that included cuts to state spending.

ECONOMY

The World Bank describes the Dominican Republic as a "lower-middle-income" developing country. In the 1990s, industrial growth that exploited the country's huge hydroelectric potential, mining, and tourism has augmented the traditional agricultural economy, though the country is far from being politically stable. Agriculture is a major activity. Leading crops include avocados, bananas, beans, mangoes, oranges, plantains, rice, sugarcane, and tobacco.

Gold and nickel are mined. Sugar refining is a major industry, with the bulk of the production exported to the United States. Leading exports are ferronickel, sugar, coffee, cocoa, and gold. Its main trading partner is the United States.

SANTO DOMINGO (formerly CIUDAD TRUJILLO, 1936–61)

Capital and chief port of the Dominican Republic, on the south coast of the island, on the River Ozama. Founded in 1496, the city is the oldest continuous European settlement in the Americas. It was the seat of the Spanish viceroys in the early 1500s, and base for the Spaniards' conquering expeditions until it was devastated by an earthquake in 1562. It houses more than a third of the country's population, many of whom work in the sugar industry.

EAST TIMOR

Yellow represents the vestiges of East Timor's colonial past. Black stands for the darkness to be overcome, while red recalls the struggle for liberation. The five-pointed star is the guiding light of peace.

The Democratic Republic of East Timor is a small island nation in Southeast Asia. It became part of Indonesia in 1975, but it became independent in 2002, ending almost 500 years of foreign domination.

The Timor Sea separates East Timor from Australia, which lies about 310 mi [500 km] away. East Timor occupies the northeastern part of the island of Timor. It also includes some small islands and the enclave of Oscùsso-Ambeno on the northwestern coast of West Timor, the Indonesian part of the island. The land is largely mountainous, rising to about 9,711 ft [2,960 m].

Area 5,743 sq mi [14,874 sq km]
Population 1,019,000
Capital (population) Dili (52,000)
Government Multiparty republic
Ethnic groups Austronesian (Malayo-Polynesian), Papuan
Languages Tetum, Portuguese (both official), Indonesian, English
Religions Roman Catholic 90%, Muslim 4%, Protestant 3%
Currency US dollar = 100 cents
Website www.gov.east-timor.org

CLIMATE

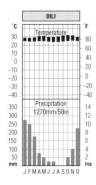

East Timor has a monsoon climate, with high temperatures throughout the year. The average temperature is more than 75°F [24°C], although the mountainous zone is cooler. The monsoon extends from November to May. Average annual rainfall is less than 59 in [1,500 mm]. The mountainous zone is wetter, with a shorter dry period of four months.

HISTORY

Europeans first visited the island of Timor in the 16th century and, following skirmishing between the Dutch and the Portuguese, East Timor came under Portuguese rule by treaties in 1859 and 1893. Japan occupied Timor between 1942 and 1945 during World War II, but the Portuguese returned after Japan was defeated. Following a coup in Portugal in 1974, the new government announced that its colonies should prepare for independence. The Portuguese withdrew from East Timor in August 1975 and Fretilin (the Revolutionary Front for an Independent East Timor) declared the territory independent, but in December 1975, Indonesia seized the territory.

The territory became the 27th province of Indonesia named Timor/Timur. Indonesian rule proved oppressive and guerrilla resistance grew. In 1999, Indonesia agreed to let the people of East Timor vote on independence or local autonomy in a UN-supervised referendum. The people voted by almost 99% for independence. East Timor declared itself independent. However, pro-Indonesian militias, who were widely believed to have the backing of the Indonesian government, caused massive destruction until an international peacekeeping force restored order. An estimated 1,400 people were killed and 250,000 were displaced.

Commission, arguing that it is an attempt to bury the past rather than providing justice. However, the governments of East Timor and Indonesia have argued that it is important for the two countries to build a strong relationship between them

ECONOMY

Agriculture is the main activity. Arable farms cover about 5% of the land and coffee is widely grown. Crops include bananas, cassava, corn, rice, soybeans, and sweet potatoes. Coffee, sandalwood, and marble are exported.

Following independence, the new country, with its shattered economy and infrastructure, was heavily dependent on foreign aid. However, offshore petroleum and natural gas deposits hold out hope for future development. Production of natural gas began in 2004. The petroleum and natural gas deposits are located under the Timor Sea and, following independence, East Timor denounced the maritime boundary given to Australia by Indonesia because it gave Australia control over what was probably the richest oil field region. In 2005, Australia agreed on a division of the expected revenues on the petroleum and natural gas deposits, while East Timor agreed to postpone discussions on the disputed maritime boundary.

POLITICS

East Timor gained full independence on May 20, 2002 and became the 191st member of the United Nations in September. The UN set up a Mission of Support in East Timor to help the Timorese authorities. The country's first president, a ceremonial head of state, was Xanana Gusmao, who took 82.6% of the votes in a presidential election in April 2002. Gusmao was the former leader of the separatist guerrilla fighters and he had been jailed by the Indonesians between 1992 and 1999.

Following independence, the new government replaced the UN-supervised transitional government and, in 2005, the UN withdrew the last of its troops. Some Australian troops remained to train an East Timorese army. East Timor has an elected National Parliament, which elects the country's head of state.

A major problem arising in independent East Timor was the bringing to trial of individuals responsible for atrocities committed during the country's violent transition. An Indonesian court tried people for human rights abuses in East Timor in 1999, while a joint Indonesian and East Timorese Truth and Friendship Commission was established to heal the wounds of the past. Some organizations have criticized the Truth and Friendship

DILI

Capital of East Timor, on the north coast of Timor. Originally settled in 1520 by the Portuguese, who made it the capital of Portuguese Timor in 1596. Occupied by the Japanese during World War II. Occupied by Indonesian forces from 1975, but finally independent in 2002. The city retains links with its Portuguese past but has rebuilt itself after the massive destruction of 1999.

ECUADOR

Ecuador's flag was created by a patriot, Francisco de Miranda, in 1806. The armies of Simón Bolívar, the South American general, won victories over Spain, and flew this flag. At the center is Ecuador's coat of arms, showing a condor over Mount Chimborazo.

The Republic of Ecuador straddles the Equator on the west coast of South America. Three ranges of the high Andes Mountains form the backbone of the country. Between the towering, snowcapped peaks of the mountains, some of which are volcanoes, lie a series of high plateaus, or basins. Nearly half of Ecuador's population lives on these plateaus.

West of the Andes lie the flat coastal lowlands, which border the Pacific Ocean and average 60 mi [100 km] in width. The eastern alluvial lowlands, often called the Oriente, are drained by headwaters of the River Amazon.

Area 109,483 sq mi [283,561 sq km]
Population 13,213,000
Capital (population) Quito (1,616,000)
Government Multiparty republic
Ethnic groups Mestizo 65%, Amerindian 25%, White 7%, Black 3%
Languages Spanish (official), Quechua
Religions Roman Catholic 95%
Currency US dollar = 100 cents
Website www.vivecuador.com

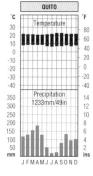

CLIMATE

The climate in Ecuador is greatly influenced by the altitude. The coastal lowlands are hot, despite the cooling effect of the cold offshore Peru Current. The Andes have spring temperatures all the year, while the eastern lowlands are hot and humid. The rainfall is heaviest in the eastern lowlands and the northern coastal lowlands.

HISTORY

The Inca people of Peru conquered much of what is now Ecuador in the late 15th century. They introduced their language, Quechua, which is widely spoken today. In 1532 a colony was founded by the Spaniards in the territory, which was then called Quito. The country became independent in 1822, following the defeat of a Spanish force by an army led by General Antonio Jose de Sucre in a battle near Quito. Ecuador became part of Gran Colombia, a confederation which also included Colombia and Venezuela. Ecuador became a separate nation in 1830.

In 1832, Ecuador annexed the volcanic Galapagos Islands, which lie 610 mi [970 km] west of Ecuador, of which they form a province. The archipelago, which contains six main islands and more than 50 smaller ones, later became world-famous through the writings of Charles Darwin, who visited the islands in 1835. His descriptions of the unique endemic flora and fauna gave him crucial evidence for his theory of natural selection.

POLITICS

The failure of successive governments to tackle the country's many social and economic problems caused great instability in Ecuador throughout the 20th century. A war with Peru in 1941 led to loss of territory and border disputes flared up again in 1995, though the two countries

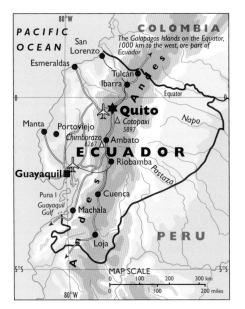

demonstrations. In 2000, economic problems made Ecuador adopt the US dollar as its sole unit of currency.

ECONOMY

The World Bank classifies Ecuador as a "lower-middle-income" developing country. Agriculture employs 10% of the populous. Bananas, cocoa, and coffee are all important export crops. Other products in the hot coastal lowlands include citrus fruits, rice, and sugarcane, while beans, corn and wheat are important in the highlands. Cattle are raised for dairy products and meat, while fishing is important in the coastal waters. Forestry is a major activity. Ecuador produces balsa wood and such hardwoods as mahogany. Mining is important and petroleum and petroleum products now play a major part in the economy. Ecuador started to export petroleum in the early 1970s and is a member of the Organization of Petroleum Exporting Countries. Manufactures include cement, Panama hats, paper products, processed food and textiles. Major exports are food and live animals, and mineral fuels. Ecuador's main trading partners are the United States and Colombia.

eventually signed a peace treaty in January 1998.

Military regimes ruled the country between 1963 and 1966 and again from 1976 to 1979. However, under a new constitution introduced by the second of these military juntas and approved by a national referendum, civilian government was restored. Civilian governments have ruled Ecuador since multiparty elections in 1979. But the volatile character of politics here was evident throughout the 1980s and 1990s. For example, a state of emergency, albeit of short duration, was declared in 1986 and, in 1995, the vice-president was forced to leave the country after accusations that he had bribed opposition deputies.

In 1996, the president was deposed on the grounds of mental incompetence and, in 1998, accusations of fraud marred the victory of President Jamil Mahaud of the center-right Popular Democracy Party. The early years of the 21st century were marked by political instability as successive presidents faced opposition from the military and public

QUITO

Capital of Ecuador, at 9,260 ft [2,850 m], Quito lies almost on the Equator. The site was first settled by Quito Native Americans, and was captured by the Incas in 1487. It was taken by Spain in 1534, and liberated in 1822 by Antonio José de Sucre. A cultural and political center, it is the site of the Central University of Ecuador (1787). Products include textiles and handcrafts.

EGYPT

A flag consisting of three bands of red, white and black, the colors of the Pan-Arab movement, was adopted in 1958. The present design has a gold eagle in the center. This symbolizes Saladin, the warrior who led the Arabs in the 12th century.

The Arab Republic of Egypt is Africa's second largest country by population after Nigeria. Most of Egypt is desert. Almost all the people live either in the Nile Valley and its fertile delta or along the Suez Canal, the artificial waterway between the Mediterranean and Red seas. This canal shortens the sea journey between the United Kingdom and India by 6,027 mi [9,700 km]. Recent attempts have been made to irrigate parts of the Western Desert.

Apart from the Nile Valley, Egypt has three other main regions. The Western and Eastern deserts are part of the Sahara. The Sinai Peninsula (Es Sina), to the east of the Suez Canal, is very mountainous and contains Egypt's highest peak, Gebel Katherina (8,650 ft [2,637 m]); few people live in this area.

Area [386,659 sq mi1,001,449 sq km]
Population 76,117,000
Capital (population) Cairo (6,801,000)
Government Republic
Ethnic groups Egyptians/Bedouins/Berbers 99%
Languages Arabic (official), French, English
Religions Islam (mainly Sunni Muslim) 94%, Christian (mainly Coptic Christian) and others 6%
Currency Egyptian pound = 100 piastres
Website www.egypt.gov.eg/english/default.asp

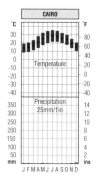

CLIMATE

Egypt has a desert climate and is one of the world's sunniest countries. The low rainfall occurs in winter, if at all. Winters are mild, but summers hot. Conditions become unpleasant when hot and dusty winds blow from the deserts into the Nile Valley.

HISTORY

Ancient Egypt, which was founded about 5,000 years ago, was one of the great early civilizations. Throughout the country, pyramids, temples, and richly deco-

rated tombs are memorials to its great achievements. After Ancient Egypt declined, the country came under successive foreign rulers. Arabs occupied Egypt in AD 639–42. They introduced the Arabic language and Islam. Their influence was so great that many Egyptians now regard themselves as Arabs.

Egypt came under British rule in 1882, but it gained partial independence in 1922, becoming a monarchy.

POLITICS

In 1952, following a military revolution led by General Muhammad Naguib, the monarchy was abolished and Egypt became a republic. Naguib became president, but he was overthrown in 1954 by Colonel Gamal Abdel Nasser. President Nasser sought to develop Egypt's economy, and he announced a major project to build a new dam at Aswan to provide electricity and water for irrigation. When Britain and the United States failed to provide finance for building the dam, Nasser seized the Suez Canal Company in July 1956. In retaliation, Israel, backed by British and French troops, invaded the Sinai Peninsula and the Suez Canal region. However, under international

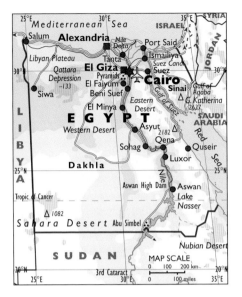

victorious in the first contested presidential elections, but members of the banned Muslim Brotherhood, standing as independents, made gains in parliament.

ECONOMY

Egypt is Africa's second most industrialized country after South Africa, but remains a developing country. The people are poor, farming employs 34% of the workers. Most *fellahin* (peasants) grow food crops such as beans, corn, rice, sugarcane, and wheat, but cotton is the chief cash crop. Egypt depends increasingly on the Nile. Its waters are seasonal, and control and storage have become essential in the last 100 years. The Aswan High Dam is the greatest Nile dam, and the water behind it in Lake Nasser makes desert reclamation possible. The electricity produced is important for industrial development.

pressure, they were forced to withdraw. Construction of the Aswan High Dam began in 1960 and it was fully operational by 1968.

In 1967, Egypt lost territory to Israel in the Six Day War and Nasser tendered his resignation, but the people refused to accept it. After his death in 1970, Nasser was succeeded by his vice-president, Anwar el-Sadat. In 1973, Egypt launched a surprise attack in the Sinai Peninsula, but its troops were finally forced back to the Suez Canal. In 1977, Sadat began a peace process when he visited Israel and addressed the Knesset (Israel's parliament). Finally, in 1979, Egypt and Israel signed a peace treaty under which Egypt regained the Sinai Peninsula. However, extremists opposed contacts with Israel and, in 1981, Sadat was assassinated. He was succeeded as president by Hosni Mubarak.

In the 1990s, attacks on foreign visitors severely damaged tourism, despite efforts to curb the activities of Islamic extremists. In 1997, terrorists killed 58 foreign tourists near Luxor. Unrest continued in the 21st century. In 2005, Mubarak was

CAIRO (AL-QAHIRAH)

Capital of Egypt and port on the River Nile. The largest city in Africa, Cairo was founded in 969 by the Fatimid dynasty and subsequently fortified by Saladin. Medieval Cairo became capital of the Mamluk empire, but declined under Turkish rule. During the 20th century it grew dramatically in population and area. Nearby are world-famous archaeological sites, the sphinx and the Pyramids of Giza. Old Cairo is a World Heritage Site containing over 400 mosques and other fine examples of Islamic art and architecture. Its five universities include the world's oldest, housed in the mosque of Al Azhar (972) and the center of Shiite Koranic study.

EL SALVADOR

This flag was adopted in 1912, replacing the earlier "Stars and Stripes." The blue and white stripes are featured on the flags of several Central American countries which gained their independence from Spain at the same time in 1821.

E l Salvador is the smallest and most densely populated country in Central America. It has a narrow coastal plain along the Pacific Ocean. The majority of the interior is mountainous with many extinct volcanic peaks, overlooking a heavily populated central plateau. Earthquakes are common; in 1854, an earthquake destroyed the capital, San Salvador. In October 1986, another earthquake killed 400 people and caused widespread damage.

Grassland and some virgin forests of original oak and pine are found in the highlands. The central plateau and valleys have areas of grass and deciduous woodland, while tropical savanna and forest cover the coastal regions.

Area 8,124 sq mi [21,041 sq km]
Population 6,588,000
Capital (population) San Salvador (473,000)
Government Republic
Ethnic groups Mestizo 90%, White 9%, Amerindian 1%
Languages Spanish (official)
Religions Roman Catholic 83%
Currency US dollar = 100 cents
Website www.elsalvadorturismo.gob.sv

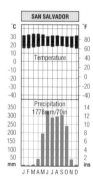

CLIMATE

The coast has a hot tropical climate. Inland, the climate is moderated by altitude. The center region has similar temperatures by day, but nights are cooler. Rain falls most afternoons between May and October.

HISTORY

In 1524–26 Spanish explorer Pedro de Alvarado conquered Native American tribes such as the Pipil, and the region became part of the Spanish Viceroyalty of Guatemala. Independence was achieved in 1821, and in 1823 El Salvador joined the Central American Federation. The federation dissolved in 1839.

El Salvador declared independence in 1841, but was continually subject to foreign interference (especially from Guatemala and Nicaragua). It was at this time that El Salvador's coffee plantations developed.

POLITICS

Following a serious collapse in the world coffee market, Maximiliano Hernández Martínez seized power in a palace coup in 1931. In 1944 a general strike overthrew his brutal dictatorship. After a period of progressive government, a military

COFFEE

Plant and the popular caffeine beverage produced from its seeds (coffee beans). The plants of the genus Coffea are evergreen with white fragrant flowers. Originally native to Ethiopia, they are now cultivated in the tropics, especially Brazil, Colombia, and the Ivory Coast. Family Rubiaceae.

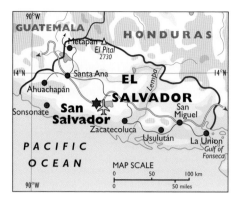

junta headed by Julio Adalberto Rivera from 1962 to 1967 and Fidel Sánchez Hernández from 1967 to 1972 seized power. Honduras' discriminatory immigration laws exacerbated tension on the border between the two countries. The "Soccer War" of 1969 broke out following an ill-tempered World Cup qualifying match. Within four days, El Salvador captured much of Honduras. A ceasefire occurred and the troops withdrew.

In the 1970s, the repressive National Republican Alliance (ARENA) regime compounded El Salvador's problems of overpopulation, unequal distribution of wealth, and social unrest. Civil war broke out in 1979 between US-backed government forces and the Farabundo Marti National Liberation Front (FMLN). The 12-year war claimed 75,000 lives and caused mass homelessness. A ceasefire held from 1992, and the FMLN became a recognized political party. In 1993 a UN Truth Commission led to the removal of senior army officers for human rights abuses, and the decommissioning of FMLN arms. Armando Calderón Sol became president in 1994 elections; Francisco Flores succeeded him in 1999.

In 2001, massive earthquakes killed about 1,200 people and left one million homeless. Tony Saca won 2004 presidential elections to become the fourth successive ARENA president.

ECONOMY

El Salvador is a lower-middle-income developing country. Farmland and pasture account for approximately 60% of land use. El Salvador is the world's 10th largest producer of coffee. Its reliance on the crop caused economic structural imbalance. Sugar and cotton grow on the coastal lowlands. Fishing is important, but manufacturing is on a small scale. The civil war devastated the economy. Between 1993 and 1995, El Salvador received more than US$100 million of credit from the International Monetary Fund.

LAGUNA EL JOCOTAL

Freshwater lagoon 7.5 mi [12 km] south of San Miguel, lying in a volcanic crater, the lagoon varies in size from 2 sq mi [5 sq km] in the dry season, to 6 sq mi [15 sq km] in the rainy season. More than 130 species of aquatic bird use the lagoon with the tree duck being especially prevalent. It has been declared a Ramsar Site

SAN SALVADOR

Capital and largest city of El Salvador in central El Salvador. Founded in 1524 near the volcano of San Salvador, which rises to 6,184 ft [1,885 m] and last erupted in 1917. The city has frequently been damaged by earthquakes. The main industry is the processing of the coffee which is grown on the rich volcanic soils of the area. Other manufactures include beer, textiles, and tobacco.

EQUATORIAL GUINEA

Green equals agriculture, white is peace and red is for the blood shed in the fight for independence. Blue is for the Atlantic Ocean. The tree on the coat of arms is the tree under which the 1843 treaty with Spain was signed. The six stars represent the mainland and the five islands.

The Republic of Equatorial Guinea is located in west-central Africa and is one of the smallest countries on the African continent. It consists of a mainland territory between Cameroon and Gabon, called Río Muni (Mbini), and five islands in the Bight of Biafra (Bonny), the largest of which is Bioko (Fernando Póo).

Bioko is a volcanic island with fertile soils and a steep rocky coast. Malabo's harbor is part of a submerged volcano. Bioko is mountainous, rising to a height of 9,869 ft [3,008 m] at the Pico de Santa Isabel. It has varied vegetation with trees such as teak, mahogany, oak, walnut and rosewood, and grasslands at higher levels.

Mainland Río Muni (90% of all land) consists mainly of hills and plateaus behind the coastal plains. Its main river, the Lolo, rises in Gabon. Dense forest covers most of Río Muni and provides a habitat for animals such as lions, gazelles and elephants .

Area 10,830 sq mi [28,051 sq km]
Population 523,000
Capital (population) Malabo (30,000)
Government Multiparty republic (transitional)
Ethnic groups Bubi (on Bioko), Fang (in Rio Muni)
Languages Spanish and French (both official)
Religions Christianity
Currency CFA franc = 100 centimes
Website www.ceiba-guinea-ecuatorial.org

HISTORY

In 1472 Portuguese navigator Fernão do Pó sighted the largest island of Bioko. In 1778, Portugal ceded the islands and commercial mainland rights to Spain in exchange for some Brazilian territories. Yellow fever hit Spanish settlers on Bioko, and they withdrew in 1781. In 1827, Spain leased bases on Bioko to Britain, and the British settled some freed slaves. Descendants of these former slaves (*Fernandinos*) remain on the island. Spain returned in the mid-19th century and developed plantations on Bioko.

In 1956, the islands became the Overseas Provinces in the Gulf of Guinea. In 1959, the territory was divided into two

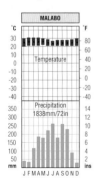

CLIMATE

Situated on the equator, Equatorial Guinea has a tropical climate. High temperatures and high humidity are the norm with an average annual temperature of 77°F [25°C]. Bioko has heavy rainfall and there is a dry season from December to February. Río Muni has a similar climate to Bioko, though rainfall diminishes inland.

COCOA

Drink obtained from the seeds of the tropical American evergreen tree Theobroma cacao. The seeds are crushed and some fatty substances are removed to produce cocoa powder. Cocoa is the basic ingredient of chocolate. The Ivory Coast is the world's largest producer. Family Sterculiaceae.

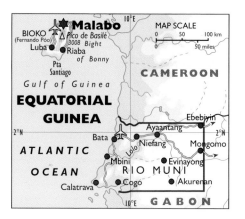

dent is seen to control all the political parties.

In 2004 a coup attempt by foreign mercenaries was foiled and the leaders were arrested.

ECONOMY

Equatorial Guinea is a poor country. Agriculture employs around 60% of the people, though many farmers live at subsistence level, making little contribution to the economy. The main food crops are bananas, cassava, and sweet potatoes. The chief cash crop is cocoa, grown on Bioko, though this has been hit by a worldwide dip in cocoa prices.

Petroleum has been produced off Bioko since 1966. By 2002 it accounted for more than 80% of exports. Despite the rapid expansion of the economy and massive increase in revenue, a UN human rights report stated that 65% of the people still live in "extreme poverty."

The government has promised that agriculture will benefit from the large amounts of revenue gained from petroleum, but this has yet to materialize. Other natural resources that have yet to be developed include titanium, iron ore, manganese and uranium. The country has forfeited much aid from the World Bank and the IMF due to corruption and mismanagement.

provinces, Fernando Póo and Río Muni, and named Spanish Guinea. The two territories reunified in 1963 and became the Autonomous Territories of Equatorial Guinea. In 1968 the territory gained independence as the Republic of Equatorial Guinea.

POLITICS

In 1969 as a result of social unrest caused by factors such as ethnic conflict and economic problems, President Francisco Macías Nguema annulled the constitution. A military dictatorship ensued with up to 100,000 refugees fleeing to neighboring countries. Nguema's dictatorship endured from 1968 to 1979 during which time more than 40,000 people were killed.

In 1979, Lieutenant-Colonel Teodoro Obiang Nguema Mbasogo deposed Nguema in a military coup. A 1991 referendum voted to set up a multi-party democracy, consisting of the ruling Equatorial Guinea Democratic Party (PDGE) and ten opposition parties. The main parties and most of the electorate boycotted elections in 1993, and the PDGE formed a government. In 1996 elections, again boycotted by most opposition parties, President Obiang claimed 99% of the vote. Human rights organizations accuse his regime of routine arrests and torture of opponents and the presi-

---

MALABO

Seaport capital of Equatorial Guinea, on Bioko island, in the Gulf of Guinea in west central Africa. Founded in 1827 as a British base to suppress the slave trade, it was known as Santa Isabel until 1973. Malabo stands on the edge of a volcanic crater that acts as a natural harbor. Industries include fish processing, hardwoods, cocoa, and coffee.

ERITREA

Based on the flag of the EPLF, the red triangle symbolizes blood shed in the fight for freedom, the blue triangle represents the Red Sea, the green triangle stands for agriculture. From 1952–9 Eritrea's flag bore a green wreath and olive branch in the center of a field of United Nations' blue.

The State of Eritrea occupies a strategic geopolitical position on the Red Sea in northeastern Africa. The coastal plain extends inland between 10 and 40 mi [16–64 km]. Inland are mountains. In the southwest, the mountains descend to the Danakil Desert. This desert contains Eritrea's lowest point at 246 ft [75 m] below sea level.

Area 45,405 sq miles [117,600 sq km]
Population 4,447,000
Capital (population) Asmara (358,000)
Government Transitional government
Ethnic groups Tigrinya 50%, Tigre and Kunama 40%, Afar 4%, Saho 3% and others
Languages Afar, Arabic, Tigre and Kunama, Tigrinya
Religions Islam, Coptic Christian, Roman Catholic
Currency Nakfa = 100 cents
Website www.shabait.com

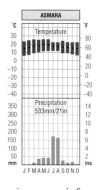

CLIMATE

The temperature ranges from 61°F [16°C] in the highlands to 81°F [27°C] on the coastal plain. Rainfall on the coastal plain is between 6 and 10 in [150–250 mm] with up to 24 in [610 mm] in the highlands. The rainy season is from June to September.

HISTORY

The first settlers were from Africa, followed by others from the Arabian peninsula. Between AD 50 and 600, Eritrea was part of the Ethiopian kingdom of Axum. The people of Axum were converted to Christianity in the 4th century, but Muslims gained control of the area in the 7th century, introducing Islam to the coastal areas. Christianity survived inland.

In the 16th century, the Ottoman Empire took over the coastal area. Italy made Eritrea a colony in 1890. In 1935, Italy also conquered Ethiopia. During World War II, a British force drove the Italians out off northeastern Africa. After the war, a British military administration ruled Eritrea.

In 1950, the UN made Eritrea a self-governing part of Ethiopia. Ethiopian rule proved unpopular and, in 1958, Eritrean nationalists formed the Eritrean Liberation Front (ELF). War broke out in 1961 between the ELF and the Ethiopians. In 1962, Ethiopia declared Eritrea to be a province, sparking off a war of independence.

The Eritrean People's Liberation Front (EPLF) was formed in 1970, replacing the ELF as the main anti-Ethiopian organization. In 1974, Ethiopian Emperor Haile Selassie was overthrown and a military government took power. EPLF victories gradually weakened Ethiopia's government. and the regime collapsed in 1991. The EPLF then formed a provisional government.

POLITICS

Eritrea declared independence in 1993, with Isaias Afewerki as president. The ruling People's Front for Democracy and Justice is the only party permitted in the country. The government has been criticized for the repression of opposition and

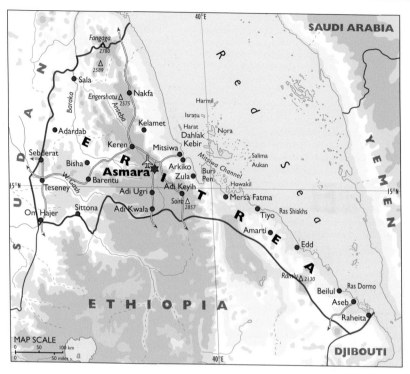

closing the private press in 2001. Eritrea's relations with Ethiopia deteriorated in 1998 over a border dispute around the town of Badme, on Eritrea's southwestern border. The conflict erupted into violence when Ethiopia bombed Asmara airport, and Eritrea attacked Mekele in northern Ethiopia. The conflict continued into 2000. A ceasefire was agreed and a peace plan drawn up. UN observers arrived to help find a settlement. In 2001, the two countries agreed to a UN-proposed mediator to demarcate the border. In 2003, the boundary commission ruled that Badme lies in Eritrea. Tension continued and, in late 2005, Eritrea ordered UN peacekeeping troops to leave the country.

ECONOMY

One of Africa's poorest countries, half the population lives below the poverty line and life expectancy is 52 years. Since 1993, the economy has been set back by droughts, border conflict, and high population increase. The main activity is farming, mostly at subsistence level. Agriculture employs 80% of the workforce.

ASMARA (ASMERA)

Capital of Eritrea, northeast Africa. Occupied by Italy in 1889 then their colonial capital and main base for the invasion of Ethiopia (1935–6). In the 1950s, the US built Africa's biggest military communications center here. Absorbed by Ethiopia in 1952, it was the main garrison in the fight against independence-seeking Eritrean rebels. In 1993 it became the capital of independent Eritrea.

ESTONIA

Estonia's flag was used between 1918 and 1940, when the country was an independent republic. It was readopted in June 1988. The blue is said to symbolize the sky, the black Estonia's black soil, and the white the snow that blankets the land in winter.

The Republic of Estonia is the smallest of the three states on the east coast of the Baltic Sea, which were formerly part of the Soviet Union, but became independent in the early 1990s. Estonia consists of a generally flat plain which was covered by ice sheets during the Ice Age. The land is strewn with moraine (rocks deposited by the ice).

The country is dotted with more than 1,500 small lakes. Water, including the large Lake Peipus (Ozero Chudskoye) and the River Narva, makes up much of Estonia's eastern border with Russia. Estonia has more than 800 islands, which together comprise about a tenth of the country. The largest island is Saaremaa (Sarema).

Farmland and pasture account for more than 33% of land use.

Area 17,413 sq mi [45,100 sq km]
Population 1,342,000
Capital (population) Tallinn (418,000)
Government Multiparty republic
Ethnic groups Estonian 65%, Russian 28%, Ukranian 2%, Belarusian 2%, Finnish 1%
Languages Estonian (official), Russian
Religions Lutheran, Russian and Estonian Orthodox, Methodist, Baptist, Roman Catholic
Currency Estonian kroon = 100 senti
Website www.riik.ee/en

CLIMATE

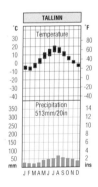

Despite its position to the north, Estonia has a fairly mild climate due to its proximity to the sea. Sea winds tend to warm the land during winter and cool it in summer. Rainfall averages from 19 to 23 in (480–580 mm).

HISTORY

The ancestors of the Estonians, who are related to the Finns, settled in the area several thousand years ago. Divided into several separate states, they were vulnerable to Viking attacks, but in the early 13th century, German crusaders, known as the Teutonic Knights, introduced Christianity. Germany took control of the southern part of Estonia and Denmark took control of the north. The Danes sold the north to the Germans in 1324 and Estonia became part of the Holy Roman Empire.

In 1561, Sweden took over northern Estonia and Poland ruled the south. Sweden controlled the entire country from 1625 until 1721 but, following the victory of Peter the Great over Sweden in the Great Northern War (1700–21), the area became part of the Russian Empire. On February 24, 1918, Estonia declared its independence. A democratic form of government was established in 1919. However, a fascist coup in 1934 ended democratic rule.

POLITICS

In 1939, Germany and the Soviet Union agreed to take over large areas of eastern Europe, and it was agreed that the Soviet Union would take over Estonia. The Soviet Union forcibly annexed the country in 1940. Germany invaded Estonia in 1941, but the Soviet Union regained con-

30% of the population. In the country's first free elections in 1992, only Estonians were permitted to vote and all Russians were excluded. Tension on this issue continued through the 1990s as dual citizenship was outlawed, while restrictions were placed on Russians applying for Estonian citizenship.

ECONOMY

Manufacturing is the most valuable activity. The timber industry is among the most important industries, alongside metalworking, shipbuilding, clothing, textiles, chemicals, and food processing, which is based primarily on extremely efficient dairy farming and pig breeding, but oats, barley and potatoes are suited to the cool climate and the average soils.

trol in 1944 when the country became the Estonian Soviet Socialist Republic. Many Estonians opposed Soviet rule and were deported to Siberia. About 100,000 Estonians settled in the West.

Resistance to Soviet rule was fueled in the 1980s when the Soviet leader Mikhail Gorbachev began to introduce reforms and many Estonians called for independence. In 1990, the Estonian parliament declared Soviet rule invalid and called for a gradual transition to full independence. The Soviet Union regarded this action as illegal, but finally the Soviet State Council recognized the Estonian parliament's proclamation of independence in September 1991, shortly before the Soviet Union itself was dissolved in December 1991.

Since independence, Estonia has sought to increase links with Europe. It was admitted to the Council of Europe in 1993; has been a member of the World Trade Organization since 1999, and a member of NATO and the European Union since 2004. But despite the fact that it had the highest standard of living among the 15 former Soviet republics, Estonia has found the change to a free-market economy hard-going.

Other problems facing Estonia include crime, rural under-development, and the status of its non-Estonian citizens, including Russians, who make up about

Like the other two Baltic states, Estonia is not rich in natural resources, though its oil shale is an important mineral deposit; enough natural gas is extracted to supply St. Petersburg, Russia's second largest city. The leading exports are mineral fuels and chemical products, followed by food, textiles and cloth, and wood and paper products. Finland and Russia are the leading trading partners.

TALLINN

Capital and largest city of Estonia, on the Gulf of Finland, opposite Helsinki. Founded in 1219 by the Danes, it became a member of the Hanseatic League in 1285. It passed to Sweden in 1561, and was ceded to Russia in 1721. Developed in the 19th century for Russia's Baltic Fleet, it remains a major port and industrial center. It was badly damaged in World War II. Tourism is increasing. Industries include machinery, cables, paper.

ETHIOPIA

The tricolor flag of Ethiopia first appeared in 1897. The central pentangle was introduced in 1996, and represents the common will of the country's 68 ethnic groups, and the present sequence was adopted in 1914.

Ethiopia is dominated by the Ethiopian Plateau, a block of volcanic mountains. Its average height is 6,000 ft to 8,000 ft [1,800 m to 2,450 m], rising in the north to 15,157ft [4620m], at Ras Dashen. The Great Rift Valley bisects the plateau. The Eastern Highlands include the Somali Plateau and the desert of the Ogaden Plateau. The Western Highlands include the capital, Addis Ababa, the Blue Nile (Abbay) and its source, Lake Tana (Ethiopia's largest lake). The Danakil Desert forms Ethiopia's border with Eritrea.

Grass, farmland, and trees cover most of the highlands. semidesert and tropical savanna cover parts of the lowlands. Dense rainforest grows in the southwest.

Area 426,370 sq mi [1,104,300 sq km]
Population 67,851,000
Capital (population) Addis Ababa (2,424,000)
Government Federation of nine provinces
Ethnic groups Oromo 40%, Amhara and Tigre 32%, Sidamo 9%, Shankella 6%, Somali 6% , others
Languages Amharic (official), many others
Religions Islam 47%, Ethiopian Orthodox 40%, traditional beliefs 12%
Currency Birr = 100 cents
Website www.mfa.gov.et

CLIMATE

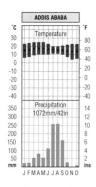

Ethiopia's climate is greatly affected by altitude. Addis Ababa, at 8,000 ft [2,450 m], has an average annual temperature of 68°F [20°C]. Rainfall is generally more than 39 in [1,000 mm], with a rainy season from April to September. The northeast and southwest lowlands are extremely hot and arid with less than 20 in [500 mm] rainfall, and frequent droughts.

HISTORY

According to legend, Menelik I, son of King Solomon and the Queen of Sheba,

founded Ethiopia in about 1000 BC. In AD 321, the northern kingdom of Axum introduced Coptic Christianity. Judaism flourished in the 6th century. The expansion of Islam led to the isolation of Axum and the kingdom fragmented in the 16th century.

In 1855, Kasa reestablished unity, and proclaimed himself Negus (Emperor) Theodore, thereby founding the modern state. European intervention marked the

ADDIS ABABA (AMHARIC, "NEW FLOWER")

Capital and largest city in Ethiopia, located on a plateau at 2,440 m [8,000 ft] in the highlands of Shewa province. Addis Ababa was made capital of Ethiopia in 1889. It is the headquarters of the African Union (AU). It is the main center for the country's vital coffee trade. Industries include food, tanning, textiles, wood products.

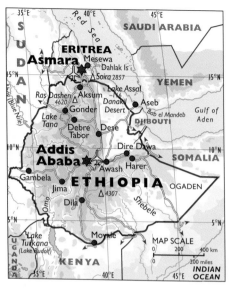

military assistance.

In 1984–5 widespread famine received global news coverage and 10,000 Falashas were airlifted to Israel. In 1987, Mengistu established the People's Democratic Republic of Ethiopia.

In 1991, the Tigrean-based Ethiopian People's Revolutionary Democratic Front (EPRDF) and the Eritrean People's Liberation Front (EPLF) brought down Mengistu. In 1995 the Federal Democratic Republic of Ethiopia was created, with Meles Zenawi as prime minister. A border war with Eritrea occurred in 1998-2000. Elections in 2005 led to protests and a crackdown on the opposition and press.

ECONOMY

Having been afflicted by drought and civil war in the 1970s and 1980s, Ethiopia is now one of the world's poorest countries. Agriculture is the main activity. In 2004, a UN report stated that Ethiopia remained on the brink of disaster, with spiraling population growth, slow economic growth and environmental degradation. Coffee is the leading cash crop and export, followed by hides, skins, and pulses.

late 19th century, and Menelik II became emperor with Italian support. He expanded the empire, made Addis Ababa his capital in 1889, and defeated an Italian invasion in 1895. In 1930, Menelik II's grandnephew, Ras Tafari Makonnen, was crowned Emperor Haile Selassie I. In 1935, Italian troops invaded Ethiopia (Abyssinia). In 1936, Italy combined Ethiopia with Somalia and Eritrea to form Italian East Africa. During World War II, British and South African forces recaptured Ethiopia, and Haile Selassie was restored as emperor in 1941.

POLITICS

In 1952, Eritrea federated with Ethiopia. The 1960s witnessed violent demands for Eritrean secession and economic equality. In 1962 Ethiopia annexed Eritrea. Following famine in northern Ethiopia, Selassie was killed in 1974. The Provisional Military Administrative Council (PMAC) abolished the monarchy. Military rule was repressive, and civil war broke out. The new PMAC leader, Mengistu Mariam, recaptured territory in Eritrea and the Ogaden with Soviet

AFRICAN UNION (AU)

Intergovernmental organization. Previously known as the Organization of African Unity (OAU). Founded in 1963 and renamed in 2001, it aims to safeguard the interests and independence of all African states, encourage the continent's development, and settle disputes among member states. Its headquarters are in Addis Ababa, Ethiopia. In 2003, the AU elected its first president, Alpha Oumar Konaré.

FIJI

A modified version of Fiji's colonial flag, it includes the UK flag and the shield from the coat of arms. The light blue represents the Pacific Ocean. Each quarter of the St. George cross features products or symbols of Fiji: sugarcane, coconut palm, bananas, dove of peace, and a lion with cocoa pod.

The Republic of Fiji Islands consists of more than 800 Melanesian islands situated in the South Pacific Ocean. The larger ones are mountainous and volcanic, and they are surrounded by coral reefs. There are also fertile coastal plains and river valleys. The rest of the islands are low, sandy coral atolls. Easily the biggest islands are Viti Levu (meaning "Big Island"), with the capital Suva on its south coast, and Vanua Levu ("Big Land"), which is just over half the size of the larger island.

Tropical forests cover more than half of the area of the islands.

Area 7,056 sq mi [18,274 sq km]
Population 881,000
Capital (population) Suva (70,000)
Government Republic
Ethnic groups Fijian 51% (predominantly Melanesian with a Polynesian admixture), Indian 44%, others
Languages English (official), Fijian, Hindustani
Religions Christian 52% (Methodist 37%, Roman Catholic 9%), Hindu 38%, Muslim 8%, others
Currency Fiji dollar = 100 cents
Website www.fiji.gov.fj

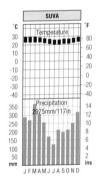

CLIMATE

Fiji has a tropical oceanic climate, with southeast trade winds blowing throughout the year, Average temperatures vary between 61°F [16°C] and 90°F [32°C]. The average annual rainfall in Suva is 118 in [3,000 mm]. Heavy rains occur, especially between November and April. Much of the rain comes in short, heavy showers often after a sunny morning. But rains may last all day during the rainy season.

HISTORY

Melanesians, possibly from Indonesia, settled on the islands thousands of years ago, while a small group of Polynesians also reached the islands about 1,900 years ago. In 1643 the Dutch navigator Abel Janszoon Tasman became the first European to reach the islands. The British Captain James Cook arrived on Vatoa, one of the southern islands in 1774. Christian missionaries began to arrive in the 1830s.

Following conflict between various Fijian tribes in the mid-19th century, a local Christian chief named Cacobau, took control of western Fiji after having helped to restore peace there, while another Christian. Ma'afu. controlled the east. In 1871, European settlers named Cacobau king of Fiji and, in 1874, Fiji, at the request of Cacobau and other chiefs, became a British Crown Colony. In the late 19th century, European traders, missionaries, and escaped convicts from Australia also settled in Fiji and, between 1879 and 1916, the British brought in more than 60,000 indentured laborers to work on the sugar plantations.

POLITICS

Fiji finally became independent on October 10, 1970, with Ratu Sir Kamisese Mara as prime minister. Fiji suffers today from its colonial past. Until the late

1980s, the Indian workers and their descendants outnumbered the native Fijians. Mixing between the two groups was minimal and the ethnic Indians were second-class citizens in terms of electoral representation, economic opportunity and land ownership. However, they played an important role in the economy. The constitution adopted on independence was intended to ease racial tension. However, in 1987, two military coups led by Lt-Colonel Sitiveni Rabuka, overthrew the elected (and first) Indian majority government, although it had been led by an ethnic Fijian, Timoci Bavadra. The leaders suspended the constitution and set up a Fijian-dominated republic outside the Commonwealth.

The country returned to civilian rule in 1990. However, in 1992, elections were held under a new constitution guaranteeing Melanesian supremacy and Rabuka became prime minister. However, thousands of ethnic Indians had already emigrated before these elections, taking their valuable skills with them and causing severe economic problems. Fiji was readmitted to the Commonwealth in 1997 after it had introduced a nondiscriminatory constitution. Peaceful elections in 1999 led to victory for the Fiji Labor Party, whose leader, an ethnic Indian, Mahendra Chaudhry, became prime minister, defeating Rabuka.

In May 2000, ethnic Fijians, led by businessman George Speight, seized parliament and held the prime minister, his cabinet, and several MPs hostage. They were eventually disarmed and arrested, but Chaudhry was dismissed as prime minister. The army appointed an ethnic

Fijian, Laisenia Qarase, leader of the nationalist Fiji United Party, as the new prime minister. His party won the elections in 2001 and the ethnic Indian Fijian Labor Party became the official opposition. Following the 2000 coup, the Commonwealth again expelled Fiji, but it was readmitted in 2002. In 2004, George Speight was sentenced to death but Fiji's president commuted the sentence to life imprisonment. In 2004, Fiji sent soldiers to Iraq for peacekeeping duties.

ECONOMY

Fiji is one of the more developed of the Pacific island states, But agriculture, which employs 70% of the population, remains the mainstay of its economy. Sugar cane, copra, and ginger are the main cash crops, and fish and timber are also exported. Other crops include bananas, cassava, coconuts, sweet potatoes, and rice.

Fiji mines gold, one of the main exports, silver, and limestone, but sugar processing makes up one-third of industrial activity. Other manufactures include beer, cement and cigarettes. Tourism is another important activity, with 300,000 to 400,000 visitors arriving annually. However, ethnic and political tensions have slowed the development of the tourist industry.

The leading markets for Fiji's exports are Australia, the United Kingdom, the United States and Japan. Imports come from Australia, New Zealand, the United States, and Japan. Fiji is heavily dependent on foreign aid.

SUVA

Seaport on the southeast coast of Viti Levu Island, in the southwest Pacific Ocean, capital of Fiji. It is the manufacturing and trade center of the islands, with an excellent harbor. Exports include tropical fruits, copra, and gold.

FINLAND

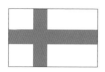

The flag of Finland was adopted in 1918, after the country had become an independent republic in 1917, following a century of Russian rule. The blue represents Finland's many lakes. The white symbolizes the blanket of snow which masks the land in winter.

The Republic of Finland (Suomi) has four geographical regions. In the south and west, on the Gulfs of Bothnia and Finland, is a low, narrow coastal strip, where most Finns live. The capital and largest city, Helsinki, is here. The Åland Islands lie in the entrance to the Gulf of Bothnia. Most of the interior is a beautiful wooded plateau, with more than 60,000 lakes. The Saimaa area is Europe's largest inland water system. A third of Finland lies within the Arctic Circle; this "land of the midnight sun" is called Lappi (Lapland).

Forests (birch, pine, and spruce) cover 60% of Finland. The vegetation becomes more and more sparse to the north, until it merges into Arctic tundra.

Area 130,558 sq mi [338,145 sq km]
Population 5,215,000
Capital (population) Helsinki (549,000)
Government Multiparty republic
Ethnic groups Finnish 93%, Swedish 6%
Languages Finnish and Swedish (both official)
Religions Evangelical Lutheran 89%
Currency Euro = 100 cents
Website www.government.fi

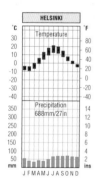

CLIMATE

Finland has short, warm summers; Helsinki's July average is 63°F [17°C]. In Lapland, the temperatures are lower, and in June the sun never sets. Winters are long and cold; Helsinki's January average is 21°F [26°C]. The North Atlantic Drift keeps the Arctic coasts free of ice.

HISTORY

In the 8th century, Finnish-speaking settlers forced the Lapps to the north. Between 1150 and 1809, Finland was under Swedish rule. The close links between the countries continue today.

Swedish remains an official language in Finland and one of the legacies of this period is a Swedish-speaking minority of 6% of the total population. In some localities on the south and west coasts, Swedish speakers are in the majority and Åland, an island closer to the Swedish coast than to Finland, is a self-governing province. Many towns use both Finnish and Swedish names. For example, Helsinki is Helsingfors, and Turku in Åbo in Swedish. Finnish bears little relation to the Swedish or any other Scandinavian language. It is closest to Magyar, the language of Hungary.

Lutheranism arrived in the 16th century. Wars between Sweden and Russia devastated Finland. Following the Northern War (1700–21), Russia gained much Finnish land. In the Napoleonic Wars, Russia conquered Finland and in 1809, it became an independent grand duchy of the Russian Empire, though the Russian tsar was its grand duke. Nationalist feelings developed during the 19th century, but in 1899 Russia sought to enforce its culture on the Finns. In 1903, the Russian governor suspended the constitution and became dictator, though following much resistance self-government was restored in 1906. Finland pro-

links with other north European countries and became an associate member of the European Free Trade Association (EFTA) in 1961. Finland became a full member of EFTA in 1986, in a decade when its economy was growing at a faster rate than that of Japan.

POLITICS

In 1992, along with most of its fellow EFTA members, Finland, which no longer needed neutrality, applied for membership of the European Union (EU). In 1994, the Finnish people voted in favor of joining the EU and the country officially joined on January 1, 1995. On January 1, 2002 the euro became Finland's sole unit of currency. Finland has also discussed the possibility of joining NATO. However, polls since the events of September 11, 2001 suggest that the majority of Finns favor nonalliance.

ECONOMY

Forests are Finland's most valuable resource. Forestry accounts for 35% of exports. The chief manufactures are wood and paper products. Post-1945 the economy has diversified. Engineering, shipbuilding, and textile industries have grown. Farming employs only 9% of workforce. The economy has slowly recovered from the recession caused by the collapse of the Soviet bloc.

claimed its independence in 1917, after the Russian Revolution and the collapse of the Russian Empire and, in 1919, it adopted a republican constitution. During World War I, the Soviet Union declared war on Finland and took the southern part of Karelia, where 12% of the Finnish people lived. Finland allied itself to Germany and Finnish troops regained southern Karelia. But at the end of the war, Russia regained southern Karelia and other parts of Finland. It also had to pay massive reparations to the Soviet Union.

After World War II, Finland pursued a policy of neutralism acceptable to the Soviet Union and this continued into the 1990s until the collapse of the Soviet Union. Finland also strengthened its

HELSINKI

Capital of Finland, in the south of the country, on the Gulf of Finland. Founded in 1550 by Gustavus I (Vasa), it became the capital in 1812. It has two universities (1849, 1908), a cathedral (1852), museums, and art galleries. The commercial and administrative center of the country, it is Finland's largest port. Industries include shipbuilding, engineering, food processing, ceramics, textiles.

FRANCE

The colors of this flag originated during the French Revolution of 1789. The red and blue are said to represent Paris, while the white represented the monarchy. The present design was adopted in 1794, and is meant to symbolize republican principles.

The Republic of France is the largest country in Western Europe. The scenery is extremely varied. The Vosges Mountains overlook the Rhine Valley in the northeast, the Jura Mountains and the Alps form the borders with Switzerland and Italy in the southeast, while the Pyrenees straddle France's border with Spain. The only large highland area entirely within France is the Massif Central between the Rhône-Saône Valley and the basin of Aquitaine. This dramatic area, covering one-sixth of the country, has peaks rising to more than 5,900 ft [1,800 m]. Volcanic activity dating back 10 to 30 million years ago appears in the form of steep-sided volcanic plugs. Brittany (Bretagne) and Normandy (Normande) form a scenic hill region. Fertile lowlands cover most of northern France, including the densely populated Paris Basin. Another major lowland area, the Aquitanian Basin, is in the southwest, while the Rhône-Saône Valley and the Mediterranean lowlands are in the southeast.

Area 212,934 sq mi [551,500 sq km]
Population 60,424,000
Capital (population) Paris (2,152,000)
Government Multiparty republic
Ethnic groups Celtic, Latin, Arab, Teutonic, Slavic
Languages French (official)
Religions Roman Catholic 85%, Islam 8%, others
Currency Euro = 100 cents
Website www.elysee.fr

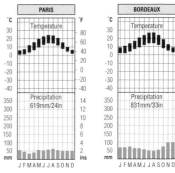

CLIMATE

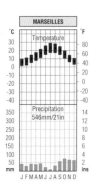

The climate varies from west to east and from north to south. The west comes under the moderating influence of the Atlantic Ocean, giving generally mild weather. To the east, summers are warmer and winters colder. The climate also becomes warmer as one

travels from north to south. The Mediterranean Sea coast experiences hot, dry summers and mild, moist winters. The Alps, Jura and Pyrenees mountains have snowy winters. Winter sports centers are found in all three areas. Large glaciers occupy high valleys in the Alps.

HISTORY

The Romans conquered France (then called Gaul) in the 50s BC. Roman rule began to decline in the 5th century AD and, in 486, the Frankish realm (as France was by then called) became independent under a Christian king, Clovis. In 800, Charlemagne, who had been king of the Franks since 768, became emperor

of the Romans. Through conquest, his empire extended from central Italy to Denmark, and from eastern Germany to the Atlantic Ocean. However, in 843, the empire was divided into three parts and the area of France contracted.

After the Norman invasion of England in 1066, large areas of France came under English rule. By 1453, after the Hundred Years' War, France drove most of the English out. In this war, the French kings lost much power to French nobles, but Louis XI, who reigned from 1461 to 1483, laid the foundations for absolute rule by French kings.

France later became a powerful monarchy, but the French Revolution (1789–99) ended absolute rule by

French kings. In 1799 Napoleon Bonaparte took power and fought a series of brilliant military campaigns before his final defeat in 1815. The monarchy was restored until 1848, when the Second Republic was founded. In 1852, Napoleon's nephew became Napoleon III, but the Third Republic was established in 1875. France was the scene of much fighting during World War I (1914–18) and World War II (1939–45), causing great loss of life and much damage to the economy.

Postwar aid from the United States started a revival in its economy, but Communist-led strikes often crippled production. France also faced growing support for independence movements in its overseas empire. After a bitter war, France withdrew from French Indochina in 1954 and then faced a long and costly struggle in Algeria, finally ending with Algeria's independence in 1962. The problems in Algeria caused considerable unrest in France in the 1950s and, in 1958, De Gaulle was recalled to power as prime minister. His government prepared a new constitution, establishing the Fifth Republic. It gave the president greater executive powers and reduced the power of parliament. The Electoral College elected De Gaulle as president for a seven-year term.

De Gaulle gave independence to many of France's overseas territories and made it a major player in an alliance of western European nations. In 1957, France became a founder member of the European Economic Community (EEC). De Gaulle opposed British membership in 1963, considering that Britain's links with the United States would give it too much influence in Europe's economy, but his popularity waned in the late 1960s when huge student demonstrations and workers' strikes paralyzed the country and he resigned as president in 1969. His successor, Georges Pompidou, changed course in foreign affairs by reestablishing closer contacts with the United States and supporting the entry of Britain into the EEC.

PARIS

The capital of France is situated on the River Seine. When the Romans took Paris in 52 BC, it was a small village on the Ile de la Cité on the Seine. Under their rule it became an important administrative center. During the 14th century Paris rebelled against the Crown and declared itself an independent commune. It suffered further civil disorder during the Hundred Years' War. In the 16th century, it underwent fresh expansion, its architecture strongly influenced by the Italian Renaiassance. In the reign of Louis XIII, Cardinal Richelieu established Paris as the cultural and political center of Europe. The French Revolution began in Paris when the Bastille was stormed by crowds in 1789. Under Emperor Napoleon I the city began to assume its present-day form. The work of modernization was continued under Napoleon III, when Baron Haussmann was commissioned to plan the boulevards, bridges, and parks. Although occupied during the Franco-Prussian War (1870–71) and again in World War II, Paris was not badly damaged. The city proper consists of the Paris department, Ville de Paris. Its many famous buildings and landmarks include the Eiffel Tower, Arc de Triomphe, Louvre, Notre Dame, and Centre Georges Pompidou. Paris is an important European cultural, commercial, and communications center and is noted for its fashion industry and for the manufacture of luxury articles.

POLITICS

Rapid urban growth has resulted in overcrowding and the growth of poorly built new districts to house immigrants, especially those from Spain and North Africa. The 4 million underprivileged workers from the Maghreb became a major political issue in the 1990s, leading to political successes in some areas for the extreme right. In France, as in most other countries, there also remains a disparity between the richer and the poorer regions. Other problems faced by France include unemployment, pollution, and the growing number of elderly people.

A socialist government under Lionel Jospin was elected in 1997. He increased the minimum wage, shortened the working week, and adopted the euro. However, in 2002 center-right parties won a resounding victory and Jean-Pierre Raffarin replaced Jospin as Prime Minister.

France has a long record of independence in foreign affairs and in 2003 it angered the US and some of its allies in the EU by opposing the invasion of Iraq, arguing that the UN weapons inspectors should be given more time to search for weapons of mass destruction in Iraq. France's stance angered some US congressmen who called for a boycott of French goods. The number of US tourist to France also fell.

A resounding "no" vote in the referendum on the European constitution in May 2005 led to the resignation of Raffarin and further decline in the relationship between Jacques Chirac and Tony Blair over the UK rebate and Common Agricultural Policy subsidies for French farmers.

ECONOMY

France is one of the world's most developed countries. It has the world's fourth largest economy. Its natural resources include its fertile soil, together with deposits of bauxite, coal, iron ore, petroleum and natural gas, and potash. France is also one of the world's top manufacturing nations and it has often innovated in bold and imaginative ways. The TGV [high speed train], Concorde and hypermarkets are all typical examples. Paris is a world center of fashion industries, but France has many other industrial towns and cities. Major manufactures include airplanes, automobiles, chemicals, electronic products, machinery, metal products, processed food, steel, and textiles.

Agriculture employs about 2% of the people, but France is the largest producer of farm products in Western Europe, producing most of the food it needs. Wheat is a leading crop and livestock farming is of major importance. The food-processing industry is well known, especially for its cheeses, such as Brie and Camembert, and its top-quality wines from areas such as Alsace, Bordeaux, Burgundy, Champagne, and the Loire valley. Fishing and forestry are leading industries. France is a popular year-round destination both for its beaches and for its mountains.

MONACO

The tiny Principality of Monaco consists of a narrow strip of coastline and a rocky peninsula on the French Riviera. Like the rest of the Riviera, it has mild, moist winters and dry, sunny summers. Average temperatures range from 50°F [10°C] in January to 75°F [24°C] in July. The average annual rainfall is about 31 in [800 mm].

Area 0.4 sq mi [1 sq km]
Population 32,000
Capital (population) Monaco (30,000)
Government Constitutional monarchy
Religions Roman Catholic 62%, Protestant 30%
Currency Euro = 100 cents
Website www.visitmonaco.com

The Genoese from northern Italy gained control of Monaco in the 12th century and, since 1297, it has been ruled for most of the time by the Genoese Grimaldi family. Monaco attracted little attention until the late 19th century when it developed into a major tourist resort. World attention was focused on Monaco in 1956 when Prince Rainier III of Monaco married the actress Grace Kelly. Their son, Prince Albert, became ruler upon Rainier's death in 2005. The country's wealth comes mainly from banking, finance, gambling, and tourism. There are three casinos, a marine museum, a zoo and botanical gardens. It also stages the Monte Carlo Rally and the Monaco Grand Prix. Manufactures include chemicals, electronic goods, and plastics. In 2001, France threatened to break its ties with Monaco unless it revised its legal system and prevented money laundering.

FRENCH GUIANA

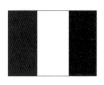

French Guiana flies the French flag. The colors originated during the French Revolution of 1789. The red and blue are said to represent Paris, while the white represented the monarchy. The present design was adopted in 1794, and is meant to symbolize republican principles.

French Guiana is a French overseas department and the smallest country in mainland South America. The coastal plain is swampy in places, but dry areas are cultivated, particularly near the capital Cayenne. The River Maroni forms the border with Suriname, and the River Oyapock its eastern border with Brazil. Inland lies a plateau, with the low Tumuchumac Mountains in the south. Most of the rivers run north toward the Atlantic Ocean.

Rainforest covers approximately 90% of the land and contains valuable hardwood species. Mangrove swamps line parts of the coast; other areas are covered by tropical savanna.

Area 90,000 sq km [34,749 sq mi]
Population 191,000
Capital (population) Cayenne (51,000)
Government Overseas department of France
Ethnic groups Black or Mulatto 66%, East Indian/Chinese and Amerindian 12%, White 12%, others 10%
Languages French (official)
Religions Roman Catholic
Currency Euro = 100 cents
Website www.guyane.pref.gouv.fr

French merchants founded Cayenne in 1637. The area became a French colony in the late 17th century, with a plantation economy dependent on African slaves. It

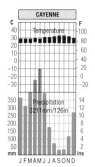

CLIMATE

French Guiana has a hot equatorial climate with high temperatures throughout the year. Rainfall is heavy, especially between December and June, but it is dry between August and October. Northeast trade winds blow across the country constantly.

HISTORY

The first people to live in what is now French Guiana were Amerindians. Today only a few of them survive in the interior. The first Europeans to explore the coast arrived in 1500 and they were followed by adventurers seeking El Dorado. The French were the first settlers in 1604 and

CENTRE SPATIAL GUYANAIS (CSG)

In 1964 the French Government chose Kourou as a satellite-launching base. An agreement in 1975 led to the newly formed European Space Agency (ESA) sharing the facility in return for financial aid. ESA now funds two-thirds of the CSG's annual budget. The CSG also counts among its clients industries from Brazil, Canada, India, Japan, and the United States. Kourou's geographical position makes it ideal for satellite launching. It is possible to launch both eastward and northward over the sea, without encountering populated areas meaning that the risks to human life are neglible.

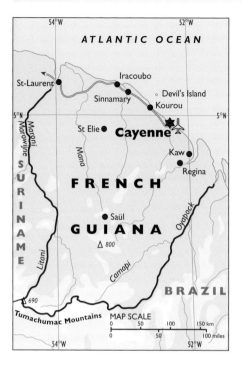

were introduced to work the land. From the time of the French Revolution, France used the colony as a penal settlement, and between 1852 and 1945 the country was notorious for the harsh treatment of prisoners. Captain Alfred Dreyfus was imprisoned on Devil's Island.

POLITICS

In 1946, French Guiana became an overseas department of France, and in 1974 it also became an administrative region. An independence movement developed in the 1980s, but most of the people want to retain links with France and continue to obtain financial aid to develop their territory.

ECONOMY

Although it has rich forest and mineral resources, such as bauxite (aluminum ore), French Guiana is a developing country with high unemployment. It depends greatly on France for money to run its services and the government is the country's biggest employer. Since 1975, Kourou has been the European Space Agency's rocketlaunching site and has earned money for France by sending communications satellites into space.

The main industries are fishing, forestry, gold mining and agriculture. Crops include bananas, cassava, rice, and sugarcane. French Guiana exports shrimp, timber, and rosewood essence.

remained French except for a brief period in the early 19th century. Slavery was abolished in 1848, and Asian laborers

DEVIL'S ISLAND
(ÎLE DU DIABLE)

The smallest island of the three Iles du Salut off the coast of French Guiana. A notorious French penal colony from 1852 to 1945 and most famous for the prison and the harsh treatment of its inmates whether they be political prisoners or murderers. Few prisoners made it off the island due to the high levels of disease and escape was almost impossible. The publicity surrounding the case of Captain Alfred Dreyfus in 1894 brought the horrors of this prison to the fore.

CAYENNE

Capital and chief port of French Guiana, on the Atlantic coast. Founded in 1643 and named after an Indian Chief. Places of interest include the cathedral, the ruined 17th-century Fort Cépérou, Hôtel de Ville and museums of local history.

GABON

Gabon's flag was adopted in 1960 when the country became independent from France. The central yellow stripe symbolizes the Equator which runs through Gabon. The green stands for the country's forests. The blue symbolizes the sea.

The Gabonese Republic lies on the Equator in west-central Africa. In area, it is a little larger than the United Kingdom, with a coastline 500 mi [800 km] long. Behind the narrow, partly lagoon-lined coastal plain, the land rises to hills, plateaus and mountains divided by deep valleys carved by the River Ogooué and its tributaries.

Dense rainforest covers about 75% of Gabon, with tropical savanna in the east and south. The forests teem with wildlife, and Gabon has several national parks and wildlife reserves.

Area 103,347 sq mi [267,668 sq km]
Population 1,355,000
Capital (population) Libreville (362,000)
Government Multiparty republic
Ethnic groups Bantu tribes: Fang, Bandjabi, Bapounou, Eshira, Myene, Nzebi, Obamba and Okande
Languages French (official), Fang, Myene, Nzebi, Bapounou/Eschira, Bandjabi
Religions Christianity 75%, animist, Islam
Currency CFA franc = 100 centimes
Website www.senat.ga

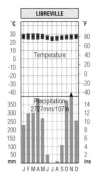

CLIMATE

Most of Gabon has an equatorial climate. There are high temperatures and humidity throughout the year. The rainfall is heavy and the skies are often cloudy.

HISTORY

Explorers from Portugal reached the Gabon coast in the 1470s, and the area later became a source of slaves.

In 1839 France established the first European settlement. In 1849 freed slaves founded Libreville. Gabon became a French colony in the 1880s and achieved full independence in 1960.

Léon Mba was Gabon's first president from 1960 to 1967. In 1964, an attempted coup was put down when French troops intervened and crushed the revolt. In 1967, following the death of Léon

Mba, Bernard-Albert Bongo, who later renamed himself El Hadj Omar Bongo, became president. He made Gabon a one-party state in 1968.

POLITICS

Free elections took place in 1990. The Gabonese Democratic Party (PDG) won a majority in the National Assembly.

RIVER OGOOUÉ (OGOWE)

The main river of Gabon, its basin drains nearly the entire country. Some of its tributaries can be found in Cameroon the Republic of Congo, and Equatorial Guinea. The Ogooué is about 560 mi [900 km] in length, rising in the northwest of the Bateke Plateau. It empties into the Gulf of Guinea, south of Port Gentil. The delta is approximately 62 mi [100 km] long and 62 mi wide. The total waterbasin is 86,431sq mi [223,856 sq km], and is largely made up of forest.

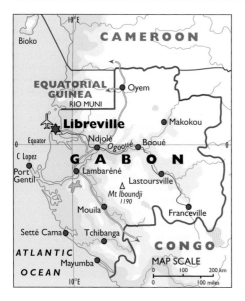

wished. In November 2005 he was again elected president and despite the preemptive protestations of the opposition, the elections were deemed free and fair by international observers.

ECONOMY

Gabon's abundant natural resources include its forests, petroleum and natural gas deposits near Port Gentil, together with manganese and uranium. These mineral deposits make Gabon one of Africa's wealthiest countries.

However, agriculture still employs about 75% of the workforce, most farmers producing little more than they need to support their families. Crops include bananas, cassava, corn, and sugarcane. Cocoa and coffee are grown for export. Other exports include petroleum, manganese, timber and wood products, and uranium.

President Bongo, of the PDG, won the presidential elections in 1993, although accusations of fraud and corruption led to riots in Libreville. The international community condemned Bongo for his harsh suppression of popular demonstrations. He was reelected in 1998. In 2003, constitutional changes enabled Bongo to stand as president as many times as he

BANTU

Group of African languages generally considered as forming part of the Benue-Congo branch of the Niger-Congo family. Swahili, Xhosa, and Zulu are among the most widely spoken of the several hundred Bantu languages spoken from the Congo Basin to South Africa, and almost all are tone languages, that is to say languages that use pitch to signal a difference in meaning between words. There are more than 70 million speakers of Bantu.

LIBREVILLE

Capital and largest city of Gabon, at the mouth of the River Gabon, on the Gulf of Guinea. Founded by the French in 1843, and named Libreville (French for Freetown) in 1849, it was initially a refuge for escaped slaves. The city expanded with the development of the country's minerals and is now also an administrative center. Places of interest include the Musée des Arts et Traditions, the French cultural center, the presidential palace, St. Marie's Cathedral, the Eglise St. Michel with its carved wooden columns, Nkembo, the Arboretum de Sibang and two cultural villages. Other industries include timber (hardwoods), palm oil, and rubber.

GAMBIA, THE

The colors represent features of the Gambian landscape. Green symbolizes the land and agricultural produce. Blue stands for the River Gambia, a vital trade route. Red represents the hot African sun. The two white bands stand for peace and unity.

The Republic of the Gambia is the smallest country in mainland Africa. It consists of a narrow strip of land bordering the River Gambia. The Gambia is almost entirely enclosed by Senegal, except along the short Atlantic coastline. The land is flat near the sea.

Mangrove swamps line the river banks. Much tropical savanna has been cleared for farming. The Gambia is rich in wildlife and has six national parks and reserves as well as several forest parks.

Area 4,361 sq mi [11,295 sq km]
Population 1,547,000
Capital (population) Banjul (42,000)
Government Military regime
Ethnic groups Mandinka 42%, Fula 18%, Wolof 16%, Jola 10%, Serahuli 9%, others
Languages English (official), Mandinka, Wolof, Fula
Religions Islam 90%, Christianity 9%, traditional beliefs 1%
Currency Dalasi = 100 butut
Website www.gambia.gm

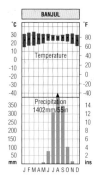

CLIMATE

The Gambia has hot and humid summers, but winter temperatures (November to May) drop to around 61°F [16°C]. In the summer, moist winds heading southwest bring rain, which is heaviest on the coast.

handed this colony over to France.

During the 1860s and 1870s, Britain and France discussed the exchange of the

HISTORY

Portuguese mariners reached Gambia's coast in 1455, when the area was part of the Mali empire. In the 16th century, Portuguese and English slave traders operated in the area. English traders bought rights to trade on the River Gambia in 1588, and in 1664 the English established a settlement on an island in the river estuary. In 1765, the British founded a colony called Senegambia, which included parts of present-day Gambia and Senegal. In 1783, Britain

MANGROVE

Common name for any one of 120 species of tropical trees or shrubs found in marine swampy areas. Also known as coastal woodland, tidal forest and mangrove forest.Its stilt-like aerial roots, which arise from the branches and hang down into the water, produce a thick undergrowth, useful in the reclaiming of land along tropical coasts. The composition varies greatly from region to region. Even within the same delta the composition of the mangrove can vary substantially according to the conditions of salinity, tidal system and substrate (soil foundation). Some species also have roots that rise up out of the water. It grows to a height of 70 ft [20 m].

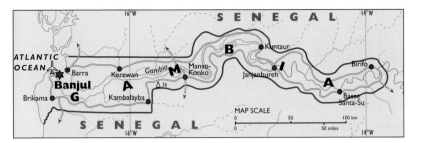

Gambia for some other French territory. No agreement was reached and Britain made the Gambia a British colony in 1888. It remained under British rule intil it achieved full independence in 1965 with Dawda Jawara as prime minister. In 1970 the Gambia became a republic.

POLITICS

Relations between the French-speaking Senegalese and the English-speaking Gambians form a major political issue. In 1981, an attempted coup in the Gambia was put down with the help of Senegalese troops. In 1982, The Gambia and Senegal set up a defense alliance, called the Confederation of Senegambia, though this alliance was dissolved in 1989.

In 1992, Jawara was reelected as president for a fifth term. In July 1994, he was overthrown in a military coup and fled into exile. The coup was led by Yahya Jammeh who was elected president in 1996. His regime faced charges of political repression. In 2001, Jammeh lifted the ban on opposition parties and was reelected, though he is still criticized for curtailing press freedom.

ECONOMY

Agriculture is the main activity, employing more than 80% of the workforce. However, the government announced in 2004 that large petroleum reserves had been discovered. The main food crops include cassava, millet, and sorghum, but peanuts and peanut products are the chief exports.

The money sent home by Gambians living abroad is important for the economy. Tourism is a growing industry.

RIVER GAMBIA

The dominant feature of the Gambia, rising in the Futa Jallon highlands in the Republic of Guinea and flowing to the Atlantic Ocean. The river valley runs along the entire length and breadth of the country. It provides precious sanctuary for many birds and fish through the creeks that adjoin it.

These areas are very important to bird watchers and lovers of nature as a whole. The river is also the habitat for much of the country's wildlife such as hippopotamus and crocodiles.

BANJUL

Capital of the Gambia, the city lies on St. Mary's Island, where the River Gambia enters the Atlantic Ocean. It was founded as a trading post by the British in 1816 and originally named Bathurst after Henry Bathurst, the secretary of the British Colonial Office. Banjul is Gambia's chief port and commercial center.

The main industry is peanut processing, though tourism is growing rapidly.

GEORGIA

The Republic of Georgia adopted a new flag in 2004. The flag had been in use some 500 years before as that of the medieval Georgian kingdom. It was subsequently used as the official symbol of the political party—United National Movement.

Georgia is located on the borders of Europe and Asia, facing the Black Sea. The land is rugged with the Caucasus Mountains forming its northern border. The highest mountain in this range, Mount Elbrus (18,506 ft [5,642 m]), lies over the border in Russia.

Lower ranges run through southern Georgia, through which pass the borders with Turkey and Armenia. The Black Sea coastal plains are in the west. In the east a low plateau extends into Azerbaijan. The main river in the east is the River Kura, on which the capital, Tbilisi, stands.

Area 26,911 sq mi [69,700 sq km]
Population 4,694,000
Capital (population) Tbilisi (1,268,000)
Government Multiparty republic
Ethnic groups Georgian 70%, Armenian 8%, Russian 6%, Azeri 6%, Ossetiam 3%, Greek 2%, Abkhaz 2%, others
Languages Georgian (official), Russian
Religions Georgian Orthodox 65%, Islam 11%, Russian Orthodox 10%, Amenian Apostolic 8%
Currency Lari = 100 tetri
Website www.parliament.ge

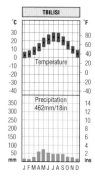

CLIMATE

The Black Sea plains have hot summers and mild winters, when the temperature seldom drops below freezing. Rainfall is heavy, but inland Tbilisi has moderate rainfall, with the heaviest rains in the spring and early summer.

HISTORY

The first Georgian state was set up nearly 2,000 years ago and, by the 3rd century BC, most of what is now Georgia was united as a single kingdom. For much of its history, Georgia was ruled by various conquerors. For example, between about 60 BC and the 11th century, the area was ruled successively by Romans, Persians, Byzantines, Arabs, and Seljuk Turks. Christianity was introduced in AD 330 and most Georgians are now members of the Georgian Orthodox Church. Georgia freed itself from foreign rule in the 11th and 12th centuries, but Mongol armies invaded in the 13th century. From the 16th to the 18th centuries, Iran and the Turkish Ottoman Empire struggled for control of the area.

In the late 18th century, Georgia sought the protection of Russia and, by the early 19th century was part of the Russian Empire. After the Russian Revolution of 1917, Georgia declared itself independent and was recognized by the League of Nations. However, Russian troops invaded in 1921, making Georgia part of the Soviet regime. From 1922, Georgia, Armenia, and Azerbaijan were linked, forming the Transcaucasian Republic. But, in 1936, the territories became separate republics within the Soviet Union. Renowned for their longevity, the people of Georgia are famous for producing Josef Stalin, who was born in Gori, 40 mi [65 km] northwest of the capital Tbilisi. Stalin ruled the Soviet Union from 1929 until his death in 1953.

which threatened its stability. In 2001, Georgia and Abkhazia signed a peace accord and agreed to the safe return of refugees. In 2002, Russian and Georgian troops attacked Chechen rebels in Pankisi Gorge in northeastern Georgia. US officials believed that Taliban fighters from Afghanistan and other Islamic terrorists had also moved into this region. In 2004, Mikhail Saakashvili was elected president, but his authority was challenged by separatists in the three minority regions.

POLITICS

A maverick among the Soviet republics, Georgia was the first to declare its independence after the Baltic states (April 1991) and deferred joining the Commonwealth of Independent States (CIS) until 1993.

In 1991, Zviad Gamsakhurdia, a non-Communist who had been democratically elected president of Georgia in 1990, found himself holed up in Tbilisi's KGB headquarters, under siege from rebel forces. They represented widespread opposition to his government's policies, ranging from the economy to the imprisonment of his opponents. In January 1992, following the breakup of the Soviet Union, Gamsakhurdia fled the country and a military council took power.

Georgia contains three regions of minority peoples: South Ossetia, in north-central Georgia, where civil war broke out in the early 1990s, with nationalists demanding the right to set up their own governments; Abkhazia in the northwest, which proclaimed its sovereignty in 1994 with fierce fighting continuing until the late 1990s; and Adjaria (or Adzharia) in the southwest, whose autonomy was recognized in Georgia's constitution in 2000.

In March 1992, Eduard Shevardnadze, former Soviet Foreign Minister, was named head of state and was elected, unopposed, later that year. Shevardnadze was reelected in 1995 and 2000, but Georgia faced mounting problems,

ECONOMY

Georgia is a developing country. Agriculture is important. Major products include barley, citrus fruits, grapes for winemaking, corn, tea, tobacco and vegetables. Food processing and silk- and perfumemaking are other important activities. Sheep and cattle are reared.

Barite (barium ore), coal, copper, and manganese are mined, and tourism is a major industry on the Black Sea coast. Georgia's mountains have huge potential for generating hydroelectric power, but most of Georgia's electricity is generated in Russia or Ukraine.

TBILISI (TIFLIS)

Largest city and capital of Georgia, on the upper River Kura. It was founded in the 5th century AD, and ruled successively by the Iranians, Byzantines, Arabs, Mongols, and Turks, before coming under Russian rule in 1801. Tbilisi's importance lies in its location on the trade route between the Black and Caspian Seas. It is now the administrative and economic focus of modern Transcaucasia. Industries include chemicals, petroleum products, locomotives, and electrical equipment.

GERMANY

This flag, adopted by the Federal Republic of Germany (West Germany) in 1949, became the flag of the reunified Germany in 1990. The red, black and gold colors date back to the Holy Roman Empire. They are associated with the struggle for a united Germany from the 1830s.

The Federal Republic of Germany lies in the heart of Europe. It is the fifth-largest country in Europe. Germany divides into three geographical regions: the north German plain, central highlands, and the south Central Alps.

The rivers Elbe, Weser, and Oder drain the fertile northern plain, which includes the industrial centers of Hamburg, Bremen, Hanover, and Kiel. In the east lies the capital, Berlin, and the former East German cities of Leipzig, Dresden, and Magdeburg. Northwest Germany (especially the Rhine, Ruhr and Saar valleys) is Germany's industrial heartland. It includes the cities of Cologne, Essen, Dortmund, Düsseldorf, and Duisburg.

The central highlands include the Harz Mountains and the cities of Munich, Frankfurt, Stuttgart, Nuremberg, and Augsburg.

Southern Germany rises to the Bavarian Alps on the border with Switzerland and Germany's highest peak, Zugspitze, (9,721 ft [2,963 m]). The Black Forest, overlooking the Rhine valley, is a major tourist attraction. The region is drained by the River Danube.

Area 137, 846 sq mi [357,022 sq km]
Population 82,425,000
Capital (population) Berlin (3,387,000)
Government Federal multiparty republic
Ethnic groups German 92%, Turkish 3%, Serbo-Croatian, Italian, Greek, Polish, Spanish
Languages German (official)
Religions Protestant (mainly Lutheran) 34%, Roman Catholic 34%, Islam 4%, others
Currency Euro = 100 cents
Website www.deutschland.de

CLIMATE

Germany has a temperate climate. The northwest is warmed by the North Sea. The Baltic lowlands in the northeast are cooler. In the south, the climate becomes more continental.

HISTORY

Around 3,000 years ago, various tribes from northern Europe began to settle in what is now Germany, occupying the valleys of the Rhine and the Danube. The Romans called this region Germania after the Germani, the name of one of the tribes. In the 5th century, the Germanic tribes attacked the Roman Empire and plundered Rome. The western part of the Roman Empire split up into several kingdoms, the largest of which was the Kingdom of the Franks.

In 486, Clovis, a Frankish king, extended his rule to include Gaul (now France) and western Germany, introducing Christianity and other Roman practices. Frankish ruler, Charlemagne, came to power in 768 and established his capital at Aachen. From 962, much of the German Empire became part of what was later known as the Holy Roman

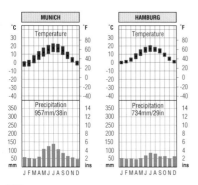

Empire under King Otto II of Germany.

In 1517, a German monk, Martin Luther, began to criticize many of the practices and teachings of the Roman Catholic Church. A Protestant movement called the Reformation soon attracted much support. By the early 17th century, the people of Germany were deeply divided by political and religious rivalries. The Thirty Years' War, which began in 1618 and lasted for 30 years, ravaged much of the country with Germany losing territory to France and Sweden, and itself being split into hundreds of states and free cities. It took almost 200 years for Germany to recover

In the 17th century, the Hohenzollern family began to assume importance in eastern Germany, gradually extending their power and building a professional civil service and army.

Prussia stayed out of the Napoleonic wars until 1806, but following defeats by Napoleon, lost its territories west of the Elbe. Prussia did help defeat Napoleon's armies at the battles of Leipzig (1813) and Waterloo (1815), and following the Napoleonic wars, gained the Rhineland, Westphalia, and much of Saxony.

In the early 1860s, the Prussian king, Wilhelm I, appointed Otto von Bismarck as prime minister. Bismarck set about strengthening Prussian power through three short wars. One conflict led to the acquisition of Schleswig-Holstein from Denmark, another led to the annexation of territory from Austria. The third was the Franco-Prussian War (1870–1), following which a victorious Germany was granted Alsace and part of Lorraine. In 1871, Wilhelm I was crowned the first Kaiser of the new German Empire and Bismarck became the Chancellor and head of government. Bismarck sought to consolidate German power and avoid conflict with Austria-Hungary and Russia, but was forced to resign in 1890 when Wilhelm II wanted to establish his own authority and extend Germany's influence in the world. Wilhelm's ambitions led Britain and France to establish the Entente Cordiale in 1904, with Britain and Russia signing a similar agreement in 1907. This left Europe divided, with Germany, Austria-Hungary and Italy forming the Triple Alliance.

After Germany and its allies were defeated in World War I, Germany became a republic and lost territories. Overseas, it lost its colonies. Germany's humiliation under the terms of the Versailles Treaty caused much resentment, made worse by the economic collapse of 1922–23. Support grew for the Nazi Party and its leader Adolf Hitler, who became chancellor in 1933. Hitler's order to invade Poland in 1939 triggered off World War II. His armies were finally defeated in 1945 and the country left in ruins. Germany was obliged to transfer the area east of the Oder and Neisse rivers to Poland and the Soviet Union. German-speaking inhabitants were expelled and the remainder of Germany was occupied by the four victorious Allied powers. In 1948, West Germany, consisting of the American, British, and French zones, was proclaimed the Federal Republic of Germany with its provisional capital at Bonn, while the Soviet zone became the German Democratic Republic with its capital in East Berlin.

POLITICS

The post-war partition of Germany together with its geographical position, made it a central hub of the Cold War, which ended with the collapse of Communism in the late 1980s/early 1990s. The reunification of Germany came on October 3, 1990. West Germany, had become a showpiece of the West through its phenomenal recovery and sustained growth, the so-called "economic miracle." It played a major part, together with France, in the revival of Western Europe through the development of the European Community (now the European Union). Although East Germany had achieved the highest standard of living in the Soviet bloc, it was short of the levels of the EU members.

Following reunification, the new country adopted the name the Federal Republic of Germany. Massive investment was needed to rebuild the East's industrial base and transport system, meaning increased taxation. In addition, the new nation found itself funnelling aid into Eastern Europe. Germany led the EU in recognizing the independence of Slovenia, Croatia, and the former Soviet republics. There were also social effects. While Germans in the West resented added taxes and the burden imposed by the East, easterners resented what many saw as the overbearing attitudes of westerners. Others feared a revival of the far right, with neo-Nazis and other right-wingers protesting against the increasing numbers of immigrant workers.

The creation of a unified state was far more complicated, expensive, and lengthy an undertaking than envisaged when the Berlin Wall came down. In 1998, the center-right government of Helmut Kohl, who had presided over reunification, was defeated by the left-of-center Social Democratic Party (SPD), led by Gerhard Schröder. Schröder led an SPD-Green Party coalition which set about tackling Germany's high unemployment and a sluggish economy. Following the attacks on the United States on September 11, 2001, Schröder announced Germany's support for the campaign against terrorism, although Germany opposed the invasion of Iraq in 2003. In 2005, Schröder was narrowly defeated in elections. A broad left-right coalition was set up. The conservative Angela Merkel became Chancellor.

ECONOMY

Despite the problems associated with reunification, Germany has the world's third largest economy after the United States and Japan. The foundation of the "economic miracle" that led to Germany's astonishing postwar recovery was manufacturing.

Germany's industrial strength was based on its coal reserves, though petroleum-burning and nuclear power plants have become increasingly important since the 1970s. Lower Saxony has oil fields, while southern Germany also obtains power from hydroelectric plants.

The country has supplies of potash and rock salt, together with smaller quantities of copper, lead, tin, uranium, and zinc. The leading industrial region is the Ruhr, which produces iron and steel, together with major chemical and textiles industries. Germany is the world's third largest producer of automobiles, while other manufactures include cameras, electronic equipment, fertilizers, processed food, plastics, scientific instruments, ships, tools, and wood and pulp products.

Agriculture employs 2.4% of the workforce, but Germany imports about a third of its food. Barley, fruits, grapes, oats, potatoes, rye, sugarbeet, vegetables, and wheat are grown. Beef and dairy cattle are raised, together with pigs, poultry, and sheep.

BERLIN

Capital and largest city of Germany, on the River Spree, northeast Germany. Berlin was founded in the 13th century. It became the residence of the Hohenzollerns and the capital of Brandenburg, and later of Prussia. It rose to prominence as a manufacturing town and became the capital of the newly formed state of Germany in 1871. Virtually destroyed at the end of World War II, Berlin was divided into four sectors: British, French, US, and Soviet. On the formation of East Germany, the Soviet sector became East Berlin, the rest West Berlin. In 1961 East Germany erected the Berlin Wall, which divided the city until 1989. On the reunification of Germany in 1990, East Berlin and West Berlin amalgamated. Sights include the Brandenberg Gate, the Kaiser Wilhelm Memorial Church, and the Victory Column in the Tiergarten. Parts of the Berlin Wall remain as a monument.

GHANA

Ghana's flag has red, green and yellow bands like the flag of Ethiopia, Africa's oldest independent nation. These colors symbolize African unity. The black star is a symbol of African freedom. Ghana's flag was adopted when the country became independent in 1957.

The Republic of Ghana faces the Gulf of Guinea in West Africa. This hot country, just north of the Equator, was formerly called the Gold Coast. Behind the thickly populated southern coastal plains, which are lined with lagoons, lies a plateau region in the southwest.

Northern Ghana is drained by the Black and White Volta Rivers, which flow into Lake Volta. This lake, which has formed behind the Akosombo Dam, is one of the world's largest artificially created lakes.

Rainforests grow in the southwest. To the north, the forests merge into savanna (tropical grassland with some woodland). More open grasslands dominate in the far north.

Area 92,098 sq mi [238,533 sq km]
Population 20,757,000
Capital (population) Accra (949,000)
Government Republic
Ethnic groups Akan 44%, Moshi-Dagomba 16%, Ewe 13%, Ga 8%, Gurma 3%, Yoruba 1%
Languages English (official), Akan, Moshi-Dagomba, Ewe, Ga
Religions Christianity 63%, traditional beliefs 21%, Islam 16%
Currency Cedi = 100 pesewas
Website www.ghana.gov.gh

Portuguese explorers reached the area in 1471 and named it the Gold Coast. The area became a center of the slave trade in the 17th century. The slave trade was ended in the 1860s and the British gradually took control of the area. The country became independent in 1957, when it was renamed Ghana.

CLIMATE

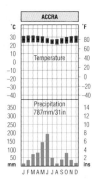

Ghana has a tropical climate. A cool offshore current reduces temperatures on the coast, and the north is hotter. The heaviest rains occur in the southwest. There are marked dry seasons in northern and eastern Ghana.

LAKE VOLTA

The largest artificial lake in Africa. Created in 1964 by the closure of the Akosombo Dam 60 mi [100 km] upstream from the mouth of the Volta. The reservoir covers 330,000 sq mi [850,000 sq km] and stretches 190 mi [320 km] from north to south and 250 mi [400 km] east to west. Built to provide hydroelectricity, the dam has rarely achieved full capacity as it takes three or four years of consecutive flood to fill the lake to its full capacity. The construction of the dam led to a substantial reduction in the fisheries of the Volta Delta.

HISTORY

Ghana was a great African empire which flourished to the northwest of present-day Ghana between the AD 300s and 1000s. Modern Ghana was the first country in the Commonwealth to be ruled by black Africans.

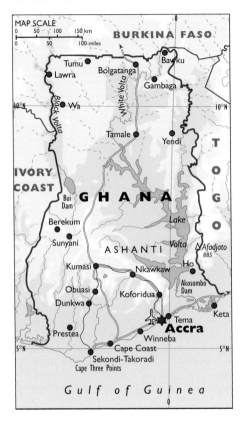

MAP SCALE
0   50   100   150 km
0   50   100 miles

in 2000. The economy expanded in the 1990s, largely because the government followed World Bank policies. When Rawlings retired, the opposition leader, John Agyekum Kufuor, leader of the New Patriotic Party, was elected president, defeating Rawlings' vice-president. He was reelected in 2004.

ECONOMY

The World Bank classifies Ghana as a "low-income" developing country. Most people are poor and farming employs 59% of the population. Food crops include cassava, peanuts, corn, millet, plantains, rice, and tropical yams. But cocoa is the most valuable export crop. Timber and gold are also exported. Other valuable crops include tobacco, coffee, coconuts, and palm kernels.

Many small factories produce goods, such as beverages, cement and clothing, for local consumption. The aluminium smelter at Tema, a port near Accra, is the country's largest factory. There are plans to construct around 378 mi [600 km] of pipeline which will form part of the West African Gas Pipeline Project. The aim is to lessen the dependence of electricity production on hydroelectric stations.

POLITICS

After independence, attempts were made to develop the economy by creating large state-owned manufacturing industries. But debt and corruption, together with falls in the price of cocoa, the chief export, caused economic problems. This led to instability and frequent coups. In 1981, power was invested in a Provisional National Defense Council, led by Flight-Lieutenant Jerry Rawlings.

The government steadied the economy and introduced several new policies, including the relaxation of government controls. In 1992, a new constitution was introduced which allowed for multiparty elections. Rawlings was reelected later that year, and served until his retirement

ACCRA

Capital and largest city of Ghana, on the Gulf of Guinea. Occupied by the Ga people since the 15th century, it became the capital of Britain's Gold Coast colony in 1875. Today it is a major port and economic center and is increasingly popular with tourists. Industries include engineering, timber, textiles, and chemicals. The principal export is cacao.

GREECE

Blue and white became Greece's national colors during the war of independence (1821–9). The nine horizontal stripes on the flag, which was finally adopted in 1970, represent the nine syllables of the battle cry "Eleutheria i thanatos" ("Freedom or Death").

The Hellenic Republic, the official name of Greece, is a rugged country lying at the southern end of the Balkan Peninsula. Olympus (Ólimbos), at 2,917 m [9,570 ft], is the highest peak. Nearly a fifth of the land area is made up of around 2,000 islands, mainly in the Aegean Sea, east of the main peninsula, but also in the Ionian Sea to the west. Only 154 are inhabited. The island of Crete is structurally related to the main Alpine fold mountain system to which the mainland Pindos Range belongs.

Area 50,949 sq mi [131,957 sq km]
Population 10,648,000
Capital (population) Athens (772,000)
Government Mulktiparty republic
Ethnic groups Greek 98%
Languages Greek (official)
Religions Greek Orthodox 98%
Currency Euro = 100 cents
Website www.culture.gr

peak in 461–431 BC, but in 338 BC Macedonia became the dominant power. In

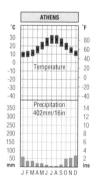

CLIMATE

Low-lying areas in Greece have mild, moist winters and hot, dry summers. The east coast has more than 2,700 hours of sunshine a year and only about half of the rainfall of the west. The mountains have a more severe climate, with snow on the higher slopes in winter.

HISTORY

Around 2,500 years ago, Greece became the birthplace of Western civilization. Crete was the center of the Minoan civilization, an early Greek culture, between about 3000 and 1400 BC. Following the end of the related Mycaenean period on the mainland (1580-100 BC), a "dark age" lasted until about 800 BC. In about 750 BC, the Greeks began to colonize the Mediterranean, creating wealth through trade. The city-state of Athens reached its

GREEK ART AND ARCHITECTURE

Greek art divides into four periods: Geometric (late 11th–late 8th century BC), Archaic (late 8th century–480 BC), Classical (480–323 BC), and Hellenistic (323–27 BC). Only a few small bronze horses survive from the Geometric period. During the Archaic period, stone sculpture appeared, vase painting proliferated, and the human figure became a common subject. Civic wealth and pride was a feature of the Classical period, and sculpture reached its peak of serene perfection. The Hellenistic period is noted for increasingly dramatic works. The earliest remaining Doric temple is the Temple of Hera at Olympia (late 7th century BC), the most outstanding is the Parthenon. Among Ionic temples, the Erechtheum is considered the most perfect. The Corinthian mausoleum at Halicarnassus (350 BC) was one of the Seven Wonders of the World.

quakes. In 2000, Greece and Turkey signed agreements aimed at improving relations between them.

ECONOMY

Manufacturing is important. Products include processed food, cement, chemicals, metal products, textiles, and tobacco. Lignite, bauxite, and chromite are mined.

Farmland and grazing land cover about 75% of the land. Major crops include barley, grapes for winemaking, dried fruits, olives, potatoes, sugarbeet and wheat. Poultry, sheep, goats, pigs and cattle are raised. The vital tourist industry is based on the warm climate, beautiful scenery, and the historical sites dating back to the days of classical Greece.

334–331 BC, Alexander the Great conquered southwestern Asia. Greece became a Roman province in 146 BC and, in AD 365, part of the Byzantine Empire. In 1453, the Turks defeated the Byzantine Empire. But between 1821 and 1829, the Greeks defeated the Turks. The country became an independent monarchy in 1830.

After World War II (1939–45), when Germany had occupied Greece, a civil war broke out between Communist and nationalist forces. This war ended in 1949, and a military dictatorship took power in 1967. The monarchy was abolished in 1973 and democratic government restored in 1974.

POLITICS

Greece joined the European Community in 1981. But despite efforts to develop the economy, Greece remains one of the poorest countries in the European Union. The euro became the sole unit of currency on January 1, 2002. Relations with Turkey have long been difficult. In 1999, the two countries helped each other when both were hit by major earth-

ATHENS (ATHÍNAI)

Capital and largest city of Greece, situated on the Saronic Gulf. The ancient city was built around the Acropolis, a fortified citadel, and was the greatest artistic and cultural center in ancient Greece, gaining importance after the Persian Wars (500–449 BC). Athens prospered under Cimon and Pericles during the 5th century BC and provided a climate in which the great classical works of philosophy and drama were created. The most noted artistic treasures are the Parthenon (438 BC), the Erechtheum (406 BC), and the Theatre of Dionysus (500 BC, the oldest of the Greek theaters). Modern Athens and its port of Piraeus form a major Mediterranean transportation and economic center. Overcrowding and severe air pollution are damaging the ancient sites and the tourist industry.

GUATEMALA

Guatemala's flag was adopted in 1871, but its origins go back to the days of the Central American Federation (1823–39), which was set up after the break from Spain in 1821. The Federation included Costa Rica, El Salvador, Guatemala, Honduras, and Nicaragua.

The Central American republic of Guatemala contains a densely populated fertile mountain region. The capital, Guatemala City, is situated here. The highlands run east–west and contain many volcanoes. Guatemala is subject to frequent earthquakes and volcanic eruptions. Tajumulco, an inactive volcano, is the highest peak in Central America, at 13,816 ft [4,211 m].

South of the highlands lie the Pacific coastal lowlands. North of the highlands is the thinly populated Caribbean plain and the vast Petén tropical forest. Guatemala's largest lake, Izabal, drains into the Caribbean Sea.

Hardwoods, such as mahogany, rubber, palm, and chicozapote (from which chicle, used in chewing gum, is obtained), grow in the tropical forests in the north, with mangrove swamps on the coast. Oak and willow grow in the highlands, with fir and pine at higher levels. Much of the Pacific plains is farmland.

Area 42,042 sq mi [108,889 sq km]
Population 14,281,000
Capital (population) Guatemala City (1,007,000)
Government Republic
Ethnic groups Ladino (mixed Hispanic and Amerindian) 55%, Amerindian 43%, others 2%
Languages Spanish (official), Amerindian languages
Religions Christianity, indigenous Mayan beliefs
Currency US dollar; Quetzal = 100 centavos
Website
www.visitguatemala.com/site/home/index_3.html

HISTORY

Between AD 300 and 900, the Quiché branch of the Maya ruled much of Guatemala, but inexplicably abandoned their cities on the northern plains. The Quiché ruins at Tikal are the tallest temple pyramids in the Americas.

In 1523–24, the Spanish conquistador Pedro de Alvarado defeated the native tribes. In 1821, Guatemala became independent. From 1823–39, it formed part of the Central American Federation.

CLIMATE

Guatemala lies in the tropics. The lowlands are hot and wet, with the central mountain region being cooler and drier. Guatemala City, at about 5,000 ft [1,500 m] above sea level, has a pleasant, warm climate, with a marked dry season between November and April.

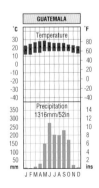

GUATEMALA CITY (CIUDAD GUATEMALA)

The capital of Guatemala lies on a plateau in the Sierra Madre. It is the largest city in Central America. Founded in 1776, the city was the capital of the Central American Federation from 1823–39. It was badly damaged by earthquakes in 1917–18 and 1976. Industries include mining, furniture, textiles, handcrafts.

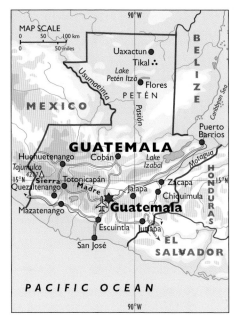

tions, and a peace agreement with the URNG ended 35 years of civil war. Alfonso Portillo became president in 1999 elections, despite admitting to killing two men.

In 2004, US$35 million was paid in damages to victims of the civil war.

ECONOMY

The World Bank classifies Guatemala as a "lower-middle-income" developing country. Agriculture employs nearly half of the workforce and coffee, sugar, bananas and beef are the leading exports. Other important crops include cardomon and cotton, while corn is the chief food crop. But Guatemala still has to import food to feed the people.

Tourism and manufacturing are growing in importance. Manufactures include processed farm products, textiles, wood products, and handcrafts.

Various dictatorial regimes interfered in the politics of other Central American states, arousing resentment and leading to the creation of the Central American Court of Justice. In 1941, Guatemala nationalized the German-owned coffee plantations. After World War II, Guatemala embarked on further nationalization of plantations.

In 1960, the mainly Quiché Guatemalan Revolutionary National Unity Movement (URNG) began a guerrilla war that claimed more than 200,000 lives. During the 1960s and 1970s, terrorism and political assassinations beset Guatemala. In 1976, an earthquake devastated Guatemala City, killing more than 22,000 people.

POLITICS

Guatemala has a long-standing claim over Belize, but this was reduced in 1983 to the southern fifth of the country. In 1985, Guatemala elected its first civilian president for 15 years. Alvaro Arzú Irigoyen became president in 1996 elec-

---

THE PETEN

The largest province of Guatemala situated in the northern third of the country. It is a vast lowland region which, until recently, had only a small population and has thereby retained much of its original forest. This small population is due, in part, to the hot, moist climate and distinct lack of surface water over much of the region during the dry season. Another reason is that for many years the region served as a base for anti-government guerrilla armies, thus making the area undesirable to the populous as a whole. The Peten's forests, along with areas of Belize and southern Mexico, make up the largest unbroken stretch of tropical forest north of the Brazilian Amazon.

GUINEA

Guinea's flag was adopted when the country became independent from France in 1958. It uses the colors of the flag of Ethiopia, Africa's oldest nation, which symbolize African unity. The red represents work, the yellow justice and the green solidarity.

The Republic of Guinea, which faces the Atlantic Ocean in West Africa, can be divided into four regions: an alluvial coastal plain, which includes the capital, Conakry; the highland region of the Fouta Djallon, the source of one of Africa's longest rivers, the Niger; the northeast savanna; and the southeast Guinea Highlands, which rise to 5,748 ft [1,752 m] at Mount Nimba.

Mangrove swamps grow along parts of the coast. Inland, the Fouta Djallon is largely open grassland. Northeastern Guinea is tropical savanna, with acacia and shea scattered across the grassland. Rainforests of ebony, mahogany, and teak grow in the Guinea Highlands.

Area 94,925 sq mi [245,857 sq km]
Population 9,246,000
Capital (population) Conakry (1,232,000)
Government Multiparty republic
Ethnic groups Peuhl 40%, Malinke 30%, Soussou 20%, others 10%
Languages French (official)
Religions Islam 85%, Christianity 8%, traditional beliefs 7%
Currency Guinean franc = 100 cauris
Website www.guinee.gov.gn

CLIMATE

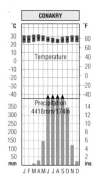

Guinea has a tropical climate. Conakry on the coast has heavy rains during its relatively cool season between May and November. Hot, dry harmattan winds blow southwestward from the Sahara in the dry season. The Fouta Djalon is cooler than the coast. The driest region is in the northeast. This region and the southeastern highlands have greater temperature variations than on the coast.

HISTORY

The northeast Guinea plains formed part of the medieval Empire of Ghana. The Malinke formed the Mali Empire, which dominated the region in the 12th century. The Songhai Empire supplanted the Malinke in the 15th century.

Portuguese explorers arrived in the mid-15th century and the slave trade began soon afterward. From the 17th century, other European slave traders became active in Guinea. In the early 18th century, the Fulani embarked on a *jihad* (holy war) and gained control of the Fouta Djallon. Following a series of wars, France won control in the mid-19th cen-

SONGHAI

West African empire, founded around AD 700. In 1468, Sonni Ali captured the market city of Timbuktu, and the Songhai Empire acquired control of most of the trade in western Africa. Askia Muhammad I succeeded Sonni, and further increased the stranglehold on trade routes. The Empire began to disintegrate because of factional in-fighting. The Songhai peoples still control much of the trans-Saharan trade.

2000, rebel incursions from these countries killed more than 1,000 people, caused massive population displacement, and threatened to destablilize Guinea. Conté was reelected in 2003, though the poll was boycotted by the opposition. His ailing health brings into question whether he will survive the full term. The president survived an assassination attempt in 2005 when shots were fired at his car.

ECONOMY

The World Bank classifies Guinea as a "low-income" developing country. It is the world's second-largest producer of bauxite, which accounts for 90% of its exports. Guinea has 25% of the world's known reserves of bauxite.

tury and, in 1891, it made Guinea a French colony. France exploited its bauxite deposits and mining unions developed.

In 1958, Guinea voted to become an independent republic and France severed all aid. Its first president, Sékou Touré (1958–84), adopted a Marxist program of reform and embraced Pan-Africanism. Opposition parties were banned, and dissent brutally suppressed. In 1970, Portuguese Guinea (now Guinea-Bissau) invaded Guinea. Conakry later acted as the headquarters for independence movements in Guinea-Bissau. A military coup followed Touré's death in 1984, and Colonel Lansana Conté established the Military Committee for National Recovery (CMRN). Conté improved relations with the West and introduced free-enterprise policies.

Other natural resources include diamonds, gold, iron ore, and uranium. Due to the mining industry, the rail and road infrastructure is improving. Agriculture (mainly at subsistence level) employs 78% of the workforce. Major crops include bananas, cassava, coffee, palm kernels, pineapples, rice, and sweet potatoes. Cattle and other livestock are raised in highland areas.

POLITICS

Civil unrest forced the introduction of a multiparty system in 1992. Elections in 1993 confirmed Conté as president, amid claims of voting fraud. In February 1996, Conté foiled an attempted military coup. He was reelected in 1998.

By 2000, Guinea was home to about 500,000 refugees from the wars in neighboring Sierra Leone and Liberia. In

CONAKRY

Capital city of Guinea. Conakry is located on Tombo Island, in the Atlantic Ocean and is connected to the mainland via a causeway. Founded in 1884, then occupied by French forces in 1887, it is Guinea's largest city. It is a major port and the administrative and commercial center of Guinea. Its economy revolves largely round the port, from which it exports aluminum and bananas. Manufactures include food products, automobiles, and beverages. The city is noted for its botanical gardens.

GUINEA-BISSAU

The red symbolizes the blood shed in the liberation struggle. Yellow stands for the hot African sun. Green represents the fertile land and hope for the future. The black star represents the African continent and pays homage to the flag of Ghana, the first African colony to gain independence.

The Republic of Guinea-Bissau is a small country in West Africa. The land is mostly low lying, with a broad, swampy, coastal plain and many flat offshore islands, including the Bijagós Archipelago.

Mangrove forests line the coasts, and dense rainforest covers much of the coastal plain.

Area 13,948 sq mi [36,125 sq km]
Population 1,388,000
Capital (population) Bissau (200,000)
Government Interim government
Ethnic groups Balanta 30%, Fula 20%, Manjaca 14%, Mandinga 13%, Papel 7%
Languages Portuguese (official),Crioulo
Religions Traditional beliefs 50%, Islam 45%, Christianity 5%
Currency CFA franc = 100 centimes
Website www.republica-da-guine-bissau.org

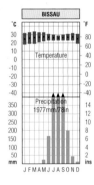

CLIMATE

The country has a tropical climate, with its dry season from December to May and its rainy season from June to November.

HISTORY

It was first visited by Portuguese navigators in 1446. From the 17th to the early 19th century, Portugal used the coast as a slave-trading base. Portugal appointed a governor to administer Guinea-Bissau and the Cape Verde Islands in 1836, but in 1879 the territories separated and Guinea-Bissau became a colony, then called Portuguese Guinea. Development was slow, partly because the territory did not attract settlers on the same scale as Portugal's much healthier African colonies of Angola and Mozambique.

In 1956, African nationalists in Portuguese Guinea and Cape Verde founded the African Party for the Independence of Guinea and Cape Verde (PAIGC). Because Portugal seemed determined to hang on to its overseas territories, the PAIGC began a guerrilla war in 1963. By 1968, it held two-thirds of the country. In 1972, a rebel National Assembly, elected by the people in the PAIGC-controlled areas, voted to make the country independent as Guinea-Bissau.

POLITICS

In 1974, it formally achieved independence (followed by Cape Verde in 1975). The independent nation faced many problems arising from its under-developed economy and its lack of trained personnel. Guinea-Bissau's leaders favored

BISSAU

Capital of Guinea-Bissau, near the mouth of the Geba, western Africa. Established by the Portuguese as a slaving center in 1687, it became a free port in 1869. It replaced Bonama as capital in 1941. The port has recently been improved. Industries include petroleum processing.

on national policies and socialist reforms.

In 1991, the PAIGC voted to introduce a multiparty system. The PAIGC won the 1994 elections, and Vieira was elected president. In 1998 an army rebellion sparked a civil war. The army rebels took power in 1999, but elections were held in 1999–2000. Kumba Ialá became president in 2000 but was overthrown in a coup in 2003. Civilian government was restored in 2004 when parliamentary elections were held.

union with Cape Verde. This objective was abandoned in 1980, when an army coup, led by Major João Vieira, overthrew the government. The Revolutionary Council which took over opposed unification with Cape Verde; it concentrated

ECONOMY

Guinea-Bissau is a poor country. Agriculture employs more than 70% of the workforce, but most farming is at subsistence level. Major crops include beans, coconuts, peanuts, corn, and rice. Fishing is also important.

BIJAGOS ARCHIPELAGO

Formed by the prehistoric delta of the Rio Grande de Buba and the Rio Geba, the Bijagos Archipelago consists of 88 islands and islets spread over 4,000 sq mi [10,000 sq km]. The rainy season brings freshwater into the coastal zone, while coastal currents from north and south meet, making the delta region vulnerable yet biologically rich. Between the islands, extensive mud flats are drained by a network of canals and creeks as the tide recedes. The characteristic vegetation of the islands are the palm groves which have replaced the once endless forest. The tidal areas remain relatively untouched, forming a unique mosaic of mangroves and tidal flats. Hippos have adapted to life in sea water and can be seen plodding along the beaches, while otters hunt for shellfish or wallow in the creeks together with manatees, for which the archipelago forms one of the most important strongholds in the region. Two species of dolphin live here. Reptiles include two species of crocodile and four species of marine turtle, including the green turtle, for which the Bijagos Archipelago is the most important breeding site in West Africa. Some 7,000 female green turtles nest here each year, and satellite tracking of turtles equipped with transmitters has shown that, after laying, they move north along the coast toward the waters of the Gulf d'Arguin in Mauritania. The archipelago is inhabited almost exclusively by the Bijagos ethnic group, 25,000 of whom live year round on 20 islands, with 20 others used at particular times of the year. The Bijagos assiduously maintain the palms from which they extract oil and palm wine; they also harvest shellfish from the mud flats and catch fish.

GUYANA

Guyana's flag was adopted in 1966 when the country became independent from Britain. The colors symbolize the people's energy in building a new nation (red), their perseverance (black), minerals (yellow), rivers (white), and agriculture and forests (green).

The Cooperative Republic of Guyana borders the Atlantic Ocean. It is the only English-speaking country in mainland South America and. The coastal plain, where the majority lives, is between 2 and 30 mi [3–48 km] wide, much of it below sea level. Dikes prevent flooding. Inland is hilly and forested and this terrain makes up Guyana's largest region. The land rises to 9,219 ft [2,810 m] in the Pakaraima Mountains, part of the Guiana Highlands on Guyana's western border. Other highlands are in the south and southwest. Guyana has impressive waterfalls, such as the King George VI Falls (1,601 ft [488 m]), the Great Falls (840 ft [256 m]), and Kaieteur Falls (741 ft [226 m]).

The coastal plain is largely farmed, but wet savanna covers some areas. Inland, rainforests, rich in plant and animal species, cover about 85% of the country. Savanna occurs in the southwest.

Area 214,969 sq km [83,000 sq mi]
Population 706,000
Capital (population) Georgetown (150,000)
Government Multiparty republic
Ethnic groups East Indian 50%, Black 36%, Amerindian 7%, others
Languages English (official), Creole, Hindi, Urdu
Religions Christianity 50%, Hinduism 35%, Islam 10%, others
Currency Guyanese dollar = 100 cents
Website www.guyana-tourism.com

Dutch West India Company began to set up armed bases and to import African slaves to work on the sugar plantations. However, between 1780 and 1813, the territory changed hands between the Dutch, French, and British. Britain occupied Guyana in 1814 during the Napoleonic Wars, and, in 1831, Britain founded the colony of British Guiana. After slavery was abolished in 1834, many former slaves set up their own farms with East Indian and Chinese laborers introduced to replace them. Gold was discovered in 1879. In 1889, Venezuela claimed part of the territory,

CLIMATE

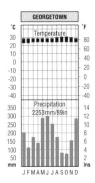

Guyana has a hot, humid equatorial climate. Rainfall ranges from 90 in [2,280 mm] on the coast to 140 in [3,560 mm] in the rainforest region. The rainfall decreases to the west and south.

HISTORY

The first inhabitants of Guyana were Arawak, Carib, and Warrau Amerindians. The Dutch founded a settlement in what is now Guyana in 1581 and, in 1620, the

GEORGETOWN

Capital and largest city of Guyana, at the mouth of the River Demerara. Founded in 1781 by the British, it was the capital of the united colonies of Essequibo and Demerara and was known as Stabroek during the brief Dutch occupation from 1784. It was renamed Georgetown in 1812. Industries, shipbuilding, brewing.

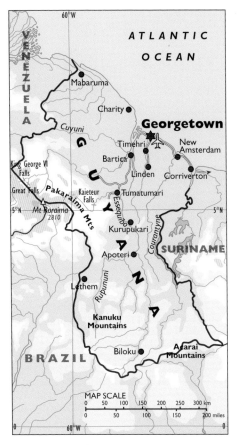

coflict broke out in the early 1960s. Elections in 1964 were won by the PNC and its ally, the United Force, and Burnham became prime minister

POLITICS

British Guiana became independent as Guyana on 26 May 1966 with Burnham as prime minister. In 1970, Guyana became a republic but remained a member of the Commonwealth. In 1980 Burnham became president and served in that post until his death in 1985. He was succeeded by the prime minister Desmond Hoyte, but, in 1992, the PPP won the elections and Ched di Jagan was elected president. On his death in 1997, he was succeeded by his wife Janet, who herself retired on health grounds in 1999. Her successor, Bharrat Jagdeo, a former finance minister, was reelected in 2001 when the PPP won both the presidential and parliamentary elections. Venezuela continues to claim Guyanese territory west of the Essequibo river, while Suriname is in dispute with Guyana over the headwaters of the Corentyne River, which forms part of the border between the two countries.

Guyana also has a long-standing dispute with Suriname over their sea boundary, which runs through a potentially important offshore oil field.

but its claims were overruled by an international arbitration tribunal. In 1953, the left-wing Guyanese Progressive People's Party (PPP) led by East Indian Cheddi Jagan won the elections. Britain then sent in troops and set up an interim administration. The constitution was restored in 1957, when the PPP split into a mostly Indian party, led by Jagan, and another group, led by a black lawyer, Forbes Burnham. Burnham's party, the more moderate People's National Congress (PNC), consisted mainly of the descendants of Africans. In 1961, British Guiana became self-governing, with Jagan as prime minister. Riots, strikes and racial

ECONOMY

Guyana is a poor developing country. Its resources include gold, bauxite (aluminum ore) and other minerals, its forests and fertile soils. Agriculture and mining are the chief activities. The leading crops are sugarcane and rice, but citrus fruits, cocoa, coffee, and plantains are also important. Farmers also produce beef, pork, poultry, and dairy products. Fishing and forestry are other important activities.

HAITI

Blue represents Haitians of African and French descent. Red represents blood shed in the struggle for independence. The coat of arms features a liberty cap on a royal palm tree and two cannons flanked by flags with a scroll reading l'Union Fait la Force (Strength in Unity).

The Republic of Haiti occupies the western third of Hispaniola, the Caribbean's second largest island. The country's culture is associated with exciting music with strong rhythms and voodoo which is practiced by around 80% of the people. The land is mainly rugged, with mountain chains forming peninsulas in the north and south. The highest peak, at 8,793 ft [2,680 m], is in Massif de la Selle in the southeast. Between the peninsulas is the Golfe de la Gonâve, which contains the large Isle de la Gonâve. Haiti's long coastline, which extends about 1,100 mi [1,770 km], is deeply indented.

Area 10,714 sq mi [27,750 sq km]
Population 7,656,000
Capital (population) Port-au-Prince (917,000)
Government Multiparty republic
Ethnic groups Black 95%, Mulatto/White 5%
Languages French and Creole (both official)
Religions Roman Catholic 80%, Voodoo
Currency Gourde = 100 centimes
Website www.haititourisme.org

CLIMATE

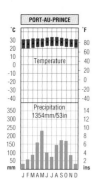

Haiti has a hot, humid tropical climate. Annual rainfall in the northern highlands is about 79 in [2,000 mm], more than twice that of the southern coast. The country is subject to tropical storms, which cause great damage.

HISTORY

In 1496, Spain established a settlement at Santo Domingo, now capital of the neighboring Dominican Republic. This was the first European settlement in the Western Hemisphere. The local Arawak Amerindians were annihilated by settlers in barely 25 years. Spain ceded the western part of Hispaniola to France in 1697. This area became Haiti. With an economy based on sugar cultivation and forestry, Haiti soon became prosperous.

In 1801, former slave Toussaint Louverture, led a revolt and proclaimed himself governor-general. A French force failed to conquer the interior of Haiti and the country became independent in 1804. Another former slave, Jean-Jacques Dessalines, declared himself emperor. In 1915, following conflict between black and mulatto Haitians, the United States invaded the country to protect its interests. The troops withdrew in 1934, although the United States maintained control over the economy until 1947.

In 1956, François Duvalier ("Papa Doc") seized power in a military coup. He was elected president in 1957. Duvalier established a brutal dictatorship. He died in 1971 and his son Jean Claude Duvalier ("Baby Doc") became president. Like his father, Baby Doc used a murderous militia, the Tontons Macoutes, to conduct a reign of terror and maintain rule. In 1986, popular unrest forced Baby Doc to flee and a military regime took over.

POLITICS

The country's first multiparty elections

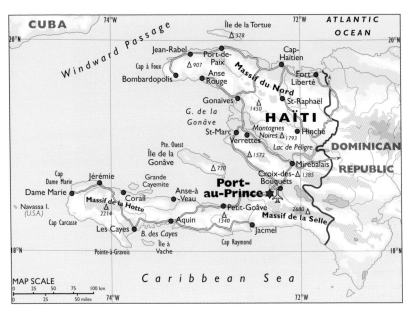

were held in December 1990: the winner was radical Roman Catholic priest, Father Jean-Bertrand Aristide, who promised sweeping reforms. Aristide was reelected in 1994, but stood down in 1995 and was succeeded by René Préval. Violence and poverty still prevailed and, in 1999, Préval dissolved parliament and declared that he would rule by decree. But in the elections of 2000, Aristide again became president amid accusations of electoral irregularities, surviving an attempted coup in 2001. In 2003, voodoo was recognized as a religion. In 2004, rebels seized several towns and cities in an antigovernment uprising. Aristide fled the country, accusing the United States of forcing him into exile. An interim president, Boniface Alexandre, was sworn in as president, an interim government, led by prime minister Gerald Latortue, a former foreign secretary and UN official, took control and a UN peace-keeping force arrived. Floods in May 2004 caused major loss of life in Haiti and the Dominican Republic, while, later that year, Hurricane Jeanne killed nearly 3,000 people in the northwest. In 2005, Hurricane Denis killed at least 45 people. Money-laundering, corruption and drug-trafficking are rife. Haiti is considered to be the transshipment point for cocaine en route to the United States.

ECONOMY

Haiti is the poorest country in the Western Hemisphere and 80% of the people live below the poverty line. Agriculture is the occupation of two-thirds of the people, but coffee is the only significant cash crop.

PORT-AU-PRINCE

Capital of Haiti, a port on the southeast shore of the Gulf of Gonâve, on the west coast of Hispaniola. Founded by the French in 1749, Port-au-Prince became the capital of Haiti in 1770. Industries include tobacco, textiles, cement, coffee, and sugar.

HONDURAS

Honduras officially adopted its present flag in 1866. It is based on the flag of the Central American Federation, set up in 1823 and consisting of Costa Rica, El Salvador, Guatemala, Honduras, and Nicaragua. Honduras left the Federation in 1838.

The Republic of Honduras is the second largest country in Central America. It has two coastlines. The north Caribbean coast extends for about 375 mi [600 km], its deep offshore waters prompted the Spanish to name the country Honduras (Spanish for "depths"); and a narrow, 50 mi [80 km] long, Pacific outlet to the Gulf of Fonseca. Along the north coast are vast banana plantations. To the east lies the Mosquito Coast. The Cordilleras highlands form 80% of Honduras, and include the capital, Tegucigalpa.

Pine forests cover 75% of Honduras. The northern coastal plains contain rainforest and tropical savanna. The Mosquito Coast contains mangrove swamps and dense forests. Mahogany and rosewood forest grow on lower mountain slopes.

Area 43,277 sq mi [112,088 sq km]
Population 6,824,000
Capital (population) Tegucigalpa (850,000)
Government Republic
Ethnic groups Mestizo 90%, Amerindian 7%, Black (including Black Carib) 2%, White 1%
Languages Spanish (official), Amerindian dialects
Religions Roman Catholic 97%
Currency Honduran lempira = 100 centavos
Website www.letsgohonduras.com

CLIMATE

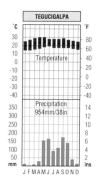

The climate is tropical, though the uplands, where the capital Tegucigalpa is situated, are cooler than the coastal plains. The heaviest rainfall occurs from November to May. The coast is often battered by hurricanes. In October 1998, Honduras and Nicaragua were hit by Hurricane Mitch, which caused floods and mudslides.

HISTORY

From AD 400 to 900, the Maya civilization flourished. The Spanish discovered the magnificent ruins at Copán in western Honduras in 1576, but these became covered in dense forest and were only rediscovered in 1839. Columbus sighted the coast in 1502. Pedro de Alvarado founded the first Spanish settlements in 1524. The Spanish gradually subdued the native population and established gold and silver mines.

In 1821 Honduras gained independence, and formed part of the Mexican

MOSQUITO COAST (MOSQUITIA)

Coastal region bordering the Caribbean Sea, 40 mi [65 km] wide, now divided between Nicaragua and Honduras. A British protectorate from 1740, it returned to its original inhabitants (the Miskito) in 1860. In 1894, it became part of Nicaragua. International arbitration awarded the northern part to Honduras in 1960. The region, which consists mainly of tropical forest, swamp and lagoons, is only thinly populated.

declaration of a state of emergency in 1988. A ceasefire was then signed in Nicaragua, after which the Contra bases were closed down.

In 1992 Honduras signed a treaty with El Salvador, settling the disputed border. Liberal Party leader Carlos Flores became president in 1997 elections. In 1998, Hurricane Mitch killed more than 5,500 people and left 14 million people homeless. Human rights organizations estimated that "death squads," often backed by the police, killed more than 1,000 street children in 2000. National Party leader Ricardo Maduro became president in 2001 elections.

Empire. From 1823 to 1838, Honduras was a member of the Central American Federation. Throughout the rest of the 19th century, Honduras was subject to continuous political interference, especially from Guatemala. Britain controlled the Mosquito Coast.

In the 1890s, American companies developed plantations in Honduras to grow bananas, which soon became the country's chief source of income. The companies exerted great political influence and the country became known as a "banana republic," a name that was later applied to several other Latin American nations.

After World War II, demands grew for greater national autonomy and workers' rights. A military coup overthrew the Liberal government in 1963. In 1969, Honduras fought the short "Soccer War" with El Salvador. The war was sparked off by the treatment of fans during a World Cup soccer series. The real reason, however, was that Honduras had forced Salvadoreans in Honduras to give up land.

POLITICS
Civilian government returned in 1982. During the 1980s, Honduras allowed US-backed Contra rebels from Nicaragua to operate in Honduras against Nicaragua's left-wing Sandinista government. Honduras was heavily dependent on US aid. Popular demonstrations against the Contras led to the

ECONOMY
Honduras is the least industrialized country in Central America, and the poorest developing nation in the Americas. It has very few mineral resources, other than silver, lead, and zinc.

Agriculture dominates the economy, forming 78% of exports and employing 38% of the workforce. Bananas and coffee are the leading exports, and corn the principal food crop. Cattle are raised in the mountain valleys and on the southern Pacific plains.

Fishing and forestry are also important. There are vast timber resources. Lack of an adequate transportation infrastructure hampers development.

TEGUCIGALPA
Capital and largest city of Honduras, located in the mountainous central Cordilleras. Founded in the 16th century by the Spanish as a silver and gold mining town, it became the national capital in 1880. Industries include sugar, textiles, chemicals, and cigarettes.

HUNGARY

Hungary's flag was adopted in 1919. A state emblem was added in 1949 and removed in 1957. The colors of red, white, and green had been used in the Hungarian arms since the 15th century. The tricolor design became popular during the 1848 rebellion against Habsburg rule.

The Republic of Hungary is a land-locked country in central Europe. The land is mostly low-lying and drained by the Danube (Duna) and its tributary, the Tisza. Most of the land east of the Danube belongs to a region called the Great Plain (Nagyalföld), which covers about half of Hungary.

West of the Danube is a hilly region, with some low mountains, called Trans-danubia. This region contains the country's largest lake, Balaton. In the northwest is a small, fertile, and mostly flat region called the Little Plain (*Kisalföld*).

Much of Hungary's original vegetation has been cleared. Large forests remain in the scenic northeastern highlands

Area 93,032 sq km [35,920 sq mi]
Population 10,032,000
Capital (population) Budapest (1,819,000)
Government Multiparty republic
Ethnic groups Magyar 90%, Gypsy, German, Serb, Romanian, Slovak
Languages Hungarian (official)
Religions Roman Catholic 68%, Calvinist 20%, Lutheran 5%, others
Currency Forint = 100 fillér
Website www.magyarorszag.hu/angol

CLIMATE

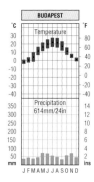

Hungary lies far from the moderating influence of the sea. As a result, summers are warmer and sunnier, and the winters colder than in Western Europe.

HISTORY

Magyars first arrived in the area from the east in the 9th century. In the 11th century, Hungary's first king, Stephen I, made Roman Catholicism the official religion. Hungary became a powerful kingdom, but in 1526 was defeated by the Turks, who occupied much of Hungary. In the late 17th century, the Austrian Habsburgs conquered Hungary. In 1867, Austria granted Hungary equal status in a "dual monarchy," called Austria-Hungary. In 1914, a Bosnian student killed Archduke Franz Ferdinand, the heir to the Austria-Hungary throne. This led to World War I, when Austria-Hungary fought alongside Germany. Defeat in 1918 led to nearly 70% of its territory being apportioned by the Treaty of Versailles to Czechoslovakia, Yugoslavia, and Romania. Some 2.6 million Hungarians live in these countries today.

The government hoped to regain these territories by siding with Hitler's Germany in World War II, but the Germans occupied the country in 1944 and later that year the Red Army invaded. Elections were held in 1945 and, in 1946, the country was declared a republic. Although the smallholders had won a clear majority of the votes in the 1945 elections, the Communists gradually took control even after failing to win a majority of the votes cast in new elections in 1947.

POLITICS

Hungary became a Communist state in 1949, with a constitution based on that

governed in coalition with the Alliance of Free Democrats. However, in elections in 1998, Victor Orbán, leader of the Fidesz-Hungarian Civic Party, became prime minister. In 2002, the Socialists and the Free Democrat coalition, led by Peter Medgyessy, won a majority in parliament. Hungary became a member of NATO and the EU in 2004.

of the Soviet Union. The first leader of the Communist government was Mathias Rákosi, who was replaced in 1953 by Imre Nagy. Nagy sought to relax Communist policies and was forced from office in 1955. He was replaced by Rákosi in 1956 and this led to a major uprising in which many Hungarians were killed or imprisoned. Nagy and his coworkers were executed for treason in 1958.

Janos Kádár came to power in the wake of the suppression, but his was a relatively progressive leadership, including an element of political reform and a measure of economic liberalism. However, in the late 1970s, the economic situation worsened and new political parties started to appear.

Kádár resigned in 1989 and the central Committee of the Socialist Workers' Party (the Communist Party) agreed to sweeping reforms, including the introduction of a pluralist system and a democratic parliament, which had formally been little more than a rubber-stamp assembly. The trial of Imre Nagy and his coworkers was declared unlawful and their bodies were reburied with honor in June 1989.

In 1990, Hungarians voted into office a center-right coalition headed by the Democratic Forum. In 1994, the Hungarian Socialist Party (made up of ex-Communists) won a majority and

ECONOMY

Under Communism the economy was transformed from agrarian to industrial. The new factories were owned by the government, as was most of the land. From the late 1980s, the government worked to increase private ownership. This change of policy caused many problems, including inflation and high rates of unemployment.

Manufacturing is the most valuable activity. The major products include aluminum made from local bauxite, chemicals, electrical and electronic goods, processed food, iron and steel, and motor vehicles. Agriculture remains important, major crops include grapes, maize, potatoes, sugarbeet, and wheat.

BUDAPEST

Capital of Hungary, on the River Danube. It was created in 1873 by uniting the towns of Buda (capital of Hungary since the 14th century) and Pest on the opposite bank. It became one of the two capitals of the Austro-Hungarian Empire. In 1918, it was declared capital of an independent Hungary. Budapest was the scene of a popular uprising against the Soviet Union in 1956. The old town contains a remarkable collection of buildings, including Buda Castle, the parliament building, the National Museum, and Roman remains.

ICELAND

Iceland's flag dates from 1915. It became the official flag in 1944, when Iceland became fully independent. The flag, which uses Iceland's traditional colors, blue and white, is the same as Norway's flag, except that the blue and red colors are reversed.

The Republic of Iceland, in the North Atlantic Ocean, though deemed part of Europe, is closer to Greenland than to Scotland. Iceland sits astride the Mid-Atlantic Ridge, the geological boundary between Europe and North America. The island is slowly getting wider as the ocean is stretched apart by the forces of plate tectonics.

Iceland has around 200 volcanoes and eruptions are frequent. An eruption under the Vatnajökull icecap in 1996 created a subglacial lake which subsequently burst, causing severe flooding. Geysers and hot springs are other volcanic features. During the thousand years that Iceland has been settled, between 150 and 200 volcanic eruptions have occurred. Icecaps and glaciers cover about one-eighth of the land. The only habitable regions are the coastal lowlands.

Vegetation is sparse or nonexistent on 75% of the land. Treeless grassland and bogs cover some areas. Deep fjords fringe the coast.

Area 39,768 sq mi [103,000 sq km]
Population 294,000
Capital (population) Reykjavik (108,000)
Government Multiparty republic
Ethnic groups Icelandic 97%, Danish 1%
Languages Icelandic (official)
Religions Evangelical Lutheran 87%, other Protestant 4%, Roman Catholic 2%, others
Currency Icelandic króna = 100 aurar
Website www.icetourist.is

CLIMATE

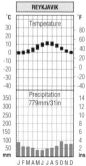

Although it lies far to the north, Iceland's climate is moderated by the warm waters of the Gulf Stream. The port of Reykjavik is ice-free all the year round.

HISTORY

Iceland was colonized by Vikings from Norway in AD 874 and the popula-

tion grew as more settlers arrived from Norway and from the Viking colonies in the British Isles. In 930, the settlers founded the oldest, and what is thought to be the world's first, parliament (the Althing). One early settler was Erik the Red, a Viking who sailed to Greenland in about 982 and founded another colony there in about 985.

Iceland was an independent country until 1262 when, following a series of civil wars, the Althing recognized the rule of the king of Norway. When Norway united with Denmark in 1380, Iceland came under the rule of Danish kingdoms. Life on Iceland was never easy. The Black Death, which swept the island in 1402, claimed two-thirds of the population, while, in the late 18th century, volcanic eruptions destroyed crops, farmland, and livestock, causing a famine. Then, during the Napoleonic Wars in the early 19th century, food supplies from Europe failed to reach the island and many people starved.

When Norway was separated from Denmark in 1814, Iceland remained under Danish rule. In the late 19th century, the invention of motorized craft, which changed the fishing industry, led

172

because of its alleged antiwhaling policy. It rejoined in 2002, but in 2003 undertook its first whale hunt for 15 years, stating that it was a scientific catch to study the impact of whales on fish stocks.

Iceland has no armed forces of its own. However, it joined NATO in 1949 and, under a NATO agreement, the United States maintains a base on the island, which remains a political issue.

ECONOMY
Iceland has few resources other than the fishing grounds which surround it. Fishing and fish processing are major industries which dominate Iceland's overseas trade. Overfishing is an economic problem. Barely 1% of the land is used to grow crops, mainly root vegetables and fodder for livestock.

However, 23% of the country is used for grazing sheep and cattle. Iceland is self-sufficient in meat and dairy products. Vegetables and fruits are grown in greenhouses heated by water from hot springs.

Manufacturing is important. Products include aluminum, cement, clothing, electrical equipment, fertilizers, and processed foods.

Geothermal power is an important energy source.

to mounting demands for self-government. In 1918, Iceland was acknowledged as a sovereign state, but remained united with Denmark through a common monarch. During World War II, when Germany occupied Denmark, British and American troops landed in Iceland to protect it from invasion by the Germans. Finally, following a referendum in which 97% of the people voted to cut all ties with Denmark, Iceland became a fully independent republic on 17 June 1944.

POLITICS
Fishing, on which Iceland's economy is based, is a major political issue. From 1975, Iceland extended its territorial waters to 200 nautical miles, causing skirmishes between Icelandic and British vessels. The issue was resolved in 1977 when Britain agreed not to fish in the disputed waters. Another problem developed in the late 1980s when Iceland reduced the allowable catches in its waters, because overfishing was causing the depletion of fishing stocks, especially of cod. The reduction of the fish catch led to a slowdown in the economy and, eventually, to a recession, though the economy recovered in the mid-1990s when the conservation measures appeared to have been successful. Iceland left the International Whaling Commission in 1992,

REYKJAVÍK
Capital of Iceland, a port on the southwest coast. Founded in 870, it was the island's first permanent settlement. It expanded during the 18th century, and became the capital in 1918. During World War II, it served as a British and US air base. Tourism is now important. Industries include food processing, fishing, textiles, metallurgy, printing and publishing, and shipbuilding.

INDIA

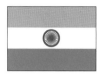

The Indian flag was adopted shortly after the country gained independence from Britain in 1947. The saffron (orange) represents renunciation, the white represents truth, and the green symbolizes mankind's relationship with nature. The central wheel represents dynamism and change.

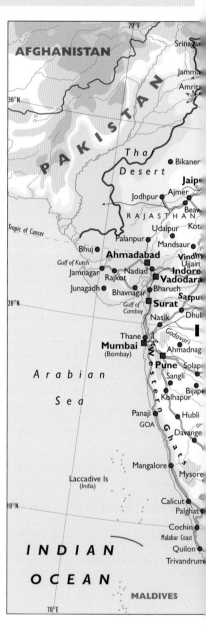

The Republic of India, the world's seventh largest country, extends from high in the Himalayas, through the Tropic of Cancer, to the warm waters of the Indian Ocean at Cape Comorin. India is the world's second most populous nation after China, and the largest democracy. The north contains the mountains and foothills of the Himalayan range. Rivers such as the Brahmaputra and Ganges (Ganga) rise in the Himalayas and flow across the fertile northern plains. Southern India consists of a large plateau called the Deccan which is bordered by two mountain ranges, the Western Ghats and the Eastern Ghats.

The Karakoram Range in the far north has permanently snow-covered peaks. The eastern Ganges delta has mangrove swamps. Between the gulfs of Kutch and Cambay are the deciduous forest habitats of the last of India's wild lions. The Ghats are clad in heavy rainfall.

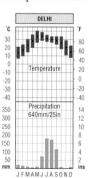

CLIMATE

India has three seasons. The weather during the cool season, from October to February, is mild in the northern plains, but southern India remains hot, though temperatures are a little lower than for

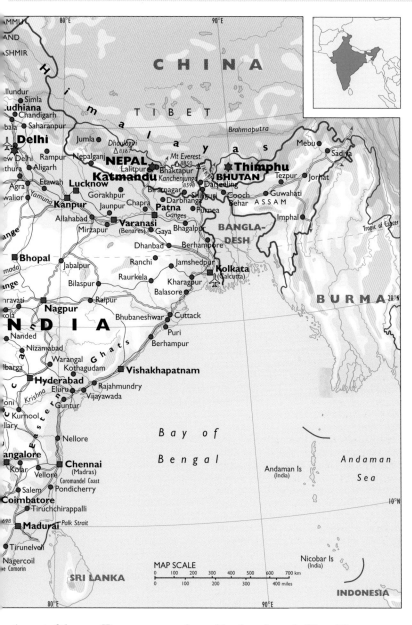

the rest of the year. Temperatures on the northern plains sometimes soar to 120°F [49°C] during the hot season from March to the end of June. The monsoon season starts in the middle of June and continues into September. At this time,

INDIA

Area 1,269,212 sq mi [3,287,263 sq km]
Population 1,065,071,000
Capital (population) New Delhi (295,000)
Government Multiparty federal republic
Ethnic groups Indo-Aryan (Caucasoid) 72%,
Dravidian (Aboriginal) 25%, others (mainly
Mongoloid) 3%
Languages Hindi, English,Telugu, Bengali,
Marathi, Tamil, Urdu, Gujurati, Malayalam,
Kannada, Oriya, Punjabi, Assamese, Kasmiri, Sindhi
and Sanskrit (all official)
Religions Hinduism 82%, Islam 12%, Christianity
2%, Sikhism 2%, Buddhism and others
Currency Indian rupee = 100 paisa
Website www.incredibleindia.org

moist southeasterly winds from the Indian Ocean bring heavy rains to India. Darjeeling in the northeast has an average annual rainfall of 120 in [3,040 mm], but parts of the Great Indian Desert in the northwest have only 2 in [50 mm] of rain a year. The monsoon rains are essential for India's farmers. If they arrive late, crops may be ruined. If the rainfall is considerably higher than average, floods may cause great destruction.

HISTORY

India's early settlers were scattered across the subcontinent in Stone Age times. The first of its many civilizations began to flourish in the Indus Valley in what is now Pakistan and western India around 4,500 years ago, and in the Ganges Valley from about 1500 BC, when Aryan people arrived in India from central Asia. The earlier, darker-skinned people, the Dravidians, moved southward, ahead of the Aryans, and their descendants are now the main inhabitants of southern India.

India was the birthplace of several major religions including Hinduism, Buddhism, and Sikhism. Islam was introduced from about AD 1000. The Muslim Mughal empire was founded in 1526. From the 17th century, Britain began to gain influence and from 1858 to 1947, India was ruled as part of the British empire.

An independence movement began in India after the Sepoy Rebellion (1857–9) and, in 1885, the Indian National Congress was founded. In 1906, Indian Muslims, concerned that Hindus formed the majority of the members of the Indian National Congress, founded the Muslim League. In 1920, Mohandas K. Gandhi, a former lawyer, became leader of the Indian National Congress which soon became a mass movement. Gandhi's policy of nonviolent disobedience proved highly effective, and in response Britain began to introduce political reforms. In the 1930s, the Muslim League called for the establishment of a Muslim state, called Pakistan.

In 1947, it was agreed that British India be partitioned into the mainly Hindu India and the Muslim Pakistan. Both countries became independent in August 1947, but the events were marred by mass slaughter as Hindus and Sikhs fled from Pakistan, and Muslims flocked to Pakistan from India. In the boundary disputes and reshuffling of minority populations that followed, some 1 million lives were lost.

POLITICS

Gandhi was assassinated in 1948 by a Hindu extremist who hated him for his tolerant attitude toward Muslims. The country adopted a new constitution in 1948 making the country a democratic republic within the Commonwealth, and elections were held in 1951 and 1952. India's first prime minister was Jawaharlal Nehru. The government sought to develop the economy and raise living standards at home, while, on the international stage, Nehru won great respect for his policy of nonalignment and neutrality. The disputed status of Kashmir was then India's thorniest security problem.

In 1966, Nehru's daughter, Indira Gandhi, took office. Her Congress Party lost support because of food shortages, unemployment, and other problems. In 1971, India helped the people of East Pakistan achieve independence from

West Pakistan to become Bangladesh. India tested its first atomic bomb in 1974, but pledged to use nuclear power for peaceful purposes only. In 1977, Mrs Gandhi lost her seat in parliament and her Congress Party was defeated by the Janata Party, a coalition led by Morarji R. Desai. Disputes in the Janata Party led to Desai's resignation in 1979 and, in 1980, Congress-I (the I standing for Indira) won the elections. Mrs Gandhi again became prime minister, but her government faced many problems. One problem was that many Sikhs wanted more control over the Punjab, and Sikh radicals began to commit acts of violence to draw attention to their cause. In 1984, armed Sikhs occupied the sacred Golden Temple, in Amritsar. In response, Indian troops attacked the temple, causing much damage and deaths. In October, 1984, two of Mrs Gandhi's Sikh guards assassinated her. Her son, Rajiv, was chosen to succeed her as prime minister, but, in 1989, Congress lost its majority in parliament and Rajiv resigned as prime minister, then during elections in 1991, he was assassinated by Tamil extremists.

India is a vast country with an enormous diversity of cultures. It has more than a dozen major languages, and many minor languages. Hindi, the national language, and the Dravidian languages of the south (Kannada, Tamil, Telugu and Malayam) are Indo-European. Sino-Tibetan languages are spoken in the north and east, while smaller groups speak residual languages in forested hill refuges.

Hinduism is all-pervasive and Buddhism is slowly reviving in the country of its origin. Jainism is strong in the merchant towns around Mount Abu in the Aravallis hills north of Ahmadabad. Islam has contributed many mosques and monuments, the Taj Mahal being the best known and India retains a large Muslim minority. The Punjab's militant Sikhs now seek separation. However, India's most intractable problem remains the divided region of Kashmir, the subject of a long conflict between India and Pakistan. However, in 2004 and 2005, both countries sought ways of easing the tension, including the opening up of cross-border transportation services.

ECONOMY

According to the World Bank, India is a "low-income" developing country. While it ranked 11th in total gross national product in 2004, its per capita GNP of US$440 placed it among the world's poorer countries. Despite initiatives, its socialist policies have failed to raise the living standards of the poor. In the 1990s, the government introduced private enterprise policies to stimulate growth. Farming employs 64% of the people. The main crops are rice, wheat, millet, sorghum, peas and beans. India has more cattle than any other country. Milk is produced but Hindus do not eat beef. India has reserves of coal, iron ore, and petroleum, and manufacturing has expanded greatly since 1947 to include high-tech goods, iron and steel, machinery, refined petroleum, textiles, jewelry, and transportation equipment.

NEW DELHI

Capital of India, in the north of the country, on the River Yamuna in Delhi Union Territory. Planned by the British architects Edwin Lutyens and Herbert Baker, it was constructed in 1912–29 to replace Calcutta (now Kolkata) as the capital of British India. Whereas Old Delhi is primarily a commercial center, New Delhi has an administrative function. Places of interest include the Coronation Durbar Site, the Crafts Museum and Humayun's Tomb. Industries include textile production, chemicals, machine tools, plastics, electrical appliances, and traditional crafts.

INDIAN OCEAN ISLANDS

CHRISTMAS ISLAND

Area 52 sq mi [135 sq km]
Population 2,771
Capital (population) The Settlement (4,000)
Government Territory of Australia
Ethnic groups Chinese 70%, European 20%, Malay 10%
Languages English (official), Chinese, Malay
Religions Buddhist 36%, Muslim 25%, Christian 18%, others
Currency Australian dollar = 100 cents
Website www.christmas.net.au

Island in the east Indian Ocean 200 mi [320 km] south of Java. Once under British domination, it was annexed to Australia in 1958. It has important lime phosphate deposits. The Australian Government has agreed to support the creation of a commercial space-launching site to aid the economy. Two-thirds of the island is national park, making it good for birdwatching.

COCOS ISLANDS

Area 5.4 sq mi [14 sq km]
Population 628
Capital (population) West Island
Government Territory of Australia
Ethnic groups Europeans, Cocos Malays
Languages Malay (Cocos dialect), English
Religions Sunni Muslim 80%, others
Currency Australian dollar = 100 cents
Website
www.dotars.gov.au/terr/cocos/index.aspx

The Territory of Cocos (Keeling) Islands is made up of 27 coral islands. Discovered in 1609 by Captain William Keeling, but uninhabited until the 19th century, the islands were annexed by the UK in 1857, and transferred to the Australian Government in 1955. Only two of the islands are inhabit-

ed and the population of the two is split along ethnic lines, with ethnic Europeans on West Island and ethnic Malays on Home Island.

Coconuts are the sole cash crop and there is a small tourist industry.

COMOROS

Area 838 sq mi [2,170 sq km]
Population 652,000
Capital (population) Moroni (30,000)
Government Independent republic
Ethnic groups Antalote, Cafre, Makoa, Oimatsaha, Sakalava
Languages Arabic and French (both official), Shikomoro
Religions Sunni Muslim 98%, Roman Catholic 2%
Currency Comoran franc = 100 centimes
Website www.arab.net/comoros

The Union des Isles Comoros, consists of three large volcanic islands and some smaller ones, at the northern end of the Mozambique Channel. The three major islands are Grande Comore (site of the capital), Anjouan, and Mohéli. They are mountainous, with a tropical climate and fertile soil.

France took over one of the islands, Mayotte, in 1843 and, in 1886, the other islands came under French protection. The Comoros became independent in 1974, but the people of Mayotte opted to remain French. In the late 1990s, separatists on Anjouan and Mohéli islands sought to secede, but in 2004 each of the large islands was granted autonomy with its own president and legislature. The Comoros is a poor country. It exports cloves, perfume oils, coconuts, copra, and vanilla.

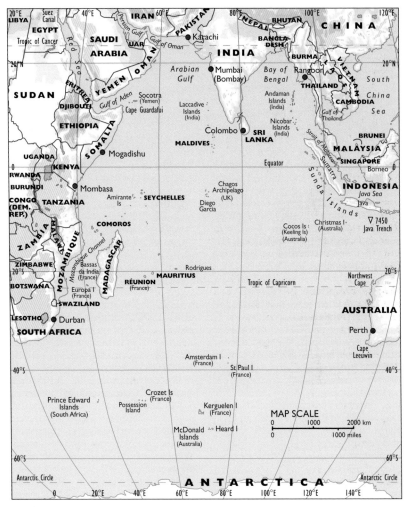

MAP SCALE
| 0 | 1000 | 2000 km |
| 0 | | 1000 miles |

MALDIVES

The Republic of the Maldives, Asia's smallest independent country, comprises some 1,200 low-lying coral islands, 202 of which are inhabited, grouped into 26 atolls. They are scattered along a broad north–south line in the Indian Ocean about 400 mi [640 km] south west of Sri Lanka.

Area 115 sq mi [298 sq km]

Population 339,000

Capital (population) Malé (74,000)

Government Republic

Ethnic groups South Indians, Sinhalese, Arabs

Languages Maldivian Dhivehi (dialect of Sinhala, script derived from Arabic), English

Religion Sunni Muslim

Currency Rufiyaa = 100 laari

Website www.maldivesinfo.gov.mv

The islands are prone to flooding. They have a tropical climate and the monsoon season lasts from April to October.

HISTORY AND POLITICS
Sri Lanka settled the islands in about 500 BC. From the 14th century, the ad-Din dynasty ruled the Maldives. In 1518, the Portuguese claimed the islands. From 1665 to 1886, the Maldives were a dependency of Ceylon (Sri Lanka).

They became a British protectorate in 1887. In 1965 they achieved independence as a sultanate and then became a republic when the sultan was deposed in 1968. Maumoon Abdul Gayoom has served as president since 1978. In 1982, the Maldives joined the Commonwealth. In 1988, Indian troops helped suppress an attempted coup.

A tsunami struck the islands in December 2004 killing 82 people.

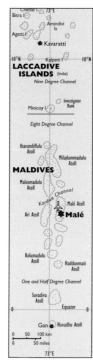

Political parties are banned, though political activity is permitted, at least nominally. The president is chosen in a yes-no referendum. The voters are presented with just one candidate, chosen on their behalf by parliament. President Gayoom came under increasing pressure from human rights groups to ease up on his autocratic style of governance. There was anti-government violence on the streets. As a result of this he announced in 2005 that he planned to introduce a multiparty democracy.

ECONOMY
The chief crops are bananas, coconuts, copra, mangoes, sweet potatoes, and spices, but much food is imported. Fishing is important and the leading export is the bonito (Maldives tuna). Since 1972 the growth in tourism has boosted foreign reserves, but the Maldives remain one of the world's poorest countries.

MAYOTTE

Area 144 sq mi [373 sq km]
Population 173,300
Capital (population) Mamoudzou (4,000)
Government Territorial collectivity of France
Ethnic groups Comorian (mixture of Bantu, Arab, and Malagasy) 92%
Languages Mahorian (a Swahili dialect), French (official)
Religions Muslim 97%, Christian (mostly Roman Catholic)
Currency Euro = 100 cents
Website http://ctt.mayotte.free.fr/anglais/Eaccueil.htm

A French administered archipelago in the Indian Ocean to the east of the Comoros. The two major islands are Grande Terre and Petite Terre (Pamanzi). Grande Terre includes the new capital, Mamoudzou. Pamanzi is the site of the old capital, Dzaoudzi.

Mayotte has a tropical climate. It is generally hot and humid. The rainy season is from November to May, during the northeastern monsoon. The dry season, from May to November, is cooler.

HISTORY AND POLITICS
Mayotte was a French colony from 1843 to 1914 when it was attached to the Comoro group and collectively achieved administrative autonomy as a French Overseas Territory. In 1974, the rest of the Comoros became independent while Mayotte voted to remain a French

dependency. In 1976 it became an over-seas collectivity of France.

ECONOMY

The economy is primarily agricultural, the chief products being bananas and mangoes.

RÉUNION

Area 969 sq mi [2,510 sq km]
Population 766,000
Capital (population) St.-Denis (122,000)
Government Overseas department of France
Ethnic groups French, African, Malagasy, Chinese, Pakistani, Indian
Languages French (official), Creole
Religions Roman Catholic 86%, Hindu, Muslim, Buddhist
Currency Euro = 100 cents
Website
www.la-reunion-tourisme.com/gb_entree.htm

The Department of Réunion is a volcanic island in the Mascarene group lying about 440 mi [700 km] east of Madagascar in the Indian Ocean. It has a mountainous, wooded center, surrounded by a fertile coastal plain. The climate is tropical, but the temperature moderates with elevation. It is cool and dry from May to November and hot and rainy from November to April.

HISTORY AND POLITICS

Discovered in 1513 by the Portuguese, but claimed by France in 1638. It became a French department in 1946. The island became part of an administrative region in 1973. There is increasing pressure on France for independence.

ECONOMY

Sugarcane dominates the economy, though vanilla, perfume oils, and tea also produce revenue. Tourism is the big hope for the future, but unemployment is high and the island relies on French aid.

SEYCHELLES

Area 176 sq mi [455 sq km]
Population 81,000
Capital (population) Victoria (24,000)
Government Republic
Ethnic groups French, African, Indian, Chinese, and Arab
Languages Creole 92%, English 5% (both official)
Religions Roman Catholic 82%, Anglican 6%, others
Currency Seychelles rupee = 100 cents
Website www.virtualseychelles.sc

The Republic of Seychelles includes a group of four large and 36 small granitic islands, plus a wide scattering of coralline islands, 14 of them inhabited, lying to the south and west. The islands experience a tropical oceanic climate.

Formerly part of the British Indian Ocean Territory (BIOT), Farquhar, Des Roches, and Aldabra (famous for its unique wildlife) were returned to the Seychelles in 1976. The BIOT now consists only of the Chagos Archipelago, with Diego Garcia, the largest island, supporting a US Navy base.

HISTORY AND POLITICS

French from 1756 and British from 1814, the islands gained their independence in 1976. A year later, a coup resulted in the setting up of a one-party socialist state that several attempts failed to remove. Multiparty elections were held in 1992 and France-Albert René, who had been elected president unopposed in 1979 and 1984, was reelected president under a new constitution adopted in 1993. René was again reelected president in 1998.

ECONOMY

The Seychelles produces copra, cinnamon, and tea, although rice is imported. Fishing and luxury tourism are the two main industries.

INDONESIA

This flag was adopted in 1945, when Indonesia proclaimed itself independent from the Netherlands. The colours, which date back to the Middle Ages, were adopted in the 1920s by political groups in their struggle against Dutch rule.

The Republic of Indonesia, in Southeast Asia, consists of about 13,600 islands, fewer than 6,000 of which are inhabited. The island of Java covers only 7% of the country's area but it contains more than half of Indonesia's population. Three-quarters of the country is made up of five main areas: the islands of Sumatra, Java, and Sulawesi (Celebes), together with Kalimantan (southern Borneo and Papua (western New Guinea). The islands are mountainous and many have extensive coastal lowlands. Indonesia contains more than 200 volcanoes, but the highest peak is Puncak Jaya, which reaches 16,503 ft [5,029 m] above sea level, is in West Papua.

Area 1,904,569 sq km [735,354 sq mi]
Population 238,453,000
Capital (population) Jakarta (9,374,000)
Government Multiparty republic
Ethnic groups Javanese 45%, Sundanese 14|%, Madurese 7%, coastal Malays 7%, approximately 300 others
Languages Bahasa Indonesian (official), many others
Religions Islam 88%, Roman Catholic 3%, Hinduism 2%, Buddhism 1%
Currency Indonesian rupiah = 100 sen
Website www.indonesiamission-ny.org

CLIMATE

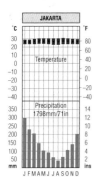

Indonesia has a hot and humid monsoon climate. Only Java and the Sunda Islands have a relatively dry season. From December to March, moist prevailing winds blow from mainland Asia. Between mid-June and October, dry prevailing winds blow from Australia.

HISTORY

From the 8th century, the empire of Sri Vijaya, which was centered on Palembang, held sway until it was replaced in the 14th century by the kingdom of Madjapahit, whose center was east-central Java. Indonesia is the world's most populous Muslim nation, though Islam was introduced as recently as the 15th century. The area came under the domination of the Dutch East India Company in the 17th century. The Dutch government took over the islands in 1799. Japan occupied the islands in World War II and Indonesia declared its independence in 1945. The Dutch finally recognized Indonesia's independence in 1949.

POLITICS

Indonesia's first president, the anti-Western Achmed Sukarno, plunged his coun-

KRAKATOA

Small, volcanic island (2,667 ft [813 m]) in western Indonesia, in the Sunda Strait between Java and Sumatra. In 1883, one of the world's largest recorded volcanic eruptions destroyed most of the island. The resulting tidal waves caused 50,000 deaths and great destruction.

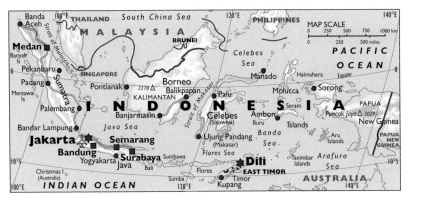

try into chaos. In 1962, Indonesia invaded Dutch New Guinea (now West Papua), and between 1963 and 1966 Sukarno sought to destabilize the Federation of Malaysia through incursions into northern Borneo. In 1967, Sukarno was toppled by General Suharto, following Sukarno's suppression of an alleged Communist-inspired uprising that cost 80,000 lives. Suharto's military regime, with US help, achieved significant economic growth, though corruption was rife. In 1975, Indonesian troops invaded East Timor, opposed by the local people. Suharto was forced to stand down in 1998 and his deputy, Bacharuddin Jusuf Habibie, succeeded him. In June 1999, Habibie's ruling Golkar Party was defeated in elections and, in October, the parliament elected Abdurrahman Wahid as president. However, Wahid, charged with corruption and general incompetence, was dismissed in 2001 and succeeded by the vice-president, Megawati Sukarnoputri (daughter of President Sukarno).

In the early 21st century, Indonesia faced many problems. East Timor seceded in 2002, while secessionist groups in Aceh province, northern Sumatra, and the Free Papua Movement in West Papua also demanded independence. Muslim-Christian clashes broke out in the Moluccas at the end of 1999, while indigenous Dyaks in Kalimantan clashed with immigrants from Madura.

In December 2004, more than 120,000 people were killed in Indonesia by a tsunami. Worst hit was Aceh, though the tragedy was followed by peace talks in 2005, ending the separatist conflict.

ECONOMY

The World Bank describes Indonesia as a "lower-middle-income" developing country. Agriculture employs more than 40% of the workforce and rice is the main food crop. Bananas, cassava, coconuts, peanuts, corn, spices, and sweet potatoes are also grown. Major cash crops include coffee, palm oil, rubber, sugarcane, tea, and tobacco. Fishing and forestry are also important.

There are important mineral reserves, including petroleum and natural gas. Bauxite, coal, iron ore, nickel and tin are also mined.

JAKARTA

Capital of Indonesia, on the northwest coast of Java. It was founded in 1619 as Batavia by the Dutch as a fort and trading post, and it became the headquarters of the Dutch East India Company. It became the capital after Indonesia gained its independence in 1949. Industries include ironworking, printing, and timber.

IRAN

Iran's flag was adopted in 1980 by the country's Islamic government. The white stripe contains the national emblem, which is the word for Allah (God) in formal Arabic script. The words Allah Akbar (God is Great) is repeated 11 times on both the green and red stripes.

The Islamic Republic of Iran contains a barren central plateau which covers about half the country. It includes the Great Salt Desert (Dasht-e-Kavir) and the Great Sand Desert (Dasht-e-Lut). The Elburz Mountains (Alborz), which border the plateau to the north, contain Iran's highest peak, Damavand (18,386 ft [5,604 m]). North of the Elburz Mountains are the fertile, densely populated lowlands around the Caspian Sea, with a mild climate and abundant rainfall. Bordering the plateau to the west are the Zagros Mountains which separate the central plateau from the Khuzistan Plain, a region of sugar plantations and oil fields, which extends to the Iraqi border.

Area 636,368 sq mi [1,648,195 sq km]
Population 69,019,000
Capital (population) Tehran (7,723,000)
Government Islamic Republic
Ethnic groups Persian 51%, Azeri 24%, Gilaki and Mazandarani 8%, Kurd 7%, Arab 3%, Lur 2%, Baluchi 2%, Turkmen 2%
Languages Persian, Turkic 26%, Kurdish
Religions Islam (Shiite Muslim 89%)
Currency Iranian rial = 100 dinars
Website www.netiran.com

CLIMATE

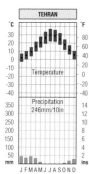

Much of Iran has a severe, dry climate, with hot summers and cold winters. Rain falls only about 30 days a year in Tehran and the annual temperature range is of more than 45°F [25°C]. The lowlands are generally milder.

HISTORY

Ancient Persia was a powerful empire. It flourished from 550 BC, when its king, Cyrus the Great, conquered the Medes, to 331 BC, when the empire was conquered by Alexander the Great. Arab armies introduced Islam in AD 641 and made Iran a great center of learning.

Britain and Russia competed for influence in the area in the 19th century, and in the early 20th century the British began to develop the country's petroleum resources. In 1925, the Pahlavi family took power. Reza Khan became shah (king) and worked to modernize the country. Persia was renamed Iran in 1935. The Pahlavi Dynasty ended in 1979 when religious leader, Ayatollah Ruhollah Khomeini, made Iran an Islamic republic.

POLITICS

Iran and Iraq fought over disputed borders from 1980–8; the war led to a great reduction in Iran's vital petroleum production, but output returned to its mid-1970s levels by 1994. Khomeini died in 1989 but his views and anti-Western attitudes continued to dominate. In 1997, the liberal Mohammad Khatami was elected president, but conservative clerics made reform difficult with spiritual leader, Ayatollah Al Khameni, retaining much power. Khatami was reelected in 2001, but the clerical establishment and institutions such as the judiciary and the Expediency Council, still blocked most

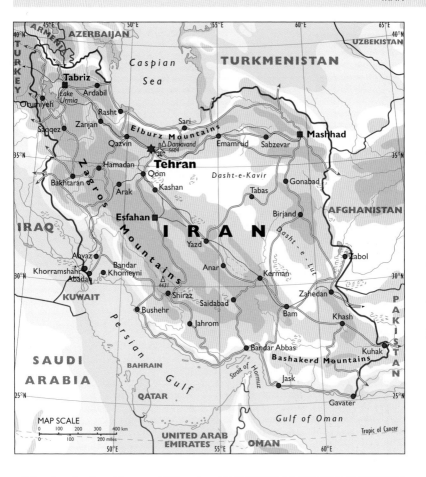

of his reformist plans. Between 2003 and 2005 the United States accused Iran of developing nuclear weapons, a charge Iran denied.

ECONOMY

Iran's prosperity is based on its petroleum which accounts for 95% of its exports. Petroleum revenues have been used to develop a manufacturing sector, but agriculture still accounts for 25% of the gross domestic product, even though farms cover only a tenth of the land. The main crops are wheat and barley. Livestock farming and fishing are also important.

TEHRAN

Capital of Iran, 65 mi [105 km] south of the Caspian Sea. Strategically placed on the edge of the plains, in the foothills of the country's highest mountains. In 1788, it replaced Isfahan as the capital of Persia. In the early 20th century, the old fortifications were demolished and a planned city established by the Shah. Now Iran's industrial, commercial, administrative, and cultural center.

185

IRAQ

Iraq's flag was adopted in 1963, when the country was planning to federate with Egypt and Syria. It uses the four Pan-Arab colors. The three green stars symbolize the three countries. Iraq retained these stars even though the union failed to come into being.

The Republic of Iraq is a southwest Asian country at the head of the Persian Gulf. Deserts cover western and southwestern Iraq, with part of the Zagros Mountains in the northeast, where farming can be practiced without irrigation. Western Iraq contains a large slice of the Hamad (or Syrian) Desert, but essentially comprises lower valleys of the rivers Euphrates (Nahr al Furat) and Tigris (Nahr Dijlah). The region is arid, but has fertile alluvial soils. The Euphrates and Tigris join south of Al Qurnah, to form the Shatt al Arab. The Shatt al Arab's delta is an area of irrigated farmland and marshes. This waterway is shared with Iran.

Area 438,317 sq km [169,234 sq mi]
Population 25,375,000
Capital (population) Baghdad (4,865,000)
Government Republic
Ethnic groups Arab 77%, Kurdish 19%, Assyrian and others
Languages Arabic (official), Kurdish (official in Kurdish areas), Assyrian, Armenian
Religions Islam 97%, Christianity and others
Currency New Iraqi dinar
Website www.un.int/iraq/homepage.htm

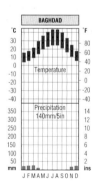

CLIMATE

The climate of Iraq varies from temperate in the north to subtropical in the south and east. Baghdad, in central Iraq, has cool winters, with occasional frosts, and hot summers. Rainfall is generally low.

HISTORY

Mesopotamia was the home of several great civilizations, including Sumer, Babylon and Assyria. It later became part of the Persian Empire. Islam was introduced in AD 637 and Baghdad became the brilliant capital of the powerful Arab Empire. However, Mesopotamia declined after the Monguls invaded it in 1258. From 1534, Mesopotamia became part of the Turkish Ottoman Empire. Britain invaded the area in 1916. In 1921, Britain renamed the country Iraq and set up an Arab monarchy. Iraq finally became independent in 1932.

By the 1950s, petroleum dominated Iraq's economy. In 1952, Iraq agreed to take 50% of the profits of the foreign petroleum companies. This revenue enabled the government to pay for welfare services and development projects. But many Iraqis felt that they should benefit more from their petroleum.

Since 1958, when army officers killed the king and made Iraq a republic, the country has undergone turbulent times. In the 1960s, the Kurds, who live in northern Iraq, Iran, Turkey, Syria and Armenia, asked for self-rule. The government rejected their demands and war broke out. A peace treaty was signed in 1975, but conflict continued.

POLITICS

In 1979, Saddam Hussein became Iraq's president. Under his leadership, Iraq invaded Iran in 1980, starting an eight-year war. During this war, Iraqi Kurds

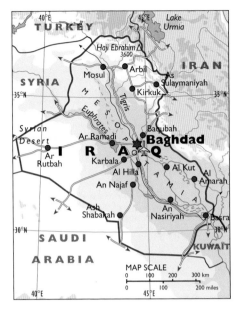

overthrow the Saddam regime in March–April 2003. The coalition forces rapidly achieved their main objectives, but violence continued even after the capture of Saddam Hussein in December 2003.

Although largely boycotted by the Sunni Arabs—a fifth of the population—elections took place in Iraq in 2005, but in an atmosphere of constant battle. Civil order was still in disarray with daily attacks on civilians, Iraqi security forces, and international agencies.

ECONOMY

Civil war and war damage in 1991 and 2003, UN sanctions, and economic mismanagement have all contributed to economic chaos. Petroleum remains Iraq's main resource, but a UN trade embargo in 1990 halted petroleum exports. Farmland covers around a fifth of the land. Products include barley, cotton, dates, fruit, livestock, wheat, and wool. Iraq still has to import food. Industries include petroleum refining and the manufacture of petrochemicals and consumer goods.

supported Iran and the Iraqi government attacked Kurdish villages with poison gas.

In 1990, Iraqi troops occupied Kuwait, but an international force drove them out in 1991. Since 1991, Iraqi troops have attacked Shiite Marsh Arabs and Kurds. In 1996, the government aided the forces of the Kurdish Democratic Party in an offensive against the Patriotic Union of Kurdistan, a rival Kurdish faction. In 1998, Iraq's failure to permit UNSCOM, the UN body charged with disposing of Iraq's deadliest weapons, access to all suspect sites led to Western bombardment of military sites. Periodic bombardment and sanctions continued, but Iraq was allowed to export a limited amount of petroleum in exchange for food and medicines.

The threat of war mounted after the terrorist attacks on the United States in 2001 and the rejection by Iraq in 2002 of the return of UN weapons inspectors. In 2002 and 2003, presure mounted on Iraq to dispose of its alleged weapons of mass destruction. Its failure to do so led to a coalition force, headed by the United States and the UK, to invade Iraq and

BAGHDAD

Capital of Iraq, on the River Tigris. Established in 762 as capital of the Abbasid caliphate, it became a center of Islamic civilization and focus of caravan routes between Asia and Europe. It was almost destroyed by the Monguls in 1258. In 1921 Baghdad became the capital of newly independent Iraq. Notable sites include the 13th-century Abbasid Palace. In 1991 and 2003 Baghdad was badly damaged by bombing in the two Gulf Wars. Industries include building materials, textiles, tanning, and bookbinding.

IRELAND

Ireland's flag was adopted in 1922 after the country had become independent from Britain, though nationalists had used it as early as 1848. Green represents Ireland's Roman Catholics, orange the Protestants, and the white a desire for peace between the two.

The Republic of Ireland consists of a large lowland region surrounded by a broken rim of low mountains. The lowlands include peat bogs. The uplands include the Mountains of Kerry with Carrauntoohill, Ireland's highest peak (3,415 ft [1,041 m]). The River Shannon is the longest in the Ireland, flowing through three large lakes, loughs Allen, Ree and Derg. Forests cover approximately 5% of Ireland. Much of the land is under pasture and a very small percentage is set aside for crops.

Area 27,132 sq mi [70,273 sq km]
Population 3,970,000
Capital (population) Dublin (482,000)
Government Multiparty republic
Ethnic groups Irish 94%
Languages Irish (Gaelic) and English (both official)
Religions Roman Catholic 92%, Protestant 3%
Currency Euro = 100 cents
Website www.irlgov.ie

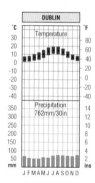

CLIMATE

Ireland has a mild, damp climate influenced by the Gulf Stream current which warms the west coast, with Dublin in the east somewhat cooler. Rain occurs throughout the year.

HISTORY

Celts settled in Ireland from about 400 BC. They were followed by the Vikings, Normans, and the English. Vikings raided Ireland from the 790s, establishing settlements in the 9th century. But Norse domination was ended in 1014 when they were defeated by Ireland's king, Brian Boru. The Normans arrived in 1169 and, gradually, Ireland came under English influence.

In 1801, the Act of Union created the United Kingdom of Great Britain and Ireland. But Irish discontent intensified in the 1840s when a potato blight caused a famine in which a million people died and nearly a million emigrated. Britain was blamed for not having done enough to help. In 1905, Arthur Griffith founded Sinn Féin ("We Ourselves"), a move-

DUBLIN (BAILE ÁTHA CLIATH)

Capital of the Republic of Ireland, at the mouth of the River Liffey on Dublin Bay. In 1014, Brian Boru recaptured it from the Danish. In 1170, it was taken by the English and became the seat of colonial government. Dublin suffered much bloodshed in nationalist attempts to free Ireland from English rule. Strikes beginning in 1913 finally resulted in the Easter Rising (1916). Dublin was the center of the late 19th-century Irish literary renaissance. It is now the commercial and cultural center of the Republic. Notable sites include Christ Church Cathedral (1053), St. Patrick's Cathedral (1190), Trinity College (1591), and the Abbey Theatre (1904).

accepted the Act, but fighting broke out in southern Ireland. In 1921, a treaty was agreed allowing southern Ireland to become a self-governing dominion, called the Irish Free State, within the British Commonwealth. With one Irish group accepting the treaty and another group, wanting complete independence, civil war occurred between 1922 and 1923.

POLITICS

Ireland became a republic in 1949 and has subsequently played an independent role in Europe, joining the EEC in 1973 and, unlike Britain, adopting the euro as its currency in 2002. The government of Ireland has worked with British governments in attempts to solve the problems of Northern Ireland. In 1998, it supported the creation of a Northern Ireland Assembly, the setting up of north–south political structures, and the amendment of the 1937 constitution removing from it the republic's claim to Northern Ireland. A referendum showed strong support for the proposals and the amendments to the constitution. The 1998 Good Friday Agreement in Northern Ireland, aimed to end the long-standing conflict, it met with much support but ran into difficulties when the underground Irish Republican Army (IRA) refused to disarm. In July 2005 the IRA issued a statement of full disarmament.

ment advocating self-government for Ireland. Another secret organization, the Irish Republican Brotherhood, was also active in the early 20th century and its supporters became known as republicans. In 1916, republicans launched what was called the Easter Rebellion in Dublin, but the uprising was crushed. The republicans took over Sinn Féin in 1918. They won most of Ireland's seats in the British parliament, but instead of going to London, they set up the Dáil Éireann (House of Representatives) in Dublin and declared Ireland an independent republic in January 1919.

In 1920, the British parliament passed the Government of Ireland Act, partitioning Ireland. The six Ulster counties

ECONOMY

Aided by EU grants, farming is now relatively prosperous and includes cattle and dairy, sheep, pigs, potatoes, and barley. Manufacturing is now the leading activity, with high-tech industries producing chemicals and pharmaceuticals, electronic equipment, machinery, paper, and textiles. Tourism and racehorses are important industries.

189

ISRAEL

Israel's flag was adopted when the Jewish state declared itself independent in 1948. The blue and white stripes are based on the tallit, a Hebrew prayer shawl. The ancient, six-pointed Star of David is in the center. The flag was designed in America in 1891.

The State of Israel is a small country in the eastern Mediterranean. Inland lie the Judaeo-Galilean highlands, from northern Israel to the northern tip of the Negev Desert in the south. To the east lie part of the Great Rift Valley, River Jordan, Sea of Galilee and Dead Sea, 1,322 ft [403 m] below sea level, the world's lowest point on land.

Area 7,954 sq mi [20,600 sq km]
Population 6,199,000
Capital (population) Jerusalem (685,000)
Government Multiparty republic
Ethnic groups Jewish 80%, Arab and others 20%
Languages Hebrew and Arabic (both official)
Religions Judaism 80%, Islam (mostly Sunni) 14%, Christianity 2%, Druze and others 2%
Currency New Israeli shekel = 100 agorat
Website www.mfa.gov.il/MFA

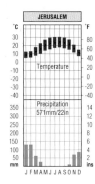

CLIMATE

Israel has hot, dry, summers. Winters are mild and moist on the coast. Annual rainfall decreases west to east and north to south with only 2.5 in [70 mm] in the Dead Sea region.

the Golan Heights (Syrian). In 1982, Israel invaded Lebanon to destroy the stronghold of the PLO (Palestine Liberation Organization), but they left in 1985.

POLITICS

In 1978 Israel signed a treaty with Egypt leading to the return of the Sinai Penin-

HISTORY

Israel is part of a region called Palestine. Most modern Israelis are descendants of immigrants who began to settle from the 1880s. Britain ruled Palestine from 1917. Large numbers of Jews escaping Nazi persecution arrived in the 1930s, provoking an Arab uprising against British rule. In 1947, the UN agreed to partition Palestine into an Arab and a Jewish state. Fighting broke out after Arabs rejected the plan. The State of Israel came into being in May 1948, but fighting continued into 1949. Other Arab-Israeli wars were fought in 1956, 1967 and 1973. The Six Day War in 1967 led to the acquisition by Israel of the West Bank and East Jerusalem. Israel also occupied the Gaza Strip, the Sinai Peninsula (Egyptian) and

WEST BANK

Region west of River Jordan, northwest of the Dead Sea. Designated an Arab district in the 1947 UN plan for the partition of Palestine. Administered by Jordan after the first Arab-Israeli War (1948), but captured by Israel in the Six Day War (1967). In 1988, Jordan passed its claim to the West Bank to the PLO. The 1994 Israeli-Palestinian Accord gave the Palestinian National Authority (PNA) limited autonomy in the West Bank. Difficulties resulting particularly from the growth of Israeli settlements and security disputes halted progress toward any Israeli withdrawal.

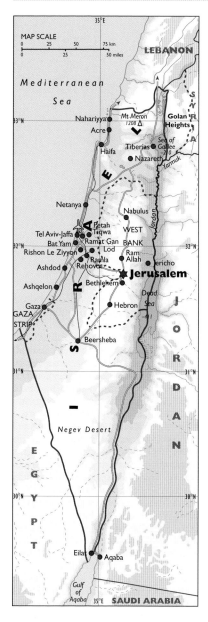

Palestinian self-rule in the Gaza Strip and in Jericho on the West Bank. The agreement was extended in 1995 to include more than 30% of the West Bank. Israel's prime minister, Yitzhak Rabin, who had been seeking a "land for peace" settlement, was assassinated in 1995 and in 1996 the right-wing hardliner Binyamin Netanyahu became prime minister.

In 1999 the left-wing Ehud Barak won elections, promising to resume the peace process. Many problems remained, particularly the extension of Jewish settlements in the occupied areas and attacks on Israel by the militant Islamic group, Hezbollah, based in southern Lebanon. In 2001, Ariel Sharon, former general and leader of the right-wing Likud, was elected prime minister adopting a hard-line policy against the Palestinians. In late 2004 the death of Palestinian leader Yasser Arafat held out hope that moderate policies might lead to the creation of a Palestinian state. Israel forcibly evicted Israeli settlers from Gaza in August 2005.

ECONOMY

The State of Israel has a high standard of living. Manufacturing is important. Israel produces potash, cotton, fruits, grain, poultry and vegetables.

JERUSALEM

Capital of Israel, it is a sacred site for Christians, Jews, and Muslims. Repeatedly occupied and destroyed through history. Jews were expelled from Jerusalem by the Romans. It was occupied by the Ottoman Turks from the Middle Ages until 1917, when it became the capital of the British-mandated territory of Palestine. In 1948, it was divided between Jordan (east) and Israel (west). In 1967, the Israeli army captured the Old City of East Jerusalem. In 1980, the united city became the capital of Israel. The UN does not recognize this status.

sula to Egypt in 1979. Conflict continued between Israel and the PLO. In 1993, the PLO and Israel agreed to establish

ITALY

The Italian flag is based on the military standard carried by the French Republican National Guard when Napoleon invaded Italy in 1796, causing great changes in Italy's map. It was finally adopted as the national flag after Italy was unified in 1861.

The Republic of Italy is bordered to the north by the Alps which overlook the northern plains, Italy's most fertile and densely populated region, drained by the River Po. The Apennines (Appennini), which form the backbone of southern Italy, reach their highest peaks (9,800 ft [3,000 m]), in the Gran Sasso Range overlooking the the central Adriatic Sea, near Pescara. Limestones are the most common rocks. Between the mountains are long, narrow basins, some with lakes.

Southern Italy contains a string of volcanoes, stretching from Vesuvius, near Naples (Nápoli), through the Lipari Islands, to Mount Etna on Sicily. Traces of volcanic activity are found throughout Italy. Ancient lava flows cover large areas and produce fertile soils. Italy is still subject to earthquakes and volcanic eruptions. Sicily is the largest island in the Mediterranean. Sardinia is more isolated from the mainland and its rugged, windswept terrain and lack of resources have set it apart.

Area 116,339 sq mi [301,318 sq km]
Population 58,057,000
Capital (population) Rome (2,460,000)
Government Multiparty republic
Ethnic groups Italian 94%, German, French, Albanian, Slovene, Greek
Languages Italian (official),German, French, Slovene
Religions Roman Catholic
Currency Euro = 100 cents
Website www.enit.it

HISTORY

Magnificent ruins throughout Italy testify to the glories of the ancient Roman Empire, which was founded in 753 BC. It reached its peak in the AD 100s and finally collapsed in the 400s, although the Eastern Roman Empire (the Byzantine Empire), survived another 1,000 years.

In the Middle Ages, Italy was split into many tiny states.and they made a huge contribution to the revival of art and learning, known as the Renaissance. Cities such as Florence and Venice testify to the artistic achievements of this period.

The struggle for unification (the Risorgimento) began early in the 19th century, but little progress was made until an alliance between France and Piedmont (then part of the Kingdom of Sardinia) drove Austria from Lombardy in 1859. Tuscany, Parma, and Modena joined Piedmont-Lombardy in 1860, and the Papal States, Sicily, Naples – including most of the southern peninsula – and Romagna were brought into the alliance. King Victor Emmanuel II was proclaimed ruler of a united Italy the following year. Venetia was acquired from Austria in 1866 and Rome was finally

CLIMATE

The north has cold, snowy winters, but warm and sunny summer months. Rainfall is plentiful, with brief but powerful thunderstorms in summer. Southern Italy has mild, moist winters and warm, dry summers.

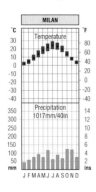

MILAN

Temperature

Precipitation
1017mm/40in

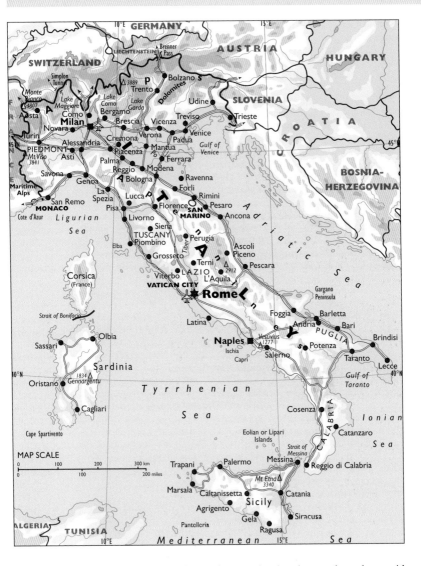

annexed in 1871. Since then, Italy has been a unified state, though the pope and his successors disputed the takeover of the Papal States. This dispute was resolved in 1929, when Vatican City was established as a fully independent state.

Since unification, the population has doubled, and though the rate of increase is notoriously slow today, the rapid growth of population, in a poor country attempting to develop its resources, forced millions of Italians to emigrate during the first quarter of the 20th century. Large numbers settled in the United States, South America, and Australia. More recently, large numbers of Italians have moved into northern Europe for similar reasons.

In 1915, Italy entered World War I alongside the Allies (Britain, France, and Russia). After the war, Italy was given nearly 9,000 sq mi [23,000 sq km] of territory that had belonged to Austria-Hungary. Benito Mussolini (Il Duce) became prime minister of Italy in 1922 and, from 1925 ruled as a dictator. In 1936, Italian forces invaded Ethiopia, while military personnel were sent to support the rebellion of General Franco in Spain. Italy agreed to fight alongside Germany in the event of war, though did not enter World War II until June 1940. During the war, Italy lost much of its colonial empire to the Allies and, in late 1943, declared war on Germany. Mussolini was captured and shot by partisans in 1945, when he tried to escape to Switzerland.

Italy became a republic in 1946 following a referendum. Allied troops left in 1947. Italy was a founder member of NATO in 1949, and of the EEC, now the European Union, in 1957. After the establishment of the EEC, Italy's economy began to expand. Much of the economic development took place in the industrialized north. Central Italy is less developed and represents a transition zone between the developed north and the poor agrarian south, which is known as the Mezzogiorno.

POLITICS

In 1992, the old political establishment was driven from office with several prominent leaders accused of links to organized crime and some imprisoned. In 1996, the left-wing Olive Tree alliance led by Romano Prodi took office, but Prodi was forced to resign in 1998 following his rejection of demands made by his Communist allies. He was replaced by Massimo D'Alemo, the first former Communist to become prime minister. His attempts to create a two-party system in Italy failed in 1999.

By the late 1990s, Italy had the world's sixth largest economy and, on 1 January 2002, the euro became its currency. In 2001, Italy moved toward the political right when a coalition of center-right parties won a substantial majority in parliament. Media tycoon Silvio Berlusconi, who had briefly served as prime minister in 1994 and who had spent several years fighting tax evasion charges, became prime minister.

ECONOMY

Fifty years ago, Italy was a mainly agricultural society. Today it is a major industrial power. It imports most raw materials used in industry. Manufactures include automobiles, chemicals, processed food, textiles, and machinery. Major crops include grapes for winemaking, olives, citrus fruits, sugarbeet and vegetables. Cattle, pigs, poultry and sheep are raised.

ROME (ROMA)

Capital of Italy, on the River Tiber (Tevere), west central Italy. Founded in the 8th century BC. The Roman Republic was founded around 500 BC. By the 3rd century BC, Rome ruled most of Italy and began to expand overseas. In the 1st century AD, the city was transformed as successive emperors built temples, palaces, public baths, arches, and columns. It remained the capital of the Roman Empire until AD 330. In the 5th century, Rome was sacked during the Barbarian invasions, and its population fell rapidly. In the Middle Ages, Rome became the seat of the papacy. In 1527, it was sacked by the army of Charles V. The city flourished once more in the 16th and 17th centuries. Italian troops occupied it in 1870, and in 1871 it became the capital of a unified Italy. The 1922 Fascist march on Rome brought Mussolini to power; he did much to turn Rome into a modern capital city.

VATICAN CITY

The State of the Vatican City, the world's smallest independent nation, is an enclave on the west bank of the River Tiber in Rome. It forms an independent base for the Holy See, the governing body of the Roman Catholic Church. The state includes St. Peter's Square, St. Peter's Basilica, and the Vatican Palace.

The treasures of the Vatican include Michelangelo's frescoes in the Sistine Chapel and attract tourists from all over the world. The Vatican Library contains a priceless collection of both Christian and pre-Christian manuscripts. The popes have lived in the Vatican almost continuously since the 5th century. Sustained by investment income and voluntary contributions, the Vatican City is all that remains of the Papal States which, until 1870, occupied most of central Italy. In 1929, Mussolini recognized the Vatican's independence, in return for papal

Area 0.17 sq mi [0.44 sq km]
Population 921
Capital (population) Vatican City (921)
Government Ecclesiastical
Ethnic groups Italian, Swiss, others
Languages Italian, Latin, French, others
Religions Roman Catholic
Currency Euro = 100 cents
Website www.vatican.va

recognition of the kingdom of Italy.

The population, including 100 Swiss Guards, the country's armed force, is entirely of unmarried males. The Commission appointed by the Pope to administer the affairs of the Vatican controls a radio station, the Pope's summer palace at Castel Gandolfo, and several churches in Rome. Vatican City has its own newspaper, police and railway stations, and issues its own stamps and coins.

SAN MARINO

The Republic of San Marino, the world's smallest republic, lies 12 mi [20 km] southwest of the Adriatic port of Rimini and is wholly surrounded by Italy. It consists largely of the limestone mass of Monte Titano (2,382 ft [725 m]), around which cluster wooded mountains, pastures, fortresses, and medieval villages. San Marino has pleasant, mild summers and cool winters.

The republic was named after St. Marinus, the stonemason saint who is said to have first established a community here in 301 AD. It has a friendship and cooperation treaty with Italy dating back to 1862 and uses the Italian euro, but issues its own stamps, which are an important source of revenue. The state is governed by an elected council and has its own legal system. San Marino has no armed forces and its police are from the

Area 24 sq mi [61 sq km]
Population 27,000
Capital (population) San Marino (2,395)
Government Republic
Ethnic groups Sanmarinese, Italian
Languages Italian (official)
Religions Roman Catholic
Currency Euro = 100 cents
Website www.visitsanmarino.com

Italian constabulary. Most of the people live in the medieval city of San Marino, which receives more than 3 million tourists a year.

Chief occupations are tourism, limestone quarrying, ceramics, textiles, and winemaking. The customs union with Italy makes San Marino an easy conduit for the illegal export of currency and certain kinds of tax evasion for Italians.

IVORY COAST

This flag was adopted in 1960 when the country became independent from France. It combines elements from the French tricolor and the Pan-African colors. The orange represents the northern savanna, the white peace and unity, and the green the forests in the south.

The Republic of the Ivory Coast, in West Africa, is officially known as Côte d'Ivoire. The southeast coast is bordered by sandbars that enclose lagoons, on one of which the former capital and chief port of Abidjan is situated. The southwestern coast is lined by rocky cliffs. Behind the coast is a coastal plain, but the land rises inland to high plains. The highest land is an extension of the Guinea Highlands in the northwest, along the borders with Liberia and Guinea. Most of the country's rivers run north–south.

Area 322,463 sq km [124,503 sq mi]
Population 17,328,000
Capital (population) Yamoussoukro (107,000)
Government Multiparty republic
Ethnic groups Akan 42%, Voltaiques 18%, Northern Mandes 16%, Krous 11%, Southern Mandes 10%
Languages French (official), many native dialects
Religions Islam 40%, Christianity 30%, traditional beliefs 30%
Currency CFA franc = 100 centimes
Website
www.afrika.no/index/Countries/C_te_d_Ivoire/index.html

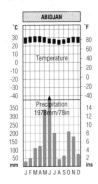

ABIDJAN

°C / °F
Temperature
Precipitation
1978mm/78in
mm / ins
J F M A M J J A S O N D

CLIMATE

Ivory Coast has a hot and humid tropical climate, with high temperatures throughout the year. There are two distinct rainy seasons in the south of the country: between May and July, and from October to November. Inland, the rainfall decreases. Northern Ivory Coast has a dry season and only one rainy season. As a result, the forests in central Ivory Coast thin out to the north, giving way to savanna.

HISTORY

The region that is now Ivory Coast came under successive black African rulers until the late 15th century, when Europeans, attracted by the chance to trade in slaves and such local products as ivory, began to establish contacts along the

coast. French missionaries reached the area in 1637 and, by the end of the 17th century, the French had set up trading posts on the coast. In 1842, France brought the Grand-Bassam area under its protection and Ivory Coast became a French colony in 1893. From 1895, it was ruled as part of French West Africa, a massive union which also included present-day Benin, Burkina Faso, Guinea, Mali, Mauritania, Niger and Senegal. In

ABIDJAN

Former capital of Ivory Coast, situated on Ebrié Lagoon, inland from the Gulf of Guinea. A chief port and commercial center, Abidjan was founded by French colonists at the end of the 19th century. Although it lost capital status to Yamoussoukro in 1983, Abidjan remains a cultural and economic center.

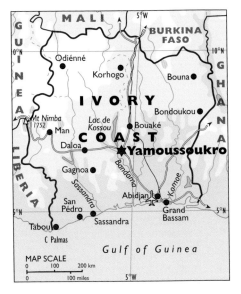

Following the death of Houphouët-Boigny in 1993, the Speaker of the National Assembly, Henri Konan Bédié, proclaimed himself president. He was reelected president in 1995. However, in December 1999, Bédié was overthrown during an army mutiny and a new administration was set up by General Robert Guei. Presidential elections, held after a new constitution was adopted in 2000, resulted in defeat for Guei by a veteran politician, Laurent Gbago. However, conflict began in 2002. By 2004 the country was divided into the government-held south and the rebel-held, mainly Muslim, north.

ECONOMY

Ivory Coast is one of Africa's more prosperous countries. Its free-market economy has proved attractive to foreign investors, especially French firms, while France has given much aid. It has an agrarian economy, which employs about three-fifths of the workforce. The chief farm products are cocoa, coffee, and cotton and make up nearly half the value of the total exports. Food crops include cassava, corn, plantains, rice, vegetables and tropical yams. Manufactures include processed farm products, timber, and textiles.

1946, Ivory Coast became a territory in the French Union. The port of Abidjan was built in the early 1950s, but the country achieved autonomy in 1958.

POLITICS

Ivory Coast became fully independent in 1960. Its first president, Félix Houphouët-Boigny, became the longest serving head of state in Africa with an uninterrupted period in office that ended with his death in 1993. Houphouët-Boigny was a paternalistic, pro-Western leader, who made his country a one-party state. In 1983, the National Assembly agreed to move the capital from Abidjan to Yamoussoukro, Houphouët-Boigny's birthplace. Visitors to Abidjan, where most of the country's Europeans live, are usually impressed by the city's general air of prosperity, but the cost of living for local people is high and there are great social and regional inequalities. Despite its political stability since independence, the country faces such economic problems as variations in the price of its export commodities, unemployment, and high foreign debt.

YAMOUSSOUKRO

Capital of Ivory Coast since 1983. Originally a small Baouké tribal village and birthplace of Ivory Coast's first president, Félix Houphouët-Boigny, it developed rapidly into the administrative and transport center of Ivory Coast. Yamoussoukro's Our Lady of Peace Cathedral (consecrated by Pope John Paul II in 1990) is the world's largest Christian church.

JAMAICA

A committee of Jamaica's House of Representatives designed the national flag. The gold represents Jamaica's sunshine and natural resources. The green symbolizes agriculture and hope for the future. The black stands for the hardships faced and overcome by its people.

Jamaica is the third largest Caribbean island. It is a parliamentary democracy, with the monarch of the UK as its head of state. The coastal plain is narrow and discontinuous. Inland are hills, plateaus and mountains. The country's central range culminates in the Blue Mountain Peak (7,402 ft [2,256 m]). The Cockpit Country in the north-west of the island is an inaccessible limestone area, known for its many deep depressions, (cockpits). Jamaica is a lush, green island.

Area 4,244 sq mi [10,990 sq km]
Population 2,713,000
Capital (population) Kingston (104,000)
Government Constitutional monarchy
Ethnic groups Black 91%, Mixed 7%, East Indian
Languages English (official), patois English
Religions Protestant 61%, Roman Catholic 4%
Currency Jamaican dollar = 100 cents
Website www.jis.gov.jm

CLIMATE

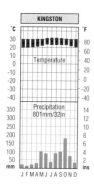

The climate is hot and humid. Temperatures range from 77°F [25°C] in January to 81°F [27°C] in July. Moist southeast trade winds bring rain to the more temperate highlands. Annual rainfall on the northern slopes may reach 200 in [5,000 mm]. But the sheltered south coast is much drier, about 30 in [750 mm] per year. The island is prone to periodic hurricanes.

HISTORY

In 1509, Spaniards occupied the island. Soon, the local Arawak Amerindian population had died out. The Spaniards imported African slaves to work the sugar plantations. The British took the island in 1655 and, with sugar as its staple product, it became a prized possession. The African slaves were the forefathers of much of the present population. But the plantations on which they worked disappeared when the sugar market collapsed in the 19th century.

In 1865, after 200 years of having their own elected body to help the British rule the island, Jamaica came under direct British rule as a Crown Colony, following the Moranty Bay rebellion. This peasant uprising, led by Baptist deacon Paul Bogle, was staged by freed slaves who were suffering acute hardship, but it was put down by British troops. In the 1930s, Jamaican leaders called for more power and riots took place in 1938, with people protesting against unemployment and Britain's racial policies. In that year, the People's National Party (PNP) was founded. In 1944, Britain granted Jamaica a new constitution, providing for an elected House of Representatives. In 1958, the island became a member of the British-sponsored Federation of the West Indies, but it withdrew in 1961.

POLITICS

Jamaica became an independent nation and a member of the British Commonwealth in 1962. It joined the Organization of American States in 1969. In the 1970s, economic problems developed.

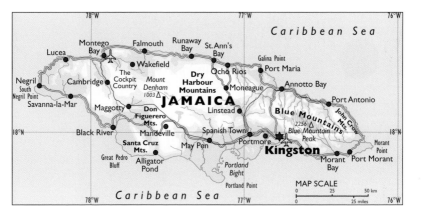

Michael Manley, leader of the PNP, became prime minister in 1972 and he pursued socialist policies, advocating a policy of non-alignment. The PNP won a second term in office in 1976. It nationalized businesses and sought closer ties with Cuba. In 1980, the Jamaica Labor Party (JLP), led by Edward Seaga, defeated the PNP in elections. Seaga privatized much state-owned business and distanced Jamaica from Cuba. His moderate policies led to increased investment and better relations with Western countries. In 1989, the PNP defeated the JLP and Manley was returned to power as prime minister. However, Manley broadly followed Seaga's moderate policies. Manley retired on health grounds in 1992 and was succeeded by Percival J Patterson. The PNP was reelected in 1993, 1998 and 2002.

Jamaica faces many problems, including drug trafficking. It has become a major transshipment point for cocaine being transported from South America to North America and Europe. Cannabis is produced in Jamaica. Corruption and money-laundering are major concerns. Price and tax increases led to riots in 1999, while in 2001 gun battles occurred with 27 people killed when the police searched for drugs in a poor district of Kingston. The murder rate in 2004 was 1,145. This figure was attributed by the police to street-gang violence.

In 2004 Hurricane Ivan destroyed thousands of homes, described as the worst natural disaster in living memory.

ECONOMY

Jamaica is a developing country. Agriculture employs about 20% of the workforce. The chief crop is sugarcane, other products include allspice, bananas, citrus fruits, cocoa, coconuts, coffee, milk, poultry, vegetables, and tropical yams. The country's chief resource is bauxite and Jamaica is one of the world's top producers. Cement, chemicals, cigars, clothing and textiles, fertilizers, machinery, molasses, and petroleum products are also produced. Service industries account for 60% of the gross domestic product. Tourism brings in vital revenue.

KINGSTON

Capital and largest city of Jamaica. It was founded in 1693. It rapidly developed into Jamaica's commercial center, based on the export of raw canesugar, bananas, and rum. In 1872 it became the island's capital. Kingston is the cultural heart of Jamaica, and the home of calypso and reggae music.

JAPAN

Japan's flag was officially adopted in 1870, though Japanese emperors had used this simple design for many centuries. The flag shows a red sun on a white background. The geographical position of Japan is also expressed in its name "Nippon" or "Nihon," meaning "source of the Sun."

Japan is an island nation in northeastern Asia containing four large islands—Honshu, Hokkaido, Kyushu and Shikoku—which make up more than 98% of the country. Thousands of small islands, including the Ryukyu island chain, make up the rest of the country.

The four main islands are mainly mountainous, while many of the small islands are the tips of volcanoes rising from the seabed. Japan has more than 150 volcanoes, about 60 of which are active. Volcanic eruptions, earthquakes and tsunamis often occur, because the islands lie on an unstable part of Earth where the continental plates are constantly moving.

Throughout Japan, complex folding and faulting has produced an intricate mosaic of landforms. Mountains and forested hills alternate with small basins and coastal lowlands, covered by alluvium deposited there by the short rivers that rise in the uplands. Most of the population lives on the coastal plains, one being the stretch from the Kanto Plain, where Tokyo is situated, along the narrow plains that border the southern coasts of Honshu, to northern Kyushu.

The pattern of landforms is further complicated by the presence of volcanic cones and calderas. The highest mountain in Japan, Fuji-san (12,388 ft [3,776 m]), is a long dormant volcano which last erupted in 1707. It is considered sacred, and is visited by thousands of pilgrims every year.

Area 145,880 sq mi [377,829 sq km]
Population 127,333,000
Capital (population) Tokyo (8,130,000)
Government Constitutional monarchy
Ethnic groups Japanese 99%, Chinese, Korean, Brazilian and others
Languages Japanese (official)
Religions Shintoism and Buddhism 84% (most Japanese consider themselves to be both Shinto and Buddhist), others
Currency Yen = 100 sen
Website http://web-japan.org

winters. At Sapporo, temperatures below 4°F [–20°C] have been recorded between December and March. Summers are warm, with temperatures often exceeding 86°F [30°C]. Rain falls throughout the year. Tokyo has higher rainfall and temperatures while the southern islands of Shikoku and Kyushu in the south have warm temperate climates with hot summers and cold winters.

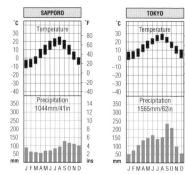

CLIMATE
The climate of Japan varies greatly. Hokkaido in the north has cold, snowy

HISTORY
Most modern Japanese are descendants of early immigrants who arrived in suc-

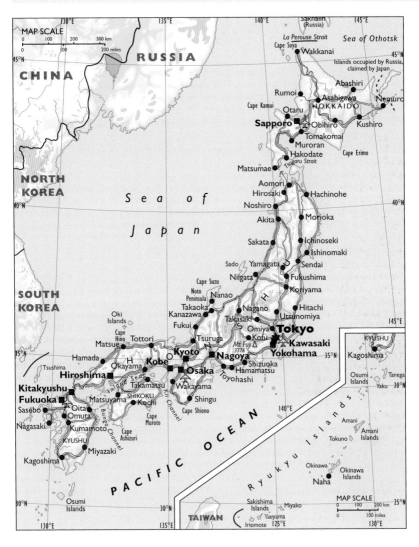

cessive waves from the Korean Peninsula and other parts of the Asian mainland. The earliest zone of settlement included the northern part of Kyushu Island and the coastlands of Setonaikai (Inland Sea). By the 5th century AD, Japan was divided among numerous clans, of which the largest and most powerful was the Yamato. The Yamato ruled from the area which now contains the city of Nara. Shinto, a polytheistic religion based on nature worship, was practiced, and the Japanese imperial dynasty established. The chiefs of the Yamato clan are regarded as ancestors of the Japanese imperial family.

The 5th century AD was a time when new ideas and technology reached Japan from China. The Japanese adopted the Chinese system of writing and their methods of calculating the calendar. Confucianism was also introduced from China and, in about 552, Buddhism reached Japan.

From the early 12th century, political power passed increasingly to military aristocrats. Government was conducted in the name of the emperor by warrior leaders called *shoguns*. Civil warfare between rival groups of feudal lords was endemic over long periods, but, under the rule of the Tokugawa *shoguns*, between 1603 and 1867, Japan enjoyed a great period of peace and prosperity. Military families (the feudal lords and their retainers, or *samurai*) formed a powerful elite. During the *shogun* era, a code of conduct called *bushido* ("the way of the warrior") was developed for the *samurai*, it stressed military skills and fearlessness, frugality, kindness, honesty, and filial piety. The *samurai's* supreme obligation was, above all, to his feudal lord.

European contact began with the arrival of Portuguese sailors in 1543, then in 1549 a Spanish missionary came to convert the Japanese to Christianity. The Japanese put an end to missionary work in the 1630s when they ordered all Christian missionaries to leave the country and forced Japanese converts to give up their faith. The only Europeans allowed to stay were Dutch traders, as they were not involved in missionary work. Japan only opened its ports to Western trade again in 1854, after American intervention.

The Meiji period from 1867 to 1912 was marked by the adoption of Western ideas and technology. An educational system and a telegraph network were set up, railroads built and modern systems of banking and taxation introduced. In addition the *samurai* were abolished and a modern army and navy established.

In 1889, Japan introduced its first constitution under which the emperor became head of state and supreme commander of the army and navy. The emperor appointed government ministers, responsible to him. The constitution also allowed for a parliament, called the Diet, with two houses.

From the 1890s, Japan began to build up an overseas empire. In 1894–5, Japan fought China over the control of Korea. Under the Treaty of Shimonoseki (1895), Japan took Taiwan. Korea was made an independent territory, leaving it open to Japanese influence. Rivalry with Russia led to the Russo-Japanese War (1904–5). Under the Treaty of Portsmouth, Japan gained the Liaodong peninsula, which Russia had leased from China, while Russia recognized the supremacy of Japan's interests in Korea. Thus Japan was established as a world power.

In World War I Japan supported the Allies. After the war Japan's foreign policy strongly supported the maintenance of world peace, becoming a founding member of the League of Nations in 1920. The army seized Manchuria in 1931 and made it a puppet state called Manchukuo, they then extended their influence into other parts of northern China. In 1933, after the League of Nations condemned its actions in Manchuria, Japan was forced to rescind its membership

During the 1930s, and especially after the outbreak of war between Japan and China in 1937, militarist control of Japan's government grew steadily. By the end of 1938, when Japan controlled most of eastern China, there was talk of bringing all of eastern Asia under Japanese control. In September 1939, Japan occupied the northern part of French Indochina and, later that month, signed an agreement with Italy and Germany, assuring their cooperation in building a "new world order," and acknowledging Japan's leadership in Asia.

In 1941 Japan launched a surprise attack on the American naval base of Pearl Harbor, in Hawaii, an action that drew the United States into World War II. On August, 6 1945, American bombers dropped the first atomic bomb on

Hiroshima. The USSR declared war on Japan and invaded Manchuria and Korea. On August 9, the Americans dropped an atomic bomb on Nagasaki. World War II ended on September 2, 1945 when Japan officially surrendered.

POLITICS

The Allies occupied Japan in August 1945. Under a new constitution, power was transferred from the emperor to the people. The army and navy were abolished and the country renounced war as a political weapon. The emperor became a constitutional monarch.

Japan signed a Treaty of Peace that took effect on April 28, 1952. The Allied occupation ended on that day. When, in 1956, the Soviet Union and Japan agreed to end the state of war between them, Japan became a member of the UN.

The conservative Liberal-Democratic Party was formed in 1955, made up of rival Japanese parties. The LDP controlled Japan's government until the 1990s, when a series of coalition governments were formed. A true opposition party emerged in the late 1990s, when the Democratic Party of Japan united with several small parties. The country underwent a serious economic crisis in 1997. In 2001, the LDP chose Junichiro Koizumi as prime minister. Koizumi promised drastic reforms to revive the economy. He won a landslide victory in September 2005 after calling a snap election when his plans to privatise Japan's postal system were defeated in the upper house. After this victory his government announced plans to continue his reform program and also to revise Japan's pacifist constitution.

ECONOMY

Japan has the world's second highest GDP after the United States. The most important sector of the economy is industry, though Japan has to import most of the raw materials and fuels it needs for its industries. Its success is based on the use of the latest technology, a skilled and hard-working labor force, vigorous export policies and a comparatively small spend on defence. Manufactures dominate its exports which include machinery, electrical and electronic equipment, vehicles and transportation equipment, iron and steel, chemicals, textiles, and ships. Japan is one of the world's top fishing nations and fish is an important source of protein. Only 15% of the land can be farmed due to its rugged nature yet the country produces about 70% of the food it requires. Rice is the chief crop, taking up about half of the farmland. Other major products include fruits, sugarbeet, tea, and vegetables.

TOKYO

Capital of Japan, on east-central Honshu, at the head of Tokyo Bay. The modern city divides into distinct districts: Kasumigaseki, Japan's administrative centre; Marunouchi, its commercial center; Ginza, its shopping and cultural center; the west shore of Tokyo Bay, its industrial center. Modern Tokyo also serves as the country's educational center with more than 100 universities. Founded in the 12th century as Edo, it became capital of the Tokugawa shogunate in 1603. In 1868, the Japanese Reformation reestablished imperial power, and the last *shogun* surrendered Edo Castle. Emperor Meiji renamed the city Tokyo, and it replaced Kyoto as the capital. In 1923, an earthquake and subsequent fire claimed more than 150,000 lives and necessitated the city's reconstruction. In 1944–5, intensive US bombing destroyed more than half of Tokyo, and another modernization and restoration program began. Industries include electronics, cameras, automobile manufacture, metals, chemicals, and textiles.

The green, white, and black on this flag are the colors of the three tribes who led the Arab Revolt against the Turks in 1917. Red is the color of the Hussein Dynasty. The star was added in 1928. Its seven points represent the first seven verses of the sacred book, the Koran.

The Hashemite Kingdom of Jordan is an Arab country in south-western Asia. The Great Rift Valley in the west contains the River Jordan and the Dead Sea. East of the Rift Valley is the Transjordan Plateau, where most Jordanians live. To the east and south lie vast areas of desert. Jordan has a short coastline on an arm of the Red Sea, the Gulf of Aqaba. The country's highest peak is Jabal Ram (5,755 ft [1,754 m]).

Area 34,495 sq mi [89,342 sq km]
Population 5,611,000
Capital (population) Amman (1,148,000)
Government Constitutional monarchy
Ethnic groups Arab 98% (Palestinians 50%)
Languages Arabic (official)
Religions Islam (mostly Sunni) 94%, Christianity (mostly Greek Orthodox) 6%
Currency Jordanian dinar = 1,000 fils
Website www.tourism.jo

CLIMATE

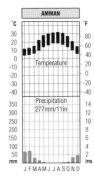

About 90% of Jordan has a desert climate, with an average annual rainfall of less than 8 in [200 mm]. Summers are hot, winters can be cold, with snow on higher areas. The north-west is the wettest area, with an average annual rainfall of 31 in [800 mm] in higher areas.

HISTORY

Jordan was first settled by Semitic peoples about 4,000 years ago, and later conquered by Egyptian, Assyrian, Chaldean, Persian, and Roman forces. The area fell to Muslim Arabs in AD 636; the Arab culture they introduced survives to this day.

By the end of the 12th century, Christian crusaders controlled parts of western Jordan, but were driven out by the great Muslim warrior Saladin in 1187. The Egyptian Mamelukes overthrew Saladin's successors in 1250 and ruled until 1517, when the area was conquered by the Ottoman Turks. Jordan stagnated under their rule, but the opening of a railroad in 1908 stimulated the economy. Arab and British forces defeated the Turks during World War I and after the war, the area east of the River Jordan was awarded to Britain by the League of Nations.

Britain created a territory called Transjordan, east of the River Jordan in 1921. It then became self-governing in 1923, but Britain retained control of its defenses, finances and foreign affairs. This territory became fully independent as Jordan in 1946.

Since the creation of the State of Israel in 1948 Jordan has suffered from instability arising from Arab-Israeli conflict. After the first Arab-Israeli War (1948-9), Jordan acquired the West Bank, which was officially incorporated into the state in 1950. This crucial area, including East Jerusalem, was lost to Israel in the war of 1967, causing many Palestinians to seek refuge in Jordan. In the 1970s, Palestinian guerrillas using Jordan as a base became a challenge to the authority of King Hussein's government. After a short civil war, the Palestinian leadership fled.

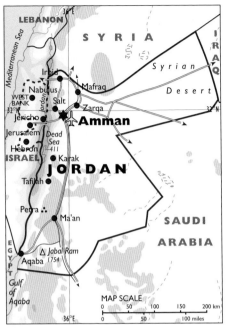

King Hussein, who had commanded great respect for his role in Middle Eastern affairs, died in 1999. He was succeeded by his eldest son who became King Abdullah II. Following the path of his father, Abdullah sought to further the Israeli-Palestinian peace process. He also worked to consolidate his country's relations with other nations in the region. Despite local opposition to the invasion of Iraq in 2003, he supported the US-led war on terrorism and worked to improve relations with Israel. However, in November 2005 suicide bombers killed 57 in Amman and terrorist group al-Qaida claimed responsibility.

ECONOMY

Jordan's economy depends substantially on aid. The World Bank classifies Jordan as a "lower-middle-income" developing country. Less than 6% of the land is farmed. Jordan has a petroleum refinery and manufactures include cement, ceramics, pharmaceuticals, processed food, fertilizers, shoes, and textiles. Service industries, including tourism, employ more than 70% of the workforce.

POLITICS

In 1988 King Hussein suddenly renounced all responsibility for the West Bank, thereby recognizing that the Palestine Liberation Organization, not Jordan, was the legitimate representative of the Palestinian people. Palestinians were still in the majority and the refugees, numbering around 900,000, placed a huge burden on an already weak economy. Jordan was further undermined by the 1991 Gulf War when, despite its official neutrality, the pro-Iraq, anti-Western stance of the Palestinians in Jordan damaged prospects of trade and aid deals with Europe and the United States, Jordan's vital economic links with Israel having already been severed. A ban on political parties was removed in 1991, and martial law lifted after 21 years. Multiparty elections were held in 1993 and, in 1994, Jordan and Israel signed a peace treaty, ending a 40-year-long state of war. The treaty restored some land in the south to Jordan.

AMMAN

Capital and largest city of Jordan. Known as Rabbath-Ammon, it was the chief city of the Ammonites in biblical times. Ptolemy II Philadelphus renamed it Philadelphia. A new city was built on seven hills from 1875, and it became the capital of Transjordan in 1921. From 1948 it grew rapidly, partly as a result of the influx of Palestinian refugees. Industries include cement, textiles, tobacco, and leather.

KAZAKHSTAN

Kazakhstan's flag was adopted on June 4, 1992, about six months after it had become independent. The blue represents cloudless skies, while the golden sun and the soaring eagle represent love of freedom. A vertical strip of gold ornamentation is on the left.

The Republic of Kazakhstan is a large country in west-central Asia. In the west, the Caspian Sea lowlands include the Karagiye Depression, which reaches 433 ft [132 m] below sea level. The lowlands extend eastward through the Aral Sea area. The north contains high plains, but the highest land is along the eastern and southern borders. These areas include parts of the Altai and Tian Shan mountain ranges.

Eastern Kazakhstan contains several freshwater lakes, the largest of which is Lake Balkhash (Balqash Köl). The water in the rivers has been used for irrigation, causing ecological problems. The Aral Sea, deprived of water, shrank from 25,830 sq mi [66,900 sq km] in 1960, to 12,989 sq mi [33,642 sq km] in 1993. Areas which once provided fish have dried up and are now barren desert.

Kazakhstan has very little woodland. Grassy steppe covers much of the north, while the south is desert or semidesert. Large, dry areas between the Aral Sea and Lake Balkhash have become irrigated farmland.

Area 1,052,084 sq mi [2,724,900 sq km]
Population 15,144,000
Capital (population) Astana (322,000)
Government Multiparty republic
Ethnic groups Kazakh 53%, Russian 30%, Ukranian 4%, German 2%, Uzbek 2%
Languages Kazakh (official), Russian, the former official language, is widely spoken
Religions Islam 47%, Russian Orthodox 44%
Currency Tenge = 100 tiyn
Website www.kz

HISTORY

From the late 15th century, the Kazakhs built up a large nomadic empire ruled by *khans*. But Kazakh power declined in the 17th century. In the early 18th century, Russia became influential in the area. In 1731, the Kazakhs in the west accepted Russian rule to gain protection from attack from neighboring peoples. By the mid-1740s, Russia ruled most of the region and, in the early 19th century, Russia abolished the *khanates*. They also

CLIMATE

The extreme climate reflects position in the heart of Asia, far from the influence of the oceans. Winters are cold and snow covers the land for about 100 days, on average, at Almaty (Alma Ata). Rainfall is generally quite low.

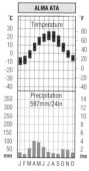

ALMA ATA

Temperature

Precipitation
597mm/24in

CASPIAN SEA

Shallow salt lake, the world's largest inland body of water. The Caspian Sea is enclosed on three sides by Russia, Kazakhstan, Turkmenistan and Azerbaijan. The southern shore forms the northern border of Iran. It has been a valuable trade route for centuries. It is fed mainly by the River Volga; there is no outlet. The chief ports are Baku and Astrakhan. It has important fisheries and an area of 143,000 sq mi [371,000 sq km].

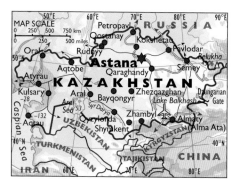

powerful entity, wealthier and more diversified than other Asian republics. It could provide the "new order" between East and West. It is the only former Soviet republic whose ethnic population is almost outnumbered by another group (the Russians), and its Muslim revival is relatively muted. Its first elected president, Nursultan Nazarbayev, a former Communist leader, introduced many reforms, including a multiparty system. However, he has been criticized for his authoritarian rule and the elections of 2004 and 2005 were widely considered to be flawed.

encouraged Russians and Ukrainians to settle in Kazakhstan.

After the Russian Revolution of 1917, many Kazakhs wanted independence, but the Communists prevailed and in 1936 Kazakhstan became a republic of the Soviet Union, called the Kazakh Soviet Socialist Republic. During and after World War II, the Soviet government moved many people from the west into Kazakhstan. From the 1950s, people were encouraged to work on a "Virgin Lands" project, which involved bringing large areas of grassland under cultivation.

POLITICS

Reforms in the Soviet Union in the 1980s led to the breakup of the country in December 1991. Kazakhstan kept contacts with Russia and most of the former Soviet republics by joining the Commonwealth of Independent States (CIS), and in 1995 Kazakhstan announced that its army would unite with that of Russia. In December 1997, the government moved the capital from Alma Ata to Aqmola (later renamed Astana), a town in the Russian-dominated north. It was hoped that this move would bring some Kazakh identity to the area.

Under Soviet rule, Kazakhstan was a dumping ground and test bed. The rocket-launching site at Baykonur (Bayqongyr) suffered great environmental damage, including the shrinking of the Aral Sea by 70%. But Kazakhstan has emerged as a

ECONOMY

The World Bank classifies Kazakhstan as a "lower-middle-income" developing country. Livestock farming, especially sheep and cattle, is an important activity, and major crops include barley, cotton, rice and wheat.

The country is rich in mineral resources, including coal and petroleum reserves, together with bauxite, copper, lead, tungsten, and zinc. Manufactures include chemicals, food products, machinery and textiles. The first major pipeline transporting petroleum direct from the Caspian opened in 2001, the pipeline runs through Russia. To reduce dependence on Russia, Kazakhstan signed an agreement in 1997 to build a new pipeline to China.

ASTANA
(formerly AQMOLA)

Capital of Kazakhstan, on the River Ishim in the steppes of north-central Kazakhstan. Under Soviet rule, Aqmola functioned as capital of the Virgin Lands. From 1961 to 1993 it was known as Tselinograd, and from 1993 to 1998 as Aqmola.

KENYA

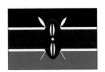

Kenya's flag dates from 1963, when the country became independent. It is based on the flag of KANU (Kenya African National Union), the political party which led the nationalist struggle. The Masai warrior's shield and crossed spears represent the defense of freedom.

The Republic of Kenya is located in East Africa straddling the Equator. Behind the narrow coastal plain on the Indian Ocean, the land rises to high plains and highlands, broken by volcanic mountains, including Mount Kenya, the highest peak at 17,057 ft [5,199 m].

Crossing the country is an arm of the Great Rift Valley with several lakes including Baringo, Magadi, Naivasha, Nakuru and, on the northern frontier, Lake Turkana (formerly Lake Rudolf).

Area 224,080 sq mi [580,367 sq km]
Population 32,022,000
Capital (population) Nairobi (2,143,000)
Government Multiparty republic
Ethnic groups Kikuyu 22%, Luhya 14%, Luo 13%, Kalenjin 12%, Kamba 11%, others
Languages Kiswahili and English (both official)
Religions Protestant 45%, Roman Catholic 33%, traditional beliefs 10%, Islam 10%
Currency Kenyan shilling = 100 cents
Website www.kenya.go.ke

CLIMATE

The coast is hot and humid, but inland the climate is moderated by the height of the land. The thickly populated south-western highlands have summer temperatures 18°F [10°C] lower than the coast. Nights can be cool, but temperatures stay above zero. The main rainy season is from April to May. Only 15% of the country has a reliable rainfall of 31 in [800 mm].

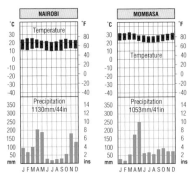

HISTORY

The Kenyan coast has been an important trading center for more than 2,000 years. Early Arab traders carried goods from eastern Asia and exchanged them for items from the local people. Portuguese explorer Vasco da Gama reached the coast in 1498. Later, the Portuguese competed with the Arabs for control of the coast.

The British took control of the coast in 1895, soon extending their influence

MOMBASA

City and port on the Indian Ocean, southwestern Kenya, partly on Mombasa Island, partly on the mainland. From the 11th to 16th centuries, Mombasa was a center of the Arab slave and ivory trades. From 1529 to 1648, the Portuguese held it. Taken by Zanzibar in the mid-19th century, the city passed to Britain in 1887, when it became capital of the British East Africa Protectorate. Kenya's chief port, Mombasa exports coffee, fruit, and grain. Industries include tourism, food processing, glass, and petroleum refining.

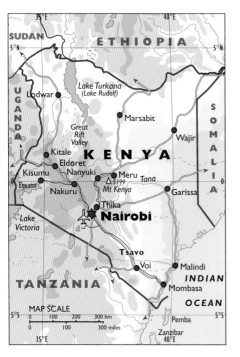

between the political and economic policies of the countries. Hopes were revived in 1999, when a new East African Community was created. Its aim was to establish a customs union, a common market, a monetary union, and, ultimately, a political union.

Jomo Kenyatta died in 1978 and was succeeded by the vice-president Daniel arap Moi, who stood down in 2002 after having been criticized for his autocratic rule, as well as corruption. The veteran Mwai Kibaki was elected as president in 2002, promising to stamp out corruption. But by 2005 he was widely criticized for failing to fulfill his election pledge.

ECONOMY

According to the United Nations, Kenya is a "low-income" developing country. Agriculture employs about 80% of the people, but many Kenyans are subsistence farmers. The chief food crop is corn. Bananas, beans, cassava and sweet potatoes are also grown. The main cash crops are coffee and tea. Manufactures include chemicals, leather, footwear, processed food, petroleum products, and textiles.

inland with many Britons setting up large farms. Opposition to British rule mounted in the 1940s, and, in 1953, a secret movement called Mau Mau launched an armed struggle. Mau Mau was eventually defeated, but Kenya finally gained independence in 1963. Kenya's first president was nationalist veteran Jomo Kenyatta.

POLITICS

Many Kenyan leaders felt that the division of the population into 40 ethnic groups might lead to instability. They argued that Kenya should have a strong central government and, as a result, Kenya has been a one-party state for much of the time since independence. Multiparty democracy was restored in the early 1990s with elections in 1992, 1997 and 2002.

In the 1960s, attempts by Kenya, Tanzania, and Uganda to collaborate collapsed due to the deep differences

NAIROBI

Capital and largest city of Kenya, in the south-central part of the country. Founded in 1899, Nairobi replaced Mombasa as the capital of the British East Africa Protectorate in 1905. Nairobi has a national park (1946), a university (1970), and several institutions of higher education. It is an administrative and commercial center. Industries: cigarettes, textiles, chemicals, food processing, furniture, and glass.

KOREA, NORTH

The flag of the Democratic People's Republic of Korea (North Korea) has been flown since Korea was split into two states in 1948. The colors are traditional ones in Korea. The design, with the red star, indicates that North Korea is a Communist country.

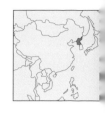

The Democratic People's Republic of Korea occupies the northern part of the Korean Peninsula extending south from northeastern China. Mountains form the heart of the country. The highest peak, Paektu-san (9,003 ft [2,744 m]) is on the northern border. East of the mountains lie the eastern coastal plains, which are densely populated, as are the coastal plains to the west which contain the capital, Pyonyang. Another small highland region in the southeast borders South Korea.

The coastal plains are mostly farmed, but some patches of chestnut, elm, and oak woodland survive on the hilltops. The mountains contain forests of such trees cedar, fir, pine, and spruce.

Area 46,540 sq mi [120,538 sq km]
Population 22,698,000
Capital (population) Pyongyang (2,725,000)
Government Single-party people's republic
Ethnic groups Korean 99%
Languages Korean (official)
Religions Buddhism and Confucianism (religious freedom now an illusion created by government-sponsored religious groups)
Currency North Korean won = 100 chon
Website www.korea-dpr.com

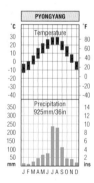

CLIMATE

North Korea has a fairly severe climate, with bitterly cold winters when winds blow from across central Asia, bringing snow. Rivers freeze over and sea ice may block harbors on the coast. In summer, moist winds from the oceans bring rain.

HISTORY

North Korea's history is described in pages 212–13 [see Korea, South]. North Korea was created in 1945, when the peninsula, a Japanese colony since 1910, was divided in two. Soviet forces occupied the north, with US forces in the south. Soviet occupation led to a Communist government being established in 1948 under the leadership of Kim Il Sung.

The Korean War began in June 1950 when North Korean troops invaded the south. North Korea, aided by China and the Soviet Union, fought with South Korea, which was supported by troops from the United States and other UN members. The war ended in July 1953. An armistice was signed but no permanent peace treaty was agreed. The war caused great destruction and loss of life, with 1.6 million Communist troops killed, wounded, or reported missing.

POLITICS

Between 1948 and his death in 1994, Kim Il Sung was a virtual dictator, ruling along similar lines to Stalin in the Soviet Union. After the war, North Korea adopted a hostile policy toward South Korea in pursuit of its aim of reunification. The situation was at times so tense as to warrant international concern.

The end of the Cold War in the late 1980s eased relations between North and South and they both joined the UN in

gram. North Korea withdrew from international talks in early 2005 stating that it had already produced nuclear weapons. However in September North Korea agreed to give up all its nuclear activities and rejoin the Nuclear Non-Proliferation Treaty. Despite reports of malnutrition North Korea formally requested an end to food aid in September 2005. It was thought that the government might be worried that taking more food aid might be perceived as a sign of weakness.

ECONOMY

North Korea's considerable resources include coal, copper, iron ore, lead, tin, tungsten, and zinc. Under Communism, North Korea has concentrated on developing heavy, state-owned industries. Manufactures include chemicals, iron and steel, machinery, processed food, and textiles. Agriculture employs about one-third of the population and rice is the leading crop. Economic decline and mismanagement, aggravated by three successive crop failures caused by floods in 1995 and 1996, and a drought in 1997, led to famine on a large scale.

1991. The two countries made several agreements, including one in which they agreed not to use force against each other. However, the collapse of Communism in the Soviet Union meant that North Korea remained isolated.

In 1993, North Korea triggered a new international crisis by announcing that it was withdrawing from the Nuclear Non-Proliferation Treaty, leading to suspicions that it, was developing its own nuclear weapons. Upon his death in 1994, Kim Il Sung was succeeded by his son, Kim Jong Il.

In the early 2000s, uncertainty surrounding North Korea's nuclear capabilities cast unease across the entire region. The United States accused North Korea of supporting international terrorism, while at the same time, talks between North and South Korea continued in an attempt to normalize relations between them. In 2003 North Korea's relations with the United States further deteriorated when the US accused the country of having a secret nuclear weapons pro-

PYONGYANG

Capital of North Korea, in the west of the country, on the River Taedong. An ancient city, it was the capital of the Choson, Koguryo, and Koryo kingdoms. In the 16th and 17th centuries, it came under both Japanese and Chinese rule. Pyongyang's industry developed during the Japanese occupation from 1910–45. It became the capital of North Korea in 1948. During the Korean War (1950–3), it suffered considerable damage.

KOREA, SOUTH

South Korea's flag, adopted in 1950, is white, the traditional symbol for peace. The central "yin-yang" symbol signifies the opposing forces of nature. The four black symbols stand for the four seasons, the points of the compass, and the Sun, Moon, Earth, and Heaven.

The Republic of Korea, as South Korea is officially known, occupies the southern part of the Korean Peninsula. Mountains cover much of the country. The southern and western coasts are major farming regions. There are many islands along the west and south coasts, the largest of which is Cheju-do, with South Korea's highest peak, Halla-san (6,398 ft [1,950 m]).

Area 38,327 sq mi [99,268 sq km]
Population 48,598,000
Capital (population) Seoul (9,888,000)
Government Multiparty republic
Ethnic groups Korean 99%
Languages Korean (official)
Religions No affiliation 46%, Christianity 26%, Buddhism 26%, Confucianism 1%
Currency South Korean won = 100 chon
Website www.kois.go.kr

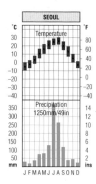

CLIMATE

South Korea is chilled in winter by cold, dry winds blowing from central Asia. Snow often covers the mountains in the east. The summers are hot and wet, especially in July and August.

HISTORY

The Chinese conquered the north in 108 BC and ruled until they were thrown out in AD 313. Mongol armies attacked Korea in the 13th century, but in 1388, a general, Yi Songgye, founded a dynasty of rulers which lasted until 1910.

From the 17th century, Korea prevented foreigners from entering the country, earning it the name the "Hermit Kingdom" until 1876, when Japan forced it to open some of its ports. Soon, the United States, Russia, and some European countries were trading with Korea. In 1910, Korea became a Japanese colony.

After Japan's defeat in World War II, North Korea was occupied by Soviet troops, while South Korea was occupied by United States forces. Attempts at reunification failed and, in 1948, a

SEOUL (KYONGSONG)

Capital of South Korea, on the River Han. The political, commercial, industrial and cultural center of South Korea, it was founded in 1392 as the capital of the Yi dynasty. It developed rapidly under the Japanese (1910–45). Following the 1948 partition, it became capital of South Korea. Seoul's capture by North Korean troops precipitated the start of the Korean War (1950–3), and the following months witnessed the city's virtual destruction. In 1951, it became the headquarters of the UN command in Korea and a rebuilding program commenced. By the 1970s, it was the hub of one of the most successful economies of Southeast Asia. In 1996, there were violent student demonstrations for reunification with North Korea. Seoul hosted the Summer Olympics in 1988 and the semifinal of the 2002 Fifa World Cup.

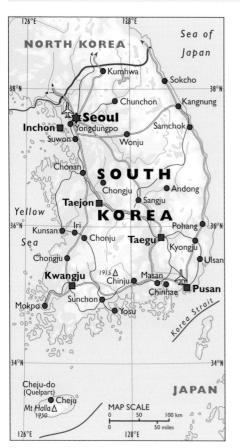

the early 1960s. Initiated by a military government and based on limited natural resources, the country used its cheap, plentiful, well-educated labor force to transform the economy. The manufacturing base of textiles remained important, but South Korea also became a world leader in footwear, shipbuilding, consumer electronics, toys, and vehicles.

In 1988, a new constitution came into force, enabling presidential elections to be held every five years. Evidence of the new spirit of democracy came in 1997 when, in presidential elections, Kim Dae-jung, leader of past pro-democracy campaigns, narrowly defeated Hoi-chang, the governing party's candidate. In foreign affairs, a major breakthrough had occurred in 1991 when both North and South Korea were admitted as full members of the United Nations. The two countries signed several agreements, including one in which they agreed not to use force against each other, but tensions between them continued. In 2000, South Korea's President Dae-jung met with North Korea's Kim Jong Il in talks aimed at establishing better relations between the countries. But the prospect of reunification seemed as distant as ever.

National Assembly was elected in South Korea. This Assembly created the Republic of Korea, with North Korea becoming a Communist state. North Korean troops invaded the South in June 1950, sparking off the Korean War (1950–3).

POLITICS

The story of South Korea after the civil war differs greatly from that of the North. Land reform based on smallholdings worked to produce some of the world's highest rice yields and self-sufficiency in food grains. The real economic miracle came with industrial expansion started in

ECONOMY

The World Bank classifies South Korea as an "upper-middle-income" developing country. It is one of the world's fastest growing industrial economies. Resources include coal and tungsten. The main manufactures are processed food and textiles. The heavy industries are chemicals, fertilizers, iron, steel, and ships. Computers, cars, and televisions are leading industrial products. Farming and fishing remain important activities. Rice is the chief food crop.

KUWAIT

The colors of Kuwait's flag are pan-Arab. The green symbolizes Kuwaiti hospitality. The white represents the commitment to peace. The red symbolizes Kuwait's determination to resist aggression. The black signifies decisiveness.

The State of Kuwait is a small, petroleum-rich, Arab country at the head of the Persian Gulf. It consists of a mainland area and several offshore islands. The capital, Kuwait City, stands on a natural harbor called Kuwait Bay. Most of the land is a flat or gently undulating plain. The highest point is about 820 ft [250 m]. There are no rivers or lakes and water supply is a problem. Water is imported, but drinking water is also produced by desalination plants. Desert scrub covers some areas, but much of Kuwait has no vegetation.

Area 6,880 sq mi [17,818 sq km]
Population 2,258,000
Capital (population) Kuwait City (29,000)
Government Constitutional monarchy
Ethnic groups Kuwaiti 45%, other Arab 35%, South Asian 9%, Iranian 4%, other 7%
Languages Arabic (official), English
Religions Islam 85%, Christianity, Hinduism
Currency Kuwaiti dinar = 1000 fils
Website www.kuwait-info.org

CLIMATE

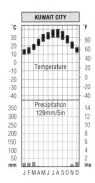

Kuwait has a hot desert climate. Annual rainfall is around 5 in [125 mm] and most occurs between November and March. Winters are mild and pleasant, but summers are hot, with average temperatures reaching 91 to 95°F [33°-35°C] between June and September. August to September are the most uncomfortable months, because the humidity is then at its highest. Periodically, conditions become most unpleasant in the interior when hot sandstorms or duststorms blow from central Arabia.

HISTORY

In the 17th century, the northwestern part of the Arabian peninsula became part of the Turkish Ottoman empire, but the area was thinly populated until about 1710, when people from Arabia settled there and built the port that later became Kuwait City. They elected the head of the Al Sabah family as their ruler, and this family still rules Kuwait today.

British interest in the area began near the end of the 18th century when Kuwait's leader, Sheikh Mubarak, feared Turkish domination. In 1899, Britain became responsible for Kuwait's defense and, in 1914, the territory became a British protectorate. Britain provided naval protection, while taking control of Kuwait's external affairs. Drilling for petroleum began in 1936 and large reserves were discovered by the US-British Kuwait Oil Company. Production was delayed by World War II (1939–45), but petroleum was produced commercially in 1946. Kuwait soon became a prosperous petroleum exporter. The country financed great improvements to the infrastructure, and Kuwaitis soon enjoyed a high standard of living.

POLITICS

Kuwait became an independent state on 19 June 1961, and the Sheikh, the head of state, became an Amir (Emir). Kuwait joined the Arab League and Iraq renewed

its claim that Kuwait was legally part of its territory. But British military intervention forced Iraq to back down. In 1963, elections were held to the National Assembly under a new constitution. However, the Amir suspended the National Assembly in 1976. The National Assembly was restored in 1981 but dissolved again in 1986.

In the Iran-Iraq War, which began in 1980, Kuwait supported Iraq. But in August 1990, Iraq invaded Kuwait, after it had accused it of taking petroleum from an Iraqi oil field near the border. The Amir and his cabinet fled to Saudi Arabia. When Iraq refused to withdraw, a United States-led and UN-supported international force began an aerial bombing campaign, called "Operation Desert Storm," in January 1991. They took Kuwait City in late February and expelled the Iraqis but not before the latter had set fire to more than 500 oil wells, causing massive pollution and destroying almost all the country's commercial and industrial installations. Kuwait's revenge was directed mainly at its huge contingent of Palestinian, Jordanian and Yemeni immigrant workers, who were seen as pro-Iraqi. In 1994, Iraq, under pressure from the UN, officially recognized Kuwait's independence and boundaries, although the countries still have no recognized maritime boundaries in the Persian Gulf.

In 1992, elections were held for a new National Assembly. In 1999, the Amir, Jabir al-Ahmad al-Jabir Al Sabah, suspended the Assembly, but liberals and Islamists predominated in the new Assembly. But the liberals did badly in elections in 2003, though Islamists again did well. In 1999, the Assembly had narrowly rejected a proposal to give women full political rights, but parliament approved these rights in May 2005. Kuwait's first woman cabinet minister was appointed in June. A recent problem faced by Kuwait is violent activity by Islamist militants, some of whom are alleged to be linked to al-Qaida. These groups have been accused of conspiring to attack Western targets.

ECONOMY

The economy is based on petroleum, which accounts for more than 90% of the exports. Kuwait has about 10% of the world's known reserves. Agriculture is practically nonexistent, though the country has a small fishing fleet. Kuwait has to import most of its food. The shortage of water has inhibited the development of industries. However, industrial products include petrochemicals, cement, food products, and construction materials. Another industry is shipbuilding and repair.

KUWAIT CITY

Capital of Kuwait, it sits on the natural harbor of Kuwait Bay in the Persian Gulf. It was first settled in the early 18th century and by the 19th century it was an important trading port. The city was invaded by Iraqi forces in the 1991 Gulf War and under Iraqi occupation renamed Saddam City, after Iraqi leader Saddam Hussein. It returned to its original name once Iraqi troops were expelled. Known as "The City" to the people of Kuwait it includes the Majlis Al-Umma (Kuwait's parliament), most governmental offices, and the headquarters of most Kuwaiti businesses.

KYRGYZSTAN

Kyrgyzstan's flag was adopted in March 1992. The flag depicts a bird's-eye view of a "yurt" (circular tent) within a radiant sun. The "yurt" recalls the traditional nomadic way of life. The 40 rays of the sun stand for the 40 traditional tribes.

The Kyrgyz Republic, is a landlocked country between China, Tajikistan, Uzbekistan and Kazakhstan. The country is mountainous, with spectacular scenery. The highest mountain, Pik Pobedy (Peak of Victory) in the Tian Shan Range, reaches 24,406 ft [7,439 m] above sea level in the east. Less than a sixth of the country is below 2,950 ft [900 m]. The largest of the country's many lakes is Lake Issyk Kul (Ysyk-Köl) in the northeast.

Area 77,181 sq mi [199,900 sq km]
Population 5,081,000
Capital (population) Bishkek (4,000)
Government Multiparty republic
Ethnic groups Kyrgyz 65%, Russian 13%, Uzbek 13%, Ukranian 1%, others
Languages Kyrgyz and Russian (both official)
Religions Islam 75%, Russian Orthodox 20%
Currency Kyrgyzstani som = 100 tyiyn
Website www.gov.kg

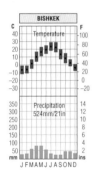

CLIMATE

The lowlands of Kyrgyzstan have warm summers and cold winters. The altitude influences the climate in the mountains, where temperatures in January plummet to 18°F [-28°C]. Far from any sea, Kyrgyzstan has a low annual rainfall.

HISTORY

The area that is now Kyrgyzstan was populated in ancient times by nomadic herders. Mongol armies conquered the region in the early 13th century. They set up areas called *khanates*, ruled by chieftains, or *khans*. Islam was introduced in the 17th century.

China gained control of the area in the mid-18th century, but, in 1876, Kyrgyzstan became a province of Russia, and Russian settlement in the area began. In 1916, Russia crushed a rebellion among the Kyrgyz, and many subsequently fled to China.

In 1922, the area became an autonomous *oblast* (self-governing region) of the newly formed Soviet Union and, in 1936, it became one of the Soviet Socialist Republics. Under Communist rule, nomads were forced to work on government-run farms, while local customs and religious worship were suppressed. However, education and health services were greatly improved.

TIAN SHAN (TIEN SHAN)

Mountain range in central Asia, 1,500 mi [2,400 km] long, forming the border between Kyrgyzstan and Xinjiang, northwest China. At their western edge, the Tian Shan ("Celestial Mountains") divide the Tarim and Junggar Basins. The range rises to 24,406 ft [7,439 m] at Pik Pobedy, on the Chinese border with Kazakhstan and Kyrgyzstan. The Issyk Kul in Kyrgyzstan is one of the world's largest mountain lakes.

POLITICS

In 1991, Kyrgyzstan became an independent country following the breakup of the Soviet Union. The Communist Party was dissolved, but the country retained ties with Russia through the Commonwealth of Independent States. Kyrgyzstan adopted a new constitution in 1994 and elections were held in 1995.

In the late 1990s, Askar Akayev, president since 1990, introduced constitutional changes and other measures which gave him greater powers and limited press freedom. In 2000, Akayev was elected to a third five-year term as president. Alleged government interference in the parliamentary elections of March 2005 sparked massive popular protest, with the people demanding a rerun of the vote and the resignation of Askar Akayev. Official buildings in the capital were seized and, with virtually no resistance from the security forces, Akayev fled to Russia. Kurmanbek Bakiev was appointed acting president and prime minister and he subsequently won a landslide victory in a presidential election in July 2005. The election was deemed to have shown clear progress in democratic standards, according to independent foreign observers.

Kyrgyzstan has the potential to be an ethnic tinderbox, with its large Russian minority (who held positions of power in Soviet days), disenchanted Uzbeks, and an influx of Chinese Muslim immigrants. In the early 2000s, many people were alarmed when Islamic guerrillas staged border raids on Kyrgyzstan as they sought to set up an Islamic state in the Fergana valley, where Kyrgyzstan borders Uzbekistan and Tajikistan.

ECONOMY

The chief activity is agriculture, especially livestock rearing. The main products include cotton, eggs, fruits, grain, tobacco, vegetables, and wool. Food is imported. Manufactures include machinery, processed food, metals, and textiles.

BISHKEK

Capital of Kyrgyzstan, central Asia, on the River Chu. Founded in 1862 as Pishpek, it was the birthplace of a Soviet general, Mikhail Frunze, after whom it was renamed in 1926 when it became administrative center of the Kyrghyz Soviet Republic. Its name changed to Bishkek in 1991, when Kyrgyzstan declared independence. The city has a university (1951). Industries include textiles, food processing, and agricultural machinery.

LAKE ISSYK KUL

At 113 mi [182 km] long, with a width of up to 38 mi [61 km] and a maximum depth of 2,303 ft [701 m], Lake Issyk Kul is one of the largest mountain lakes in the world. Its name derives from a word for "hot lake," as it does not freeze, despite being at an altitude of 5,278 ft [1,609 m]. This saltwater lake is very clear with visibility of up to 65 ft [20 m].

LAOS

Since 1975, Laos has flown the flag of the Pathet Lao, the Communist movement which won control of the country after a long struggle. The blue stands for the River Mekong, the white disc for the Moon, and the red for the unity and purpose of the people.

The Lao People's Democratic Republic is a landlocked country in Southeast Asia. Mountains and plateaus cover much of the country. The highest point is Mount Bia in central Laos, which reaches 9,242 ft [2,817 m].

Most people live on the plains bordering the River Mekong and its tributaries. This river, one of Asia's longest, forms much of the country's northwestern and southwestern borders. A range of mountains called the Annam Cordillera (Chaîne Annamatique) runs along the eastern border with Vietnam.

Area 91,428 sq mi [236,800 sq km]
Population 6,068,000
Capital (population) Vientiane (528,000)
Government Single-party republic
Ethnic groups Lao Loum 68%, Lao Theung 22%, Lao Soung 9%
Languages Lao (official), French, English
Religions Buddhism 60%, traditional beliefs and others 40%
Currency Kip = 100 at
Website www.un.int/lao

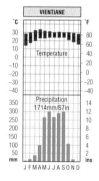

VIENTIANE

CLIMATE

Laos has a tropical monsoon climate. Winters are dry and sunny, with winds blowing in from the northeast. The temperatures rise until April, when the wind directions are reversed and moist southwesterly winds reach Laos, heralding the start of the wet monsoon season.

HISTORY

From the 9th century AD, Lao and Tai peoples set up a number of small states ruled by princes. In 1353 the area that is now Laos was united in a kingdom called Lan Xang ("land of a million elephants"). Apart from a period of Burmese rule between 1574 and 1637, the Lan Xang ruled Laos until the early 18th century. The region was divided into three separate kingdoms, Champasak, Vientiane, and Louangphrabang, which became vassals of Siam (now Thailand).

In the 19th century, Chao Anou, the king of Vientiane, united his kingdom with Vietnam in an attempt to break Siamese domination, but he was defeated and Vientiane became a Siamese province. In the late 19th century, France gradually gained control of all Siamese territory east of the River Mekong and made it a protectorate, ruling it as part of French Indochina, a region which also included Cambodia and Vietnam. After France's surrender to Germany in 1940, Japanese forces moved into Indochina. They allowed the French to continue as puppet rulers until 1945, when they interned all French authorities and military units. A Free Laos movement set up a government, but it collapsed when the French returned in 1946.

POLITICS

Under a new constitution, Laos became a monarchy in 1947 and, in 1949, the country became a self-governing state within the French Union. After full inde-

Laotian Politburo embarked upon its own *perestroika*, opening its doors to tourists and opening trade links with its neighbors, notably China and Japan. Laos became a member of the Association of Southeast Asian Nations (ASEAN) in 1997.

The economy deteriorated from the 1980s and latterly opposition has appeared with sporadic bombings occurring in Vientiane. These have been attributed to rebels in the minority Hmong tribe. Any dissent is dealt with harshly by the authorities.

ECONOMY

Laos is one of the world's poorest countries. Agriculture employs about 76% of the people, compared with 7% in industry and 17% in services. Rice is the main crop, and timber and coffee are both exported. The most valuable export is electricity, which is produced at hydroelectric power stations on the River Mekong and exported to Thailand. Laos also produces opium and in the early 1990s was thought to be the world's third biggest source of this illegal drug. Most enterprises are now outside state control. The government is working to develop alternative crops to opium.

pendence in 1954, Laos suffered from instability caused by a power struggle between royalist government forces and a pro-Communist group called the Pathet Lao. The Pathet Lao took power in 1975 after two decades of chaotic civil war in which the royalist forces were supported by American bombing and Thai mercenaries, while the Pathet Lao was assisted by North Vietnam. The king, Savang Vatthana, abdicated in 1975, and the People's Democratic Republic of Laos was proclaimed. Over 300,000 Laotians, including technicians and other experts, as well as farmers and members of ethnic minorities, fled the country. Many opponents of the government who remained were sent to reeducation camps.

Communist policies brought isolation and stagnation under the domination of the Vietnamese government in Hanoi, which had used Laos as a supply line in their war against the US. In 1986, the

VIENTIANE (VIANGCHAN)

Capital and chief port of Laos, on the River Mekong, close to the Thai border, north central Laos. It was the capital of the Lao kingdom (1707–1828). It became part of French Indochina in 1893, and in 1899 became the capital of the French Protectorate. It is a major source of opium for world markets. Industries include textiles, brewing, cigarettes, hides, wood products.

LATVIA

The burgundy and white Latvian flag, which dates back to at least 1280, was revived after Latvia achieved its independence in 1991. According to one legend, the flag was first made from a white sheet which had been stained with the blood of a Latvian hero.

The Republic of Latvia is one of three states on the southeastern corner of the Baltic Sea, known as the Baltic States. Latvia consists mainly of flat plains separated by low hills, composed of moraine (ice-worn rocks) that was dumped there by ice sheets during the Ice Age. The country's highest point is only 1,020 ft [311 m] above sea level. Small lakes and peat bogs are common. The country's main river, the Daugava, is also known as the Western Dvina.

Area 24,942 sq mi [64,600 sq km]
Population 2,306,000
Capital (population) Riga (793,000)
Government Multiparty republic
Ethnic groups Latvian 58%, Russian 30%, Belarusian, Ukranian, Polish, Lithuanian
Languages Latvian (official), Lithuanian, Russian
Religions Lutheran, Roman Catholic, Russian Orthodox
Currency Latvian lat = 100 santimi
Website www.lv

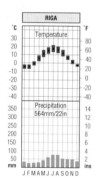

CLIMATE

Air masses from the Atlantic Ocean influence the climate of Latvia, bringing warm and rainy conditions in summer. Winters are cold. The average temperature range is 61–64°F [16-18°C] in July, and 19-27°F [–7- –3°C] in January.

HISTORY

Between the 9th and 11th centuries, the region was attacked by Vikings from the west and Russians from the east. In the 13th century, German invaders took over, naming the country Livland.

In 1561, Latvia was partitioned and most of the land came under Polish or Lithuanian rule. A Germany duchy was also established there. In 1621, the Swedish king Gustavus II Adolphus took over Riga. In 1629, the greater part of the country north of the Daugava River was

ceded to Sweden, with the southeast remaining under Lithuanian rule. But, in 1710, Peter the Great took control of Riga and, by the end of the 18th century, all of Latvia was under Russian control, although the German landowners and merchants continued to exercise considerable power. The 19th century saw the rise of Latvian nationalism and calls for independence became increasingly frequent.

After the Russian Revolution of March 1917, the Latvian National Political Conference demanded independence, but Germany occupied Riga in September. However, after the October Revolution, the Latvian National Political Conference proclaimed the country's independence in November 1918. Russia and Germany finally recognized Latvia's independence in 1920. In 1922, Latvia adopted a democratic constitution and the elected government introduced land reforms. However, a coup in May 1934 ended this period of democratic rule. In 1939, Germany and the Soviet Union agreed to divide up much of eastern Europe. Soviet troops invaded Latvia in June 1940 and Latvia was made a part of

dants. This meant that about 34% of Latvian residents were unable to vote. In 1994, Latvia restricted the naturalization of non-Latvians, denying them the vote and land ownership. In 1998, the government agreed that all children born since independence should have automatic citizenship, regardless of their parents' status.

Latvia became a member of NATO and the EU in 2004.

ECONOMY

The World Bank classifies Latvia as a "lower-middle-income" country. The country's only natural resources are land and forests, so many raw materials have to be imported.

Its industries include electronic goods, farm machinery, fertilizers, processed food, plastics, radios, washing machines, and vehicles. Latvia produces only about a tenth of its electricity needs. The rest has to be imported from Belarus, Russia and Ukraine. Farm products include barley, dairy, beef, oats, potatoes, and rye.

the Soviet Union. But German forces invaded the area in 1941 and held it until 1944, when Soviet troops reoccupied the country. Many Latvians opposed to Russian rule were killed or deported.

POLITICS

Under Soviet rule, many Russians settled in Latvia leading Latvians to fear that the Russians would become the dominant ethnic group. From the mid-1980s, when Mikhail Gorbachev was introducing reforms in the Soviet Union, Latvian nationalists campaigned against Soviet rule. In the late 1980s, the Latvian government ended absolute Communist rule and voted to restore the banned national flag and anthem. It also proclaimed Latvian the official language.

In 1990, Latvia established a multiparty political system. In elections in March, candidates in favor of separation from the Soviet Union won two-thirds of parliamentary seats. The parliament declared Latvia independent on May 4, 1990, though the Soviet Union declared this act illegal. However, the Soviet government recognized Latvia's independence in September 1991, shortly before the Soviet Union itself was dissolved. Latvia held its first free elections to its parliament (the Saeima) in 1993. Voting was limited only to those who were citizens on June 17, 1940 and their descen-

RIGA

Capital of Latvia, on the River Daugava, Gulf of Riga. Founded at the beginning of the 13th century, it joined the Hanseatic League in 1282, growing into a major Baltic port. Tsar Peter the Great took the city in 1710. In 1918, it became capital of independent Latvia. In 1940, when Latvia was incorporated into the Soviet Union, thousands of its citizens were deported or executed. Under German occupation from 1941, the city reverted to Soviet rule in 1944 and subsequently suffered further deportations and an influx of Russian immigrants. In 1991, it reassumed its status as capital of an independent Latvia.

LEBANON

Lebanon's flag was adopted in 1943. It uses the colors of Lebanese nationalists in World War I (1914–18). The cedar tree on the white stripe has been a Lebanese symbol since Biblical times. Because of deforestation, only a few of Lebanon's giant cedars survive.

The Republic of Lebanon is a country on the eastern shores of the Mediterranean Sea. Behind the coastal plain are the rugged Lebanon Mountains (Jabal Lubnán), which rise to 10,131 ft [3,088 m]. Another range, the Anti-Lebanon Mountains (Al Jabal ash Sharqi), form the eastern border with Syria. Between the two ranges is the Bekaa (Beqaa) Valley, a fertile farming region.

Area 4,015 sq mi [10,400 sq km]
Population 3,777,000
Capital (population) Beirut (1,148,000)
Government Multiparty republic
Ethnic groups Arab 95%, Armenian 4%, others
Languages Arabic (official),French, English
Religions Islam 70%, Christianity 30%
Currency Lebanese pound = 100 piastres
Website www.lebanon-tourism.gov.lb

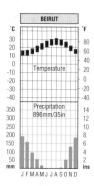

CLIMATE

The Lebanese coast has hot, dry summers and mild, wet winters. Inland, onshore winds bring heavy rain to the western slopes of the mountains in the winter months, with snow on the western slopes of the mountains.

HISTORY

There were waves of invaders from 800 BC – Egyptians, Hittites, Assyrians, Babylonians and Persians. The armies of Alexander the Great seized the area in 332 BC and the Romans took control in 64 BC. Christianity was introduced in AD 325 and in 395 the area became part of the Byzantine Empire. Muslim Arabs occupied the area in the early 7th century, converting many people to Islam.

European Crusaders arrived in Lebanon in about 1100 and the area became a battlefield between Christian and Muslim armies. The Muslim

BEIRUT (BAYRUT)

Capital and chief port of Lebanon, on the Mediterranean. From AD 635 Beirut was under Arab rule. Christian crusaders made it part of the Latin Kingdom of Jerusalem from 1110–1291. In 1516, under Druze control, it became part of the Ottoman Empire and remained so until World War I. In 1920 it became capital of Lebanon under French mandate. With the creation of Israel, thousands of Arabs sought refuge there. The outbreak of civil war in 1976 saw Beirut rapidly fracture along religious lines. In 1982, Israel devastated West Beirut in the war against the PLO. Israel began a phased withdrawal in 1985. Syrian troops entered in 1987 as part of an Arab peacekeeping force and in 1990, they dismantled the "Green Line" separating Muslim West from Christian East Beirut. All militias withdrew by 1991and restoration began. The infrastructure, economy, and culture of Beirut suffered terribly during the civil war.

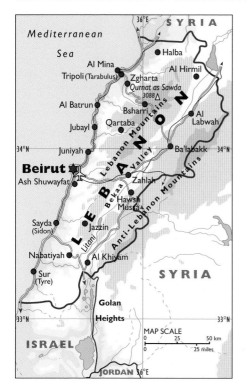

In March 1975, fierce civil war broke out between Christians, Muslims, and Druzes. Lebanon sank into a state of chaos. Assassinations, bombings, and kidnappings became routine as numerous factions fought for control.

The situation was complicated by interventions by Palestinian refugees, the Syrian army, Western and then UN forces as the country became a patchwork of occupied zones and "no-go areas." The core religious confrontation has deep roots. In 1860, thousands of Maronites, who are aligned to the Roman Catholic Church, were murdered by Druzes, who are so tangential to other Islamic sects that they are not now regarded as Muslims.

Although not directly involved, Lebanon was destabilized by the Arab-Israel War of 1967 and by the exile of the PLO leadership to Beirut in 1970. By 1990, the Syrian army had crushed the two-year revolt of Christian rebels against the Lebanese government, but peace proved fragile and a solution elusive. In 1996, Israeli forces launched a sustained attack on the pro-Iranian Hezbollah positions in southern Lebanon, with heavy civilian casualties. Sporadic fighting continued in southern Lebanon in 1997 and again flared up in early 2000. In 2005, former prime minister Rafik Hariri, a critic of Syria's military presence in Lebanon, was assassinated. Following demonstrations, Syria withdrew its forces in April.

Mamelukes of Egypt drove the last of the Crusaders out of the area around 1300. In 1516, Lebanon was taken over by the Turkish Ottoman Empire. Turkish rule continued until World War I, when British and French forces defeated the Ottoman Turks. France took over Lebanon's political affairs from 1923 until 1944, with Lebanon becoming independent in 1946.

POLITICS

The Muslims and Christians agreed to share power and Lebanon made rapid economic progress. But from the late 1950s, development was slowed by periodic conflict between Sunni and Shiite Muslims, Druzes, and Christians. The situation was further complicated by the presence of Palestinian refugees who used bases in Lebanon to attack Israel.

ECONOMY

Civil war almost destroyed valuable trade and financial services which, together with tourism, had been Lebanon's chief source of income. Manufacturing was also hit. Manufactures include chemicals, electrical goods, processed food, and textiles. Fruits, vegetables, and sugarbeet are farmed.

LESOTHO

Based on the national motto: white for peace, blue for rain and green for prosperity. The brown animal-skin shield is supported by an *assegai* (stabbing spear), a plumed spine and a bludgeon, signifying the nation's traditional peace safeguards.

The Kingdom of Lesotho is a land-locked country, surrounded by South Africa on all sides. The scenic Drakensberg Range covers most of the country and forms Lesotho's north east-ern border with KwaZulu Natal. It includes Lesotho's highest peak Thabana Ntlenyana, at 11,424 ft [3,482 m].

Most people live in the western low-lands, site of Maseru, or in the southern valley of the River Orange, which rises in northeast Lesotho and flows through South Africa to the Atlantic Ocean. Grassland covers much of Lesotho. The King holds all land in Lesotho, in trust for the Sotho nation.

Area 11,720 sq mi [30,355 sq km]
Population 1,865,000
Capital (population) Maseru (109,000)
Government Constitutional monarchy
Ethnic groups Sotho 99%
Languages Sotho and English (both official)
Religions Christianity 80%, traditional beliefs 20%
Currency Loti = 100 lisente
Website www.lesotho.gov.ls

CLIMATE

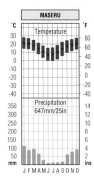

The climate is greatly affected by altitude, with 66% of the country lying above 4,921 ft [1,500 m]. Maseru has warm summers, but the temperatures fall below freezing in the winter and the mountains are colder. Rainfall varies, averaging around 28 in [700 mm].

HISTORY

The early 19th-century tribal wars dis-persed the Sotho. The Basotho nation was founded in the 1820s by King Moshoeshoe I, who united various groups fleeing from tribal wars in south-ern Africa. Moshoeshoe I was forced to yield to the British and Britain made the area a protectorate in 1868. In 1871, it became part of the British Cape Colony in South Africa. However, in 1884, Basu-toland, as the area was called, was recon-stituted as a British protectorate, where whites were not allowed to own land.

POLITICS

In 1966, Sotho opposition to incorpora-tion into the Union of South Africa saw the creation of the independent Kingdom of Lesotho, with Moshoeshoe II, great-

SOTHO

Major cultural and linguistic group of southern Africa. It includes the Northern Sotho of Transvaal, South Africa, the Western Sotho (better known as the Tswana) of Botswana, and the Southern Sotho (Basotho or Basuto) of Lesotho. Although dominating the rural territories they inhabit, the 4 million Sotho share those areas with people of other Bantu-speaking tribes. Many work and live in the urban areas and surrounding townships.

won the 1992 multiparty elections, and the military council dissolved. In 1994, Letsie III attempted to overthrow the government. In 1995, Moshoeshoe II returned to the throne. But after his death in a car crash in 1996, Letsie III again became king. In 1997, a majority of BCP politicians formed a new governing party, the Lesotho Congress for Democracy (LCD).

In 1998, an army revolt, following an election in which the ruling party won 79 out of the 80 seats, caused much damage to the economy, despite the intervention of a South African force intended to maintain order. In 2004, the government declared a state of emergency following three years of drought

ECONOMY

Lesotho is a "low-income" developing country. It lacks natural resources except diamonds. Agriculture employs two-thirds of the workforce, but most farmers live at subsistence level. Livestock farming is important. Major crops include corn and sorghum. Tourism is developing. Other sources of income include the products of light manufacturing and remittances sent home by Basotho working abroad, mainly in the mines of South Africa.

grandson of Moshoeshoe I, as its king.

In 1970, Leabua Jonathan suspended the constitution and banned opposition parties. Civil conflict between the government and Basuto Congress Party (BCP) forces characterized the next 16 years. In 1986, a military coup led to the reinstatement of Moshoeshoe II. In 1990, he was deposed and replaced by his son, Letsie III, as monarch. The BCP

ORANGE

Longest river of South Africa. It rises in the Drakensberg Mountains in northern Lesotho and flows generally west, forming the boundary between Free State and Cape Province. It continues west through the Kalahari and Namib deserts, forming South Africa's border with Namibia then emptying into the Atlantic Ocean at Oranjemund. It is 1,300 mi [2,100 km] in length.

MASERU

Capital of Lesotho, on the River Caledon, near the western border with South Africa. Originally a small trading town, it was capital of British Basutoland protectorate (1869–71, 1884–1966). It remained the capital when the Kingdom of Lesotho achieved independence in 1966. It is a commercial, transportation, and administrative center.

LIBERIA

Liberia was founded in the early 19th century as an American colony for freed slaves who wanted to return to Africa. Its flag was adopted in 1847, when Liberia became independent. The 11 red and white stripes represent the 11 men who signed Liberia's Declaration of Independence.

The Republic of Liberia is located on the Atlantic coast of west Africa. Behind the coastline 311 mi [500 km] long, lies a narrow coastal plain. Beyond, the land rises to a plateau region, with the highest land along the border with Guinea. The most important rivers are the Cavally, which forms the border with Ivory Coast, and the St. Paul.

Mangrove swamps and lagoons line the coast, while inland, forests cover nearly 40% of the land. Liberia also has areas of tropical savanna. Only 5% of the land is cultivated.

Area 43,000 sq mi [111,369 sq km]
Population 3,391,000
Capital (population) Monrovia (421,000)
Government Multiparty republic
Ethnic groups Indigenous African tribes 95% (including Kpelle, Bassa, Grebo, Gio, Kru, Mano)
Languages English (official), ethnic languages
Religions Christianity 40%, Islam 20%, traditional beliefs and others 40%
Currency Liberian dollar = 100 cents
Website
www.un.org/Depts/dpko/missions/unmil/index.html

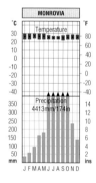

CLIMATE

Liberia has a tropical climate. There are high temperatures and humidity throughout the year. Rainfall is abundant all year round, but there is a particularly wet period from June to November. The rainfall generally increases from east to west.

HISTORY

In the late 18th century, some white Americans in the United States wanted to help freed black slaves to return to Africa. They set up the American Colonization Society in 1816, which bought land in what is now Liberia.

In 1822, the Society landed former slaves at a settlement on the coast which they named Monrovia. In 1847, Liberia became a fully independent republic with

a constitution much like that of the United States. For many years, the Americo-Liberians controlled the government. US influence remained strong and the American Firestone Company, which ran Liberia's rubber plantations covering more than 1 million acres [400,000 ha], was especially influential. Foreign countries were also involved in exploiting Liberia's mineral resources, including its huge iron-ore deposits.

POLITICS

Under the leadership (1944–71) of William Tubman, Liberia's economy grew and it adopted social reforms. In 1980, a military force composed of people from the local population killed the Americo-Liberian president William R. Tolbert, Tubman's successor. An army sergeant, Samuel K. Doe, became president. In 1985, Doe's brutal and corrupt regime won a fraudulent election.

Civil war broke out in 1989, and the Economic Community of West African States (ECOWAS) sent a five-nation peacekeeping force. Doe was assassinated and an interim government, led by

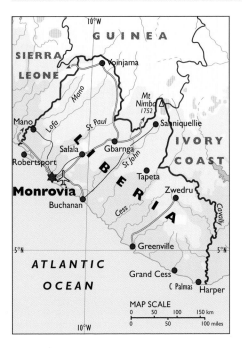

In 2001, the UN imposed an arms embargo on Liberia for trading weapons for diamonds with rebels in Sierra Leone. In 2002, Taylor imposed a state of emergency as fighting intensified with rebels. In 2003, the fighting largely ended; Taylor went into exile. The UN helped to restore order and, in 2005, Ellen Johnson-Sirleaf was elected president.

ECONOMY

Liberia's civil war devastated the economy. Agriculture employs 75% of the workforce, but many families live at subsistence level. Food crops include cassava, fruits, rice, and sugarcane. Rubber is grown on plantations and cash crops include cocoa and coffee.

Liberia's natural resources include its forests and iron ore, while gold and diamonds are also mined. Liberia has a petroleum refinery, but manufacturing is small-scale. Exports include rubber, timber, diamonds, gold, and coffee.

Revenue is also obtained from its "flag of convenience," which is used by about one-sixth of the world's commercial shipping, exploiting low taxes.

Amos Sawyer, took office. Civil war raged on, claiming 150,000 lives and leaving hundreds of thousands of people homeless by 1994. In 1995, a ceasefire occurred and the former warring factions formed a council of state. Former warlord Charles Taylor of the National Patriotic Council secured a resounding victory in 1997 elections.

AMERICAN COLONIZATION SOCIETY

Group founded in 1817 by Robert Finley to return free African-Americans to Africa for settlement. More than 11,000 African-Americans were transported to Sierra Leone and, after 1821, Monrovia. Leading members of the society included James Monroe, James Madison, and John Marshall.

MONROVIA

Capital and chief port of Liberia, West Africa, on the estuary of the River St. Paul. In 1822 freed US slaves settled Monrovia, on a site chosen by the American Colonization Society.

The city is named after James Monroe, US president from 1817–25. Monrovia is Liberia's largest city and the administrative, commercial, and financial center of the country, though it suffered much damage in the civil war. It exports latex and iron ore. Industries—bricks and cement.

LIBYA

Libya's flag was adopted in 1977. It replaced the flag of the Federation of Arab Republics which Libya left in that year. Libya's flag is the simplest of all world flags. It represents the country's quest for a green revolution in agriculture.

The Great Socialist People's Libyan Arab Jamahiriya (Libya's official name) is located in North Africa. The majority live on the Mediterranean coastal plains in the northeast and northwest. The Sahara, the world's largest desert, occupies 95% of Libya, reaching the Mediterranean coast along the Gulf of Sidra (Khalīj Surt). The Sahara is virtually uninhabited except around scattered oases.

The land rises toward the south, reaching 7,500 ft [2,286 m] at Bette Peak (Bikku Bitti) on the border with Chad. Shrubs and grasses grow on northern coasts, with some trees in wetter areas. Few plants grow in the desert, except at oases where date palms provide protection from the hot sun.

Area 679,358 sq mi [1,759,540 sq km]
Population 5,632,000
Capital (population) Tripoli (1,500,000)
Government Single-party socialist state
Ethnic groups Libyan Arab and Berber 97%
Languages Arabic (official), Berber
Religions Islam (Sunni Muslim) 97%
Currency Libyan dinar = 1000 dirhams
Website www.libyana.org

Islam. From 1551, Libya was part of the Ottoman empire. Italy took control in 1911, but lost the territory in World War II. Britain and France then jointly ruled Libya until 1951, when it became an independent kingdom.

POLITICS

In 1969, a military group headed by Colonel Muammar Gaddafi deposed the king and set up a military government. Under Gaddafi, the government took control of the economy and used money from petroleum exports to finance welfare services and development projects. However, although Libya appears to be democratic, political parties are not permitted.

CLIMATE

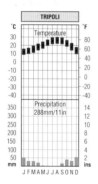

The coastal plains experience hot summers. Winters are mild with some rain. Inland, the average yearly rainfall drops to around 4 in [100 mm] or less. Daytime temperatures are high but nights are cool.

HISTORY

Libya's first known inhabitants were the Berbers. From the 7th century BC to the 5th century AD, Libya came under the Carthaginians, Greeks, and Romans. The Romans left superb ruins, but the Arabs, who invaded the area in AD 642, imposed their culture, including their religion,

---

BENGHAZI (BANGHAZI)

City on the northeastern shore of the Gulf of Sidra, Libya. Founded by the Greeks in the 6th century BC, Benghazi was captured by the Italians in 1911. Libya's second largest city, Benghazi is a commercial and industrial center for Cyrenaica province.

ECONOMY

Libya is Africa's richest country, per capita, but remains a developing one because of its dependence on petroleum, which accounts for nearly all its export revenues.

Agriculture is important, although Libya still imports food. Crops include barley, citrus fruits, dates, olives, potatoes, and wheat. Cattle, sheep, and poultry are raised. Libya has petroleum refineries and petrochemical plants. It also manufactures cement, processed food, and steel.

The "Great Man-Made River" is an ambitious project involving the tapping of subterranean water from rocks beneath the Sahara and piping it to the dry, populated areas in the north, but the water in the aquifers is nonrenewable and will eventually run dry.

Gaddafi has attracted international criticism for his support for radical movements, such as the PLO (Palestine Liberation Organization) and various terrorist groups. In 1986, his policies led the United States to bomb installations in the capital and in Benghazi. In 1994, the International Court of Justice ruled against Libya's claim to an area in northern Chad.

In 1999, Gaddafi sought to restore good relations with the outside world by surrendering for trial two Libyans suspected of planting a bomb on a PanAm airplane, which exploded over the Scottish town of Lockerbie in 1988. In addition Libya agreed to pay compensation to victims of the bombing. Gaddafi also accepted Libya's responsibility for the shooting of a British policewoman in London in 1984 and diplomatic relations with Britain were restored. In 2004 it was announced that Libya was abandoning programs to produce weapons of mass destruction, an initiative that was rewarded by visits to Libya by many Western leaders.

TRIPOLI (TARABULUS)

Capital and chief port of Libya, on the Mediterranean Sea. Founded as Oea in the 7th century BC by the Phoenicians and developed by the Romans. From the 7th century AD, the Arabs developed Tripoli as a market center for the trans-Saharan caravans. In 1551, it was captured by the Ottoman Turks. It was made capital of the Italian colony of Libya in 1911, and was an important base for Axis forces during World War II. After intensive Allied bombing in 1941–2, Britain captured the city. In 1986, the US Air Force bombed Tripoli in retaliation for Libya's alleged support of worldwide terrorism. The city is the commercial, industrial, transportation, and communications center of Libya. Its oases are the most fertile agricultural area in northern Africa.

LITHUANIA

This flag was created in 1918 when Lithuania became an independent republic. After the Soviet Union annexed Lithuania in 1940, the flag was suppressed. It was revived in 1988 and again became the national flag when Lithuania became fully independent in 1991.

The Republic of Lithuania is the southernmost of the three Baltic states. The land is essentially flat with the highest point a hill, northeast of Vilnius (945 ft [288 m]). From the southeast, the land slopes down to the fertile central lowland. In the west is an area of forested sandy ridges, dotted with lakes. South of Klaipeda, sand dunes separate a large lagoon from the Baltic Sea.

Most of the land is covered by moraine deposited by ice sheets during the Ice Age. Hollows in the moraine contain about 3,000 lakes. The longest river is the Neman, which rises in Belarus and flows through Lithuania to the Baltic Sea.

Area 25,174 sq mi [65,200 sq km]
Population 3,608,000
Capital (population) Vilnius (578,000)
Government Multiparty republic
Ethnic groups Lithuanian 80%, Russian 9%, Polish 7%, Belarusian 2%
Languages Lithuanian (official), Russian, Polish
Religions Mainly Roman Catholic
Website www.lietuva.lt

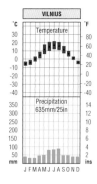

CLIMATE

Winters are cold with temperatures averaging 27°F [–3°C] in January. But summers are warm, with average temperatures in July of 63°F [17°C]. The average rainfall in the west is 25 in [630 mm]. Inland areas are drier.

HISTORY

The Lithuanian people were united into a single nation in the 12th century. The first great ruler was Mindaugas who became king in 1251. By the 14th century, Lithuania's territory extended nearly to Moscow in the east and the Black Sea in the south. Lithuania and Poland became a single state in 1569. This state collapsed in the 18th century and, by

1795, Lithuania was under Russian control. Despite rebellions, Lithuania failed to regain its independence.

In 1905, a conference of elected representatives called for self-government, Russia refused. German troops occupied Lithuania during World War I and, in February 1918, Lithuania declared its independence from Germany and Russia. Lithuania established a democratic form of government, and in 1920, Russia and Lithuania signed a peace treaty. Poland occupied Vilnius from 1920 until 1939, having incorporated it into Poland in 1923. In 1926, a coup overthrew Lithuania's democratic regime.

In 1939, Germany and the USSR agreed to divide up much of eastern Europe. Lithuania and Vilnius were ceded to the USSR in 1940 and a government was set up. German forces invaded in 1941 and held it until 1944, when Soviet troops reoccupied the country. Many Lithuanian guerrillas fought against Soviet rule between 1944 and 1952. Thousands of Lithuanians were killed and many sent to labor camps.

POLITICS

From 1988, Lithuania led the way among the Baltic states in the drive to

Democratic Party. In 1998, an independent, Valdas Adamkus, a Lithuanian-American who had fled in 1944, was elected president. Lithuania had better relations with Russia than the other two Baltic states, partly because ethnic Russians make up a lower proportion of the population than in Estonia and Latvia. Lithuania became a member of NATO and of the EU in 2004.

ECONOMY

The World Bank classifies Lithuania as a "lower-middle-income" developing country. Manufacturing is the most valuable activity. Products include chemicals, electronic goods, processed food, and machine tools. Dairy and meat farming are important, as also is fishing.

shed Communism and regain nationhood. In 1989, the parliament in Lithuania declared Soviet laws invalid unless approved by the Lithuanian parliament and that Lithuanian should be the official language. Religious freedom and the freedom of the press were restored, abolishing the monopoly of power held by the Communist Party and establishing a multiparty system.

Following parliamentary elections in February 1990, in which pro independence candidates won more than 90% of the seats, Lithuania declared itself independent in March 1990, a declaration that was rejected by the Soviet leaders. Most of the capital was then occupied by Soviet troops and a crippling economic blockade put in place. After negotiations to end the sanctions failed, Soviet troops moved into Lithuania and 14 people were killed when the troops fired on demonstrators. Finally, on 6 September 1991, the Soviet government recognized Lithuania's independence.

Parliamentary elections in 1992 were won by the Lithuanian Democratic Labor Party (former Communists). Russian troops withdrew from the country in 1993. In 1996, following new parliamentary elections, a coalition government was set up by the conservative Homeland Union and the Christian

VILNIUS

Capital of Lithuania, on the River Nerisr. Founded in 1323 as the capital of the Grand Duchy of Lithuania, the city declined after the union of Lithuania-Poland. Vilnius was captured by Russia in 1795. After World War I, it was made capital of an independent Lithuania. In 1939, Soviet troops occupied the city and, in 1940, Lithuania became a Soviet republic. During World War II, the city was occupied by German troops, and its Jewish population was all but exterminated. In 1944, it reverted to its Soviet status. In 1990, Lithuania unilaterally declared independence, leading to clashes with Soviet troops and battles on the streets of Vilnius. In 1991, the Soviet Union recognized Lithuanian independence. The old city has many historic synagogues, churches, and civic buildings, and the ruins of a 14th-century castle.

LUXEMBOURG

Luxembourg's tri-color flag derives from its coat of arms which shows a red lion on front of blue and white horizontal stripes. The first recorded use of the coat of arms is on the banner of Earl Heinrich VI in 1228.

The Grand Duchy of Luxembourg is one of the smallest and oldest countries in Europe. The north belongs to an upland region which includes the Ardennes in Belgium and Luxembourg, and the Eiffel Highlands in Germany. This scenic region contains the country's highest point, a hill in the north which reaches 1,854 ft [565 m] above sea level.

The southern two-thirds of Belgium, which is geographically part of French Lorraine, is a hilly or rolling plateau called the Bon Pays or Gut Land ("Good Land"). This region contains rich farmland, especially in the fertile Alzette, Moselle and Sûre (or Sauer) river valleys in the south and east.

Forests cover about a fifth of Luxembourg, mainly in the north, where deer and wild boar are found. Farms cover about 25% of the land and pasture another 20%.

Area 998 sq mi [2,586 sq km]
Population 463,000
Capital (population) Luxembourg (77,000)
Government Constitutional monarchy (Grand Duchy)
Ethnic groups Luxembourger 71%, Portuguese, Italian, French, Belgian, Slav
Languages Luxembourgish (official), French, German
Religions Roman Catholic 87%, others 13%
Currency Euro = 100 cents
Website www.luxembourg.lu

state in AD 963 and a duchy in 1354. In the 1440s, Luxembourg came under the House of Burgundy and, in the early 16th century, under the rule of the Habsburgs. From 1684, it came successively

CLIMATE

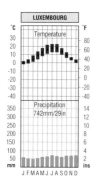

Luxembourg has a temperate climate. In the south of the country summers and falls are warm. This is when grapes ripen in the sheltered southeastern valleys. Winters are sometimes severe, particularly in the Ardennes region, where snow can cover the land for some weeks.

LUXEMBOURG

Capital of the Grand Duchy of Luxembourg, at the confluence of the Alzette and Pétrusse rivers. Luxembourg was a Roman stronghold. The walled town developed around a 10th-century fortress. The Treaty of London (1867) dismantled the fortress and demilitarized the city. It is the seat of the European Court of Justice, the Secretariat of the Parliament of the European Union, the European Monetary Fund, the European Investment Bank, and the European Coal and Steel Union. Industries include iron and steel, chemicals, textiles, and tourism.

HISTORY

Luxembourg became an independent

Germany occupied Luxembourg in both World Wars. In 1944–5, northern Luxembourg was the scene of the Battle of the Bulge. Following World War II, the economy recovered rapidly.

POLITICS
In 1948, Luxembourg joined Belgium and the Netherlands in a union called Benelux and, in the 1950s, was one of the six founders of what is now the European Union. The country's capital, a major financial center, contains the headquarters of several international agencies, including the European Coal and Steel Community and the European Court of Justice.

ECONOMY
Luxembourg has iron-ore reserves and is a major steel producer. It also has many high-technology industries, producing electronic goods and computers. Steel and other manufactures, including chemicals, glass, and rubber products, are exported. Other activities include tourism and financial services. Half the land area is farmed, but agriculture employs only 3% of workers. Crops include barley, fruits, oats, potatoes, and wheat. Cattle, sheep, pigs, and poultry are reared.

under France (1684–97), Spain (1697–1714), and Austria until 1795, when it reverted to French rule.

In 1815, following the defeat of France, Luxembourg became a Grand Duchy under the Netherlands. This was due to the Grand Duke also being the king of the Netherlands. In 1890, when Wilhelmina became queen of the Netherlands, Luxembourg broke away as its laws did not permit a woman monarch. The Grand Duchy then passed to Adolphus, Duke of Nassau-Weilburg. But, in 1912, Luxembourg's laws were changed to allow Marie Adélaïde of Nassau to become the ruling grand duchess. Her sister Charlotte succeeded in 1919, but she abdicated in 1964 in favor of her son, Jean. In 2000, Grand Duke Jean handed over the role as head of state to his son, Prince Henri.

ARDENNES

Sparsely populated wooded plateau in southeastern Belgium, northern Luxembourg, and the Ardennes department of northern France. The capital is Charleville-Mézières. It was the scene of heavy fighting in both World Wars, notably in the Battle of the Bulge (1944). In the well-preserved forest regions wild game is abundant and cleared areas support arable and dairy farming.

MACEDONIA

Macedonia's flag was introduced in August 1992. The emblem in the center of the flag was the device from the war-chest of Philip of Macedon; however, the Greeks claimed this symbol as their own. In 1995, Macedonia agreed to redesign their flag, as shown here.

The Republic of Macedonia is in southeastern Europe. This landlocked country is largely mountainous or hilly, the highest point being Mount Korab (9,068 ft [2,764 m]) on the border with Albania. Most of the country is drained by the River Vardar and its many tributaries. In the southwest, Macedonia shares two large lakes – Ohrid and Prespa – with Albania and Greece. Forests of beech, oak and pine cover large areas, especially in the west.

Area 9,928 sq mi [25,713 sq km]
Population 2,071,000
Capital (population) Skopje (430,000)
Government Multiparty republic
Ethnic groups Macedonian 64%, Albanian 25%, Turkish 4%, Romanian 3%, Serb 2%
Languages Macedonian and Albanian (official)
Religions Macedonian Orthodox 70%, Islam 29%
Currency Macedonian denar = 100 paras
Website www.vlada.mk/english/index_en.htm

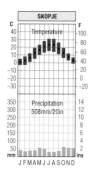

CLIMATE

Summers are hot, though highland areas are cooler. Winters are cold and snowfall is often heavy. The climate is fairly continental with rainfall throughout the year. Average temperatures in Skopje range from 34°F [1°C] in January to 75°F [24°C] in July. The average annual rainfall in the city is 21 in [550 mm].

HISTORY

Until the 20th century, Macedonia's history was closely tied to that of a larger area, also called Macedonia, which covered of northern Greece and southwestern Bulgaria. This region reached its peak in power at the time of Philip II (382–336 BC) and his son Alexander the Great (336–323 BC), who conquered an empire that stretched from Greece to India. The area became a Roman province in the 140s BC and part of the Byzantine Empire from AD 395. In the 6th century, Slavs from eastern Europe attacked and settled in the area, followed by Bulgars from central Asia in the late 9th century. The Byzantine Empire regained control in 1018. Serbia took Macedonia in the early 14th century.

The area was conquered by the Ottoman Turks in 1371 and was under their rule for more than 500 years. The Ottoman Empire began to collapse in the late 19th century and in 1913, at the end of the Balkan Wars, the area was divided between Bulgaria, Greece and Serbia.

POLITICS

As a result of the division of the area known as Macedonia, Serbia took the north and center of the region, Bulgaria took a small area in the southeast, and Greece gained the south. At the end of World War I, Serbian Macedonia became part of the Kingdom of the Serbs, Croats and Slovenes, which was renamed Yugoslavia in 1929. Yugoslavia was conquered by Germany during World War II, but when the war ended in 1945 the Communist partisan leader Josip Broz Tito set up a Communist government.

threatened in 1999 when Albanian-speaking refugees flooded into Macedonia from Kosovo. In 2001, Albanian-speaking Macedonians in northern Macedonia launched an armed struggle. The uprising ended when the government introduced changes that gave Albanian-speakers increased rights, including the recognition of Albanian as an official language. In 2004, the USA recognized the name Republic of Macedonia instead of FYROM, with other countries expected to follow this lead, despite objections from Greece.

Tito maintained unity among the diverse peoples of Yugoslavia but, after his death in 1980, the ethnic and religious differences began to reassert themselves. Yugoslavia broke apart into five sovereign republics with Macedonia declaring its independence on September 18, 1991, thereby avoiding the civil war that shattered other parts of the former Yugoslavia.

However, Macedonia ran into problems concerning recognition. Greece, worried by the consequences for its own Macedonian region, vetoed any acknowledgement of an independent Macedonia on its borders. It considered Macedonia to be a Greek name. It also objected to a symbol on Macedonia's flag, which was associated with Philip of Macedon, and a reference in the country's constitution to the desire to reunite the three parts of the old Macedonia.

Macedonia adopted a new clause in its constitution rejecting all claims on Greek territory and, in 1993, joined the United Nations under the name of The Former Yugoslav Republic of Macedonia (FYROM). In late 1993, all EU countries, except Greece, established diplomatic relations with the FYROM. Greece barred Macedonian trade in 1994, but this ban was subsequently lifted in 1995. Macedonia's stability was

ECONOMY

According to the World Bank, Macedonia ranks as a "lower-middle-income" developing country. Macedonia mines coal, chromium, copper, iron ore, lead, manganese, uranium, and zinc. Manufactures include cement, chemicals, cigarettes, cotton fabric, footwear, iron and steel, refrigerators, sulfuric acid, tobacco products, and wool yarn.

Agriculture employs 9% of the workforce, as compared with 23% in manufacturing and mining. About a quarter of the land is farmed and major crops include cotton, fruits, corn, potatoes, tobacco, vegetables, and wheat. Cattle, pigs, poultry, and sheep are also raised. Forestry is another important activity in some areas.

SKOPJE

Capital of Macedonia, on the River Vardar. Founded in Roman times, it became capital of the Serbian Empire in the 14th century, fell to the Ottoman Turks in 1392, and incorporated into Yugoslavia in 1918. In 1963 an earthquake destroyed most of the city. Industries include metals, textiles, chemicals, glassware.

MADAGASCAR

The colors on this flag are those used on historic flags in Southeast Asia. It was from this region that the ancestors of many Madagascans came around 2,000 years ago. This flag was adopted in 1958, when Madagascar became a self-governing republic under French rule.

The Republic of Madagascar, lies 240 mi [390 km] off the southeast coast of Africa and is the world's fourth largest island. In the west, a wide coastal plain gives way to a central highland region, mostly between 2,000ft to 4,000ft [600 m and 1,220 m]. This is Madagascar's most densely populated region and home of the capital, Antananarivo. The land rises in the north to the volcanic peak of Tsaratanana, at 9,436 ft [2,876 m]. The land slopes off in the east to a narrow coastal strip.

Grass and scrub grow in the south. Forest and tropical savanna once covered much of Madagascar, but farming cleared large areas, destroying natural habitats and seriously threatening the island's unique and diverse wildlife.

Area 226,657 sq mi [587,041 sq km]
Population 17,502,000
Capital (population) Antananarivo (1,250,000)
Government Republic
Ethnic groups Merina, Betsimisaraka, Betsileo, Tsmihety, Sakalava and others
Languages Malagasy and French (both official)
Religions Traditional beliefs 52%, Christianity 41%, Islam 7%
Currency Malagasy franc = 100 centimes
Website www.madagascar.gov.mg

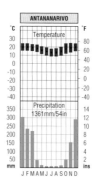

CLIMATE

Altitude moderates temperatures in the highlands. Winters (April to September) are dry, but heavy rains occur in summer. Coastlands to the east are warm and humid. The west is drier, and the south and southwest are hot and dry.

HISTORY

People from Southeast Asia began to settle on Madagascar around 2,000 years ago. Subsequent influxes from Africa and Arabia added to the island's diverse heritage, culture, and language. The Malagasy language is of Southeast Asian origin, though it included words from Arabic, Bantu languages, and European languages.

The first Europeans to reach Madagascar were Portuguese missionaries who in the early 17th century vainly sought to convert the native population. The 17th century saw the creation of small kingdoms and later the French

ANTANANARIVO (TANANARIVE)

Capital and largest city of Madagascar. Founded around 1625, the city became the residence for Imerina rulers in 1794 and the capital of Madagascar. Antananarivo was taken by the French in 1895, and became part of a French protectorate. It is the seat of the University of Madagascar (1961). A trade center for a rice-producing region, it has textile, tobacco, and leather industries.

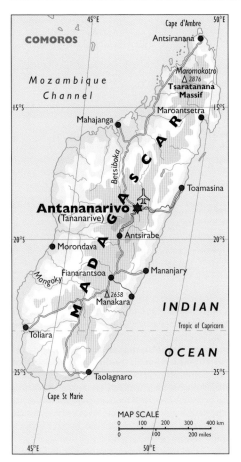

French colony. In 1942, the British overthrew Vichy colonial rule and the Free French reasserted control. In 1946–8 France brutally crushed a rebellion against colonial rule, killing perhaps as many as 80,000 islanders.

POLITICS

In 1960, it achieved full independence as the Malagasy Republic. In 1972, the military took control of government. In 1975, Malagasy was renamed Madagascar, and Lieutenant Commander Didier Ratsiraka became president. He proclaimed martial law, banned opposition parties, and nationalized many industries.

In 1992 Ratsiraka bowed to political pressure and approved a new, democratic constitution. In 1993 multiparty elections, Albert Zafy became president. Zafy was impeached in 1996, and Ratsiraka regained the presidency in 1997 elections. In 2000, floods and tropical storms devastated Madagascar.

Madagascar came to the brink of civil war in 2002 when Ratsiraka and his opponent, Marc Ravalomanana, both claimed victory in presidential elections. Ravalomanana was eventually recognized as president and Ratsiraka went into exile.

ECONOMY

Madagascar is one of the world's poorest countries. The land has been eroded because of the cutting down of the forests and overgrazing of the grasslands. Farming, fishing and forestry employ about 80% of the people.

The country's food crops include bananas, cassava, rice, and sweet potatoes. Coffee is the leading export. Other exports include cloves, sisal, sugar and vanilla. There are few manufacturing industries.

established trading posts along the east coast. The island became a haven for pirates from the late 18th to early 19th century. From the late 18th century a major part of the island was was under Merina rule. In 1817, the Merina ruler and the British governor of Mauritius agreed to the abolition of the slave trade, as a result of which the island received British military and financial assistance. British influence remained strong for several decades. France made contacts with the island in the 1860s. Finally, French troops defeated a Malagasy army in 1895 and Madagascar became a

MALAWI

The colors in Malawi's flag come from the flag of the Malawi Congress Party, which was adopted in 1953. The symbol of the rising sun was added when Malawi became independent from Britain in 1964. It represents the beginning of a new era for Malawi and Africa.

The Republic of Malawi in southern Africa is a small, landlocked country, which is nowhere more than 100 mi [160 km] wide. Its dominant physical feature is Lake Malawi, which is drained in the south by the River Shire, a tributary of the Zambezi. The land is mostly mountainous, the highest point being Mulanje, in the southeast, which reaches 9,843 ft [3,000 m].

Area 45,747 sq mi [118,484 sq km]
Population 11,907,000
Capital (population) Lilongwe (440,000)
Government Multiparty republic
Ethnic groups Chewa, Nyanja, Tonga, Tumbuka, Lomwe, Yao, Ngoni and others
Languages Chichewa and English (both official)
Religions Protestant 55%, Roman Catholic 20%, Islam 20%
Currency Malawian kwacha = 100 tambala
Website www.malawi.gov.mw

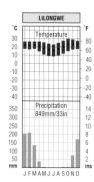

CLIMATE

The low-lying areas of Malawi are hot and humid all year round; the uplands have a pleasant climate. Lilongwe, at about 3,609 ft [1,100 m] above sea level, has a warm and sunny climate. Frosts sometimes occur in July and August, the middle of the long dry season. The wet season extends from November to May.

Wooded savanna and tropical grasslands cover much of the country, with swampy vegetation in many river valleys.

HISTORY

The Bantu-speaking ancestors of the people of Malawi first reached the area around 2,000 years ago, introducing an Iron Age culture and developing kingdoms in the region. In the first half of the 19th century, two other Bantu-speaking groups, the Ngoni (or Angoni) and the Yao invaded the area. The Yao took slaves and sold them to Arabs who traded along the coast. In 1859, the British missionary-explorer David Livingstone reached the area and was horrified by the cruelty of the slave trade. The Free Church of Scotland established a mission in 1875, while Scottish businessmen worked to found businesses to replace the slave trade. The British made treaties with local chiefs on the western banks of what was then called Lake Nyasa and, in 1891, the area was made the British Protectorate of Nyasaland.

The Federation of Rhodesia and Nyasaland was established by Britain in 1953. This included Northern Rhodesia (Zambia) and Southern Rhodesia (Zimbabwe). The people of Nyasaland opposed the creation of the federation, fearing domination by the white minority community in Southern Rhodesia. In 1958, Dr Hastings Banda took over leadership of the opposition to the federation and also to the continuance of British rule. Faced with mounting protests, Britain dissolved the federation in 1963. During 1964, Nyasaland became fully independent as Malawi. Banda became the country's first prime minister and, in 1966, after

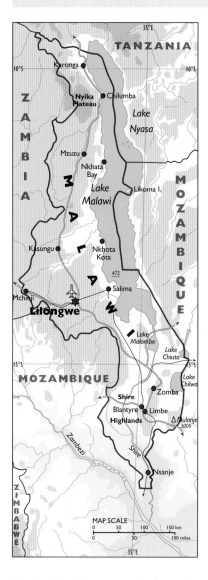

from most of black Africa in being conservative and pragmatic, hostile to its socialist neighbors, but friendly with South Africa. His austerity program and agricultural policies seemed to have wrought an economic miracle, but a swift decline in the 1980s, combined with the arrival of a million refugees from war-torn Mozambique, led to a return to poverty, despite massive aid packages. Another immediate and ongoing problem was the high incidence of HIV putting pressure on the country's limited welfare services. Political dissent led to the restoration of a multiparty system in 1993. Banda and his party were defeated in the elections of 1994 with Bakili Muluzi becoming president. Banda was arrested and charged with murder, but he died in 1997.

ECONOMY

The overthrow of Banda led to a restoration of political freedoms. The abolition of school fees and school uniforms nearly doubled school enrolment. Malawi remains one of the world's poorest countries. Reforms in the 1990s included encouraging small farmers to diversify production, but free enterprise and privatization angered some farmers who have suffered from the ending of subsidies.

Although fertile farmland is limited, agriculture dominates the economy employing more than 80% of the workforce. Tobacco is the leading export, followed by tea, sugar, and cotton. The main food crops include cassava, peanuts, corn, rice, and sorghum. Many farmers raise cattle, goats, and other livestock.

LILONGWE

Capital of Malawi, southeast Africa, in the center of the country, 50 mi [80 km] west of Lake Malawi. It replaced Zomba as the capital in 1975, and rapidly became Malawi's second-largest city.

the adoption of a new constitution, making the country a single-party republic, Banda became the first president.

POLITICS

Banda declared himself president for life in 1971. His autocratic regime differed

MALAYSIA

This flag was adopted when the Federation of Malaysia was set up in 1963. The red and white bands date back to a revolt in the 13th century. The star and crescent are symbols of Islam. The blue represents Malaysia's role in the Commonwealth.

Malaysia consists of two main parts, Peninsular Malaysia (the Malay peninsula) and northern Borneo. Peninsular Malaysia is made up of 11 states and two of the three components of the federal territory (Kuala Lumpur and Putrajaya). Northern Borneo comprises two states and one component of the federal territory (Labuan)

Peninsular Malaysia is dominated by fold mountains with a north–south axis. The most important is the Main Range, which runs from the Thai border to the southeast of Kuala Lumpur, reaching 7,159 ft [2,182 m] at its highest point, Gunong Kerbau. South of the Main Range lie the flat, poorly drained lowlands of Johor. The short rivers have built up a margin of lowlands around the coast.

Northern Borneo has a mangrove-fringed coastal plain, backed by hill country, with east–west fold mountains in the interior. The most striking mountain, and Malaysia's highest point, is the granite peak of Mount Kinabalu, in Sabah, at 13,455 ft [4,101 m].

Area 127,320 sq mi [329,758 sq km]
Population 23,522,000
Capital (population) Kuala Lumpur (1,145,000), Putrajaya (administrative center)
Government Constitutional monarchy
Ethnic groups Malay and other indigenous groups 58%, Chinese 24%, Indian 8%, others
Languages Malay (official), Chinese, English
Religions Islam, Buddhism, Daoism, Hinduism, Chirstianity, Sikhism
Currency Ringgit = 100 cents
Website www.tourism.gov.my

CLIMATE

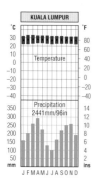

Malaysia has a hot equatorial climate. Temperatures are high all year, though the mountains are much cooler than lowland areas. Rainfall affects the whole country and is heavy throughout the year.

HISTORY

The Malay peninsula has long been a crossroads for sea traders from China and India. Hinduism and Buddhism were introduced from India in the 9th century AD. An early golden age of Malay political power came in the 15th century with the rise of the Kingdom of Malacca (now Melaka), on the southwestern coast of the Malay peninsula. Malacca controlled the important sea routes and attracted traders from all parts of Asia. Arab traders introduced Islam and, in 1414, Malacca's ruler became a Muslim. Many of the people on the peninsula soon embraced Islam, which remains the official religion of Malaysia today.

The first Europeans to reach the area were the Portuguese, and Malacca became a Portuguese possession in 1511. The Dutch, who had been trading in the area during the early 17th century, took Malacca in 1641, and many people from the Dutch-controlled Sulawesi and Sumatra settled in the peninsula, adding to the region's complex ethnic mix. The British, who had been seeking a suitable trading post in Southeast Asia, took over

Malacca in 1794 and though Malacca was returned to the Dutch in 1814, it reverted to British rule in 1824. Through the activities of Stamford Raffles, an agent for the British-owned East India Company, Singapore was occupied by the British in 1819 and made a British territory in 1824. The Straits Settlement, consisting of Penang (now Pinang), Malacca and Singapore, was founded by the British in 1826. In 1867, the Straits Settlement became a British colony. British rule was gradually extended, with Sabah and Sarawak becoming a British protectorate in 1888. In 1896, Negeri Sembilan, Penang, Perak and Selangor became the Federated Malay States. Under British rule, the economy devel-

oped and thousands of Chinese and Indian workers came to work on the rubber plantations.

Japan occupied the area that is now Malaysia and Singapore during World War II, but British rule was restored in 1945 following Japan's defeat. In the late 1940s and 1950s, inspired by the Chinese revolution, Communists fought the British, but guerrilla warfare ended with the independence of the Federation of Malaya in 1957. In 1963, Malaya joined with Singapore, and what is now Sabah and Sarawak, to form the nation of Malaysia, with Tunku Abdul Rahman of the Alliance Party as prime minister. Brunei was invited to join, but no agreement was achieved on entry terms. Arguments between Singapore and the Malaysian government occurred from the outset, causing Singapore to withdraw in 1965, and become an independent sovereign state.

One of the problems faced by the nation has been its great ethnic and religious diversity, with Malays of both Chinese and Indian origin, many brought in by the British to work the tin mines and rubber plantations. There are also a number of Eurasians, Europeans, and aboriginal peoples, notably in Sabah and Sarawak. This patchwork has caused tensions, especially between the Muslim Malays and the politically dominant, mainly Buddhist, Chinese. But while riots did break out in 1969, there was

SOUTH CHINA SEA

Part of the Pacific Ocean, surrounded by southeast China, Indochina, the Malay Peninsula, Borneo, the Philippines, and Taiwan; connected to the East China Sea by the Formosa Strait. The world's largest sea, its chief arms are the Gulf of Tonkin and Gulf of Thailand. The Si, Red, Mekong, and Chao Phraya rivers flow into it. It has an area of 848,000 sq mi [2.3 million sq km] and an average depth of 3,740 ft [1,140 m].

never any escalation into serious armed conflict, nor was economic development affected.

Malaysia faced attacks by Indonesia, which objected to Sabah and Sarawak joining Malaysia. Indonesia's policy of "confrontation" forced Malaysia to increase its defense expenditure. Malaysia was also reluctant to have dealings with Communist countries, but at the same time was keen to remain independent of the Western bloc and aware of the need for Southeast Asian nations to work together. From 1967, it was playing a major part in regional affairs, especially through its membership of ASEAN (Association of Southeast Asian Nations), together with Indonesia, the Philippines, Singapore and Thailand. (Later members of ASEAN include Brunei in 1984, Vietnam in 1995, Laos and Burma (Myanmar) in 1997, and Cambodia in 1999.)

POLITICS

From the 1970s, Malaysia achieved rapid economic progress, especially under the leadership of Dr Mahathir Mohamad, who became prime minister in 1981. Mahathir encouraged the development of industry in order to diversify the economy and reduce the country's reliance on agriculture and mining. The first Malaysian automobile, the Proton Saga, went into production in 1985 and by the early 1990s, manufacturing accounted for about 20% of the gross domestic product. By 1996 its share of the GDP had risen to nearly 35%. However, as with many of the economic "tigers" in Asia's eastern rim, Malaysia was hit by a recession in 1997–8. In response to the crisis, the government ordered the repatriation of many temporary foreign workers and initiated a series of austerity measures aimed at restoring confidence and avoiding the chronic debt problems affecting some other Asian countries. In 1998, the economy shrank by about 5%.

During the economic crisis, differences developed between Mahathir Mohamad and his deputy prime minister and finance minister, Anwar Ibrahim. Anwar wanted Malaysia to work closely with the International Monetary Fund (IMF) to promote domestic reforms and strict monetary and fiscal policies. By the summer of 1998, he had gone further, attacking corruption and nepotism in government. Mahathir, who was suspicious of international "plots" to undermine Malaysia's economy, put much of the blame for the crisis on foreign speculators. He sacked Anwar from the government and also from the ruling United Malays National Organization (UMNO). Anwar was later convicted of conspiracy and charged with sexual misconduct. He was jailed for six years.

In late 1999, Mahathir called a snap election to consolidate his power and strengthen his mandate to deal with the economy. With the economy appearing to be rebounding from recession, Mahathir's coalition retained its two-thirds majority in parliament. But many Malays voted for the conservative Muslim Parti Islam. This meant that Mahathir had to rely more on the Chinese and Indian parties in his coalition. The opposition also gained strength by forming a united front at the 1999 elections. In 2003, Mahathir was succeeded by Abdullah Ahmad Badawi, who won a landslide victory in 2004.

KUALA LUMPUR (MALAY, "ESTUARY MUD")

Capital of Malaysia, southern Malay Peninsula. Founded in 1857, Kuala Lumpur became the capital of the Federated Malay States in 1895, capital of the Federation of Malaya in 1957, and capital of Malaysia in 1963. A commercial city, its striking modern architecture includes one of the world's tallest buildings, the twin Petronas Towers, at 1,483 ft [452 m]. Industries include tin and rubber.

ECONOMY

The World Bank classifies Malaysia as an "upper-middle-income" developing country. Manufacturing is the most important sector of the economy and accounts for a sizeable proportion of the exports. The manufacture of electronic equipment is now a major industry, and, by 1994, Malaysia ranked second in the world in producing radios and fifth in television receivers. Other electronic products include clocks, semiconductors for computers, stereo equipment, tape recorders, and telephones. Other major industrial products include chemicals, petroleum products, plastics, processed food, textiles and clothing, rubber and wood products. Partly because of industrialization, Malaysia is becoming increasingly urbanized. By 2000, about 57% of the population lived in cities and towns.

Malaysia leads the world in the production of palm oil, and, in the mid-1990s, it ranked third in producing natural rubber. Malaysia also ranked fifth in the production of cocoa beans. Other important crops include apples, bananas, coconuts, pepper, pineapples, and many other tropical fruits, rice (Malaysia's chief food crop), sugarcane, tea, and tobacco. Some farmers raise livestock, including cattle, pigs, and poultry. The country's rainforests contain large reserves of timber, and wood and wood products, including plywood and furniture, play an important part in the economy.

The mining of tin is important with Malaysia the eighth largest producer of tin ore in the world. There is also bauxite, copper, gold, iron ore, and ilmenite (an ore containing titanium). Since the 1970s, the production of petroleum and natural gas has steadily increased.

By the mid-1990s, the country's leading exports were machinery and transportation equipment, which accounted for about 55% of the value of the exports. Other exports included manufactures, mineral fuels, animal and vegetable oils, inedible raw materials, and food.

BRUNEI

Negara Brunei Darussalam (as Brunei is offically known), is a sultanate located in north Borneo, southeast Asia.

Bounded in the northwest by the South China Sea, Brunei consists of humid plains with forested mountains running along its southern border with Malaysia.

CLIMATE

Brunei has an equatorial climate. Temperatures range from 73-90°F [23-32°C]. There is high humidity. Rainfall varies from 98 in [2,500 mm] on the coast to 295 in [7,500 mm] inland, but there is no defined rainy season.

HISTORY AND POLITICS

During the 16th century, Brunei ruled over the whole of Borneo and parts of the Philippines. Brunei gradually lost its influence in the region. This was caused by problems regarding royal succession combined with European colonialism. It

Area 2,226 sq mi [5,765 sq km]
Population 358,000
Capital (population) Banda Seri Begawan (55,000)
Government Constitutional sultanate
Ethnic groups Malay 67%, Chinese 15%, others
Languages Malay (official), Chinese, English
Religions Islam (official) 67%, Buddhist 13%, Christian 10%, indigenous beliefs and others
Currency Bruneian dollar = 100 cents
Website www.brunei.gov.bn/index.htm

became a British protectorate in 1888.

Brunei achieved independence in 1983. Brunei has been ruled by the same family for over six centuries. The Sultan has executive authority.

ECONOMY

Petroleum and natural gas are the main source of income, accounting for 70% of GDP. Recently, attempts have been made to increase agricultural production.

243

MALI

The colors on Mali's flag are those used on the flag of Ethiopia, Africa's oldest independent nation. They symbolize African unity. This flag was used by Mali's African Democratic Rally prior to the country becoming independent from France in 1960.

The Republic of Mali is the largest country in west Africa. It is mainly flat, with the highest land in the Adrar des Iforhas on the border with Algeria. Saharan Mali contains many wadis (dry river valleys). The old trading city of Timbuktu lies on the edge of the desert. The only permanent rivers are in the south, the main rivers being the Séné- gal, which flows westward to the Atlantic Ocean to the north of Kayes, and the Niger, which makes a large turn, called the Niger Bend, in south-central Mali.

More than 70% of Mali is desert or semidesert with sparse vegetation. Central and southeastern Mali is a dry grass- land region known as the Sahel. In prolonged droughts, the northern Sahel dries up and becomes part of the Sahara. Fertile farmland and tropical savanna covers southern Mali, the most densely populated region.

Area 478,838 sq mi [1,240,192 sq km]
Population 11,957,000
Capital (population) Bamako (1,016,000)
Government Multiparty republic
Ethnic groups Mande 50% (Bambara, Malinke, Soninke), Peul 17%, Voltaic 12%, Songhai 6%, Tuareg and Moor 10%, others
Languages French (official) and many African languages
Religions Islam 90%, Traditional beliefs 9%, Christianity 1%
Currency CFA franc = 100 centimes
Website www.officetourisme-mali.com

CLIMATE

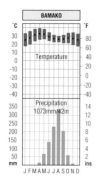

Northern Mali is part of the Sahara, with a hot, practi- cally rainless cli- mate. But the south has enough rain for farming. In the southwest of the country, unpleasant weath- er is experienced when dry and dusty harmattan winds blow from the Sahara Desert.

HISTORY

From the 4th to the 16th centuries, Mali was part of three major black African cul- tures – ancient Ghana, Mali and Song- hai. Reports on these empires were made by Arab scholars who crossed the Sahara to visit them. One major center was Tim- buktu (Tombouctou), in central Mali. In the 14th century, this town was a great center of learning in history, law, and the Muslim religion. It was also a trading center and stopping point for Arabs and their camel caravans. At its height, the Mali Empire was West Africa's richest and most powerful state. However, fol- lowing the defeat of the Songhai empire by Morocco in 1591, the area was divid- ed into small kingdoms.

In 1893, the region became known as French Sudan, and was incorporated into the Federation of West Africa in 1898. Nationalist movements grew more vocal in their opposition to colonialism. In 1958, French Sudan voted to join the French Community as an autonomous republic. In 1959, it joined with Senegal to form the Federation of Mali.

Shortly after gaining independence, Senegal seceded and, in 1960, Mali became a one-party republic. Its first

special administration for Tuaregs in northern Mali. Konaré was reelected in 1997. In 1999, he commuted Traoré's death sentence for corruption to life imprisonment. General Amadou Toumani Toure succeeded Konaré as president in 2002 elections.

ECONOMY

Mali is one of the world's poorest countries and 70% of the land is desert or semidesert. Only about 2% of the land is used for growing crops, while 25% is used for grazing animals. Despite this, agriculture employs nearly 85% of the workforce, many of whom still subsist by nomadic livestock rearing. Farming is hampered by water shortages, and the severe droughts in the 1970s and 1980s led to a great loss of animals and much human suffering. The farmers in the south grow millet, rice, sorghum, and other food crops to feed their families.

The chief cash crops are cotton (the main export), peanuts, and sugarcane. Many of these crops are grown on land which is irrigated with river water. Only a few small areas in the south are worked without irrigation, while the barren deserts in the north are populated only by a few poor nomads.

Fishing is an important economic activity. Mali has vital mineral deposits of gold and salt.

president, Modibo Keita, committed Mali to nationalization and pan-Africanism. Mali adopted its own currency in 1962 and in 1963, it joined the Organization of African States (OAS). Economic crisis forced Keita to revert to the franc zone, and permit France greater economic influence. Opposition led to Keita's overthrow in a military coup in 1968.

POLITICS

The military formed a National Liberation Committee and appointed Moussa Traoré as prime minister. During the 1970s, the Sahel suffered a series of droughts that contributed to a devastating famine in which thousands of people died.

In 1979, Mali adopted a new constitution, and Traoré was elected president. In 1991, a military coup overthrew Traoré, and a new constitution (1992) saw the establishment of a multiparty democracy. Alpha Oumar Konaré, leader of the Alliance for Democracy in Mali (ADEMA), won the ensuing presidential election. A political settlement provided a

BAMAKO

Capital of Mali, on the River Niger, 90 mi [145 km] northeast of the border with Guinea. Once a center of Muslim learning (11th–15th centuries), it was occupied by the French in 1883 and became capital of the French Sudan (1908). Industries include shipping, peanuts, meat, and metal products.

MALTA

A red flag with a white cross was used by the Knights of Malta. In 1943 a George Cross was added after the award bestowed by King George VI of Britain. Upon independence, in 1964 a red edge was added to the cross.

The Republic of Malta is an archipelago republic in the Mediterranean Sea, 60 mi [100 km] south of Sicily. Malta consists of two main islands, Malta (95 sq mi [246 sq km]) and Gozo (26 sq mi [67 sq km]), the small island of Comino, lies between the two large islands. There are also two tiny islets.

The islands are low-lying. Malta island is composed mostly of limestone. Gozo is largely covered by clay, and as a result its landscapes are less arid. Malta has no forests, and 38% of the land is arable.

Area 122 sq mi [316 sq km]
Population 397,000
Capital (population) Valetta (9,000)
Government Multiparty republic
Ethnic groups Maltese 96%, British 2%
Languages Maltese and English (both official)
Religions Roman Catholic 98%
Currency Maltese lira = 100 cents
Website www.gov.mt

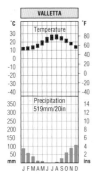

CLIMATE

Malta is typically Mediterranean. Summers are hot and dry. Winters are mild and wet. The sirocco, a hot wind from North Africa, can raise temperatures considerably in the spring.

HISTORY

Malta has evidence of Stone Age settlement dating back 4,000 years. In 850 BC, the Phoenicians colonized Malta. The Carthaginians, Greeks, and Romans followed. In AD 395, Malta became part of the Eastern Roman (Byzantine) Empire. In 870, the Arab invasion brought Islam, but Roger I, Norman King of Sicily, restored Christian rule in 1091. In 1530, the Holy Roman Emperor gave Malta to the Knights Hospitallers. In 1565, the Knights held Malta against a Turkish siege. In 1798, the French captured Malta but, with help from Britain, they

were driven out in 1800. In 1814, Malta became a British colony and a strategic military base.

During World War I, Malta was an important naval base. In World War II, Italian and German airplanes bombed the islands. In recognition of the bravery of the Maltese resistance, the British King

KNIGHTS HOSPITALLERS

(Order of the Hospital of St. John of Jerusalem) Military Christian order recognized in 1113 by Pope Paschal II. In the early 11th century, a hospital was established in Jerusalem for Christian pilgrims. They adopted a military role in the 12th century to defend Jerusalem. After the fall of Jerusalem (1187), they moved to Acre, then Cyprus, then Rhodes (1310), from where they were expelled by the Ottoman Turks (1522). The Pope then gave them Malta, where they remained until driven out by Napoleon in 1798. The order still exists as an international humanitarian charity.

George VI awarded the George Cross to Malta in 1942. Malta became a base for NATO in 1953.

POLITICS

Malta became independent in 1964, and a republic in 1974. Britain's military agreement with Malta expired in 1979, with Malta then ceasing to be a military base. In the 1980s, the people declared Malta neutral. Malta applied to join the European Union in the 1990s, but the application was scrapped when the Labor Party won the elections in 1996. But, following its election defeat in 1998, the bid for EU membership was renewed and Malta finally became a member in 2004.

ECONOMY

The World Bank classifies Malta as an "upper-middle income" developing country, although it lacks natural resources. Most of the workforce is employed in commercial shipbuilding, manufacturing, and the tourist industry. Machinery and transportation equipment account for more than 50% of exports. Manufacturing industries include chemicals, electronic equipment, and textiles. The rocky soil makes farming difficult, Malta produces only 20% of its food. It has a small fishing industry.

MALTESE (MALTI)

Maltese is a Semitic language, that is, one of a group of languages spoken by peoples native to North Africa and the Middle East and forming one of the five branches of the Afro-Asiatic language family. It is spoken by about 340,000 people in Malta and Gozo, and is the only Semitic tongue officially written in the Latin alphabet. The modern language is closely related to western Arabic dialects, but it also shows the strong influence of the Latin that was spoken in Malta. The language developed from the Arabic spoken by the Arabs who invaded Malta in 870. French-speaking Roger I, Norman King of Sicily, ruled from 1091. In 1530 the Knights Hospitallers, who spoke Italian and Latin, were given Malta by the Holy Roman Emperor and it remained under their rule until 1798. In 1814, Malta became a British colony and the British endeavoured to make English the local language. After independence in 1964 Maltese became the national language.

VALLETTA

Port and capital of Malta, on the northeast coast of the island. Founded in the 16th century, it was named after Jean Parisot de la Valette, Grand Master of the Order of the Knights of St. John, who organized the reconstruction of the city after repelling the Turks' Great Siege of 1565. Notable sights include the Royal University of Malta (1592), and the Cathedral of San Giovanni (1576).

MAURITANIA

The Islamic Republic of Mauritania adopted its flag in 1959, the year before it became fully independent from France. It features a yellow star and crescent. These are traditional symbols of the national religion, Islam, as is the color green.

The Islamic Republic of Mauritania in northwestern Africa is nearly twice the size of France, though France's population is more than 28 times that of Mauritania. Over two-thirds of the land is barren, most of it being part of the Sahara. Apart from a small nomadic population, most Mauritanians live in the south, either on the plains bordering the Senegal River in the southwest or on the tropical savanna in the southeast. The highest point is Kediet Ijill (3,002 ft [915 m]). It is an area rich in haematite (high-quality iron ore).

Area 395,953 sq mi [1,025,520 sq km]
Population 2,999,000
Capital (population) Nouakchott (735,000)
Government Multiparty Islamic republic
Ethnic groups Mixed Moor/Black 40%, Moor 30%, Black 30%
Languages Arabic and Wolof (both official), French
Religions Islam
Currency Ouguiya = 100 5 khoums
Website www.mauritania.mr

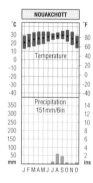

CLIMATE

The amount of rain and the length of the rainy season increases north to south. The desert has dry northeast and easterly winds throughout the year. Southwesterly winds bring summer rain to the south.

HISTORY

From the 4th to the 16th centuries, parts of Mauritania belonged to two great African empires—ancient Ghana and Mali. Portuguese explorers arrived in the 1440s.

European contact did not begin in the area until the 17th century when trade in gum arabic, a substance obtained from an acacia tree, became important, with Britain, France and the Netherlands to the fore. France set up a protectorate in

Mauritania in 1903, attempting to exploit the trade in gum arabic. In 1920 the country became a French colony and

BANC D'ARGUIN

Lying on the coast of Mauritania, the Banc d'Arguin is the most important area of intertidal flats on the coast of Africa. In winter it is home to over 2 million shorebirds, while some 40,000 waterbirds, including great white pelicans, greater flamingos, spoonbills, and several species of tern also breed here. The Banc d'Arguin is also an important fishing area and the Imraguen fishing community and their predecessors have lived from the local fishery for the past 500 years or more. The sight of their lanteen sailed boats working their way up the channels between the mudflats, backed by the dunes of the Sahara, is a unique spectacle in Africa. The Banc d'Arguin is accessible from Nouakchott 93 mi [150 km] to the south.

when the people voted to create a multiparty democracy. In 1992, an army colonel, Maaouiya Ould Sidi Ahmed Taya, who had served as leader of a military administration since December 1984, was elected president. However, subsequent legislative elections in 1992 were boycotted by opposition parties. Taya was reelected in 1997 and 2003.

ECONOMY

The World Bank classifies Mauritania as a "low-income" developing country. Agriculture employs over half the workforce, with the majority living at subsistence level. Many are still cattle herders who drive their herds from the Senegal River through the Sahelian steppelands, coinciding with the seasonal rains. However, droughts in the 1980s greatly reduced the domestic animal populations, forcing many nomadic farmers to seek help in urban areas. Plagues of locusts in 2004 also caused severe damage. Farmers in the southeast grow such crops as beans, dates, millet, rice, and sorghum. Rich fishing grounds lie off the coast. The country's chief natural resource is iron ore and the vast reserves around Fderik provide a major source of revenue.

a territory of French West Africa (an area that included present-day Benin, Burkina Faso, Guinea, Ivory Coast, Mali, Niger, and Senegal, as well as Mauritania). Mauritania became a self-governing territory in the French Union in 1958, achieving full independence in 1960.

POLITICS

In 1961, Mauritania's political parties were merged into one by the president, Mokhtar Ould Daddah, who made the country a one-party state. Upon the withdrawal by Spain from Spanish (now Western) Sahara, a territory bordering Mauritania to the north, in 1976, Morocco occupied the northern two-thirds of the territory, while Mauritania took the rest. Saharan guerrillas belonging to POLISARIO (the Popular Front for the Liberation of Saharan Territories) then began an armed struggle for independence. In 1979, Mauritania withdrew from the southern part of Western Sahara, then occupied by Morocco.

From 1978, Mauritania was ruled by a series of military regimes. In 1991, the country adopted a new constitution

NOUAKCHOTT

Capital of Mauritania, northwest Africa, in the southwestern part of the country, 5 mi [8 km] from the Atlantic Ocean. Originally a small fishing village, it was chosen as capital when Mauritania became independent in 1960. Nouakchott now has an international airport and is the site of modern storage facilities for petroleum. Light industries have been developed and handcrafts are important.

MAURITIUS

The flag is unique in having four equal horizontal bands. The colors derive from Mauritius' coat of arms. Red represents the blood shed in the liberation struggle, blue the Indian Ocean, yellow the bright future afforded by independence, and green the islands' lush vegetation.

The Republic of Mauritius consists of the large island of Mauritius, which is situated 500 mi [800 km] east of Madagascar. This island makes up just over 90% of the country, which also includes the island of Rodrigues, about 348 mi [560 km] east of Mauritius, and several small islands. The main island is fringed by coral reefs, lagoons and sandy beaches. The land in the interior rises to a high lava plateau (2,717 ft [828 m]) enclosed by rocky peaks.

Area 788 sq mi [2,040 sq km]
Population 1,220,000
Capital (population) Port Louis (148,000)
Government Multiparty republic
Ethnic groups Indo-Mauritian 68%, Creole 27%, Sino-Mauritian 3%, Franco-Mauritian 2%
Languages English (official), Creole, French, Hindi, Urdu, Hakka, Bojpoori
Religions Hindu 50%, Roman Catholic 27%, Muslim (largely Sunni) 16%, Protestant 5%
Currency Mauritian rupee = 100 cents
Website www.gov.mu

CLIMATE

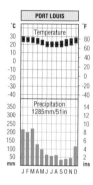

Mauritius has a tropical climate, with heavy rains in the winter. Southeast winds bring rain to the interior plateau of the main island, which is also occasionally hit by destructive tropical cyclones in summer. Average annual rainfall on the interior plateau may reach 200 in [5,100 mm]. The southwest is much drier, with about 35 in [890 mm]. Temperatures range from 72°F [22°C] in the winter (June to October) to 79°F [26°C] in the summer (November to April).

HISTORY

In 1498, Vasco da Gama's fleet accidentally saw the island and, in 1510, a Portuguese navigator, Pedro Mascarenhas, arrived and named it Cimé. Later Portuguese navigators used the island as a port of call, but no permanent settlement

was established. In 1598, the Dutch became the first nation to claim the island and they renamed it after Maurice, Prince of Orange and Count of Nassau. However, an attempt at settlement failed in the 1650s. A second attempt was abandoned in 1710, by which time the famous dodo, which was unique to Mauritius, had become extinct. Following the Dutch withdrawal, the island became a haven for pirates.

The French East India Company claimed Mauritius for France in 1715 and named it the Isle de France. The French developed the economy and imported African slaves. In 1767, control of the island passed to the French government, although the settlers revolted in 1796 when the French government tried to abolish slavery.

British forces landed on Mauritius in 1810, starting a period of British rule. In 1834, Britain abolished slavery on the island which they renamed Mauritius. Most former slaves refused to work on the sugar plantations and so Britain introduced indentured laborers from India, recruiting around 450,000 workers

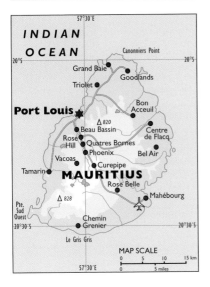

INDIAN OCEAN

Canonniers Point

Grand Baie
Goodlands
Triolet
Bon Acceuil
Port Louis
△ 820
Beau Bassin
Rose Hill
Quatres Bornes
Centre de Flacq
Vacoas
Phoenix
Bel Air
Curepipe
Tamarin
MAURITIUS
Rose Belle
△ 828
Mahébourg
Pte. Sud Ouest
Chemin Grenier
Le Gris Gris

MAP SCALE
0   5   10   15 km
0   5 miles

which lasted until 1976. The MMM won the 1992 elections and Aneerood Jugnauth became prime minister.

Mauritius became a republic in 1992 and Caseem Uteem of the MMM was elected president. In 1995, an alliance led by Navin Ramgoolam and Paul Berenger won the elections and Ramgoolam became prime minister. But Jugnauth was returned to power in 2000. He served as prime minister until 2003, when he handed over to Paul Berenger, who had been his deputy. Berenger was the first non-Hindu to become prime minister. Navin Ramgoolam again became prime minister after elections in July 2005.

Despite tensions between the Indian and Creole communities, Mauritius is a stable democracy with free elections and a good human rights record. As a result, it has attracted foreign investment.

between 1837 and 1910, when indentured labor was ended.

In 1926, the first Indo-Mauritians were elected to the government council while, in 1936, a Creole politician, Dr Maurice Cure, founded the Mauritian Labor Party (MLP). In 1942, representatives from all Mauritian communities were invited to serve on consultative committees and, in 1948, many Indian and Creole were given the vote in elections to a new, enlarged legislature. Internal self-government was introduced in 1957. Elections were held under universal adult suffrage following the introduction of a new constitution in 1958. The MLP, led by Dr Seewoosagur Ramgoolam, won a majority. However, ethnic riots between Indians and Creoles occurred in 1964.

POLITICS
Mauritius became independent in 1968. In 1969, Paul Berenger founded the socialist Mauritian Militant Movement (MMM). In 1971, the MMM supported strikers and organized opposition to the government. In response, the government declared a state of emergency

ECONOMY
Mauritius has become one of Africa's success stories. It is now a middle-income country with a diversified economy. Arable land covers more than half of the country and sugarcane plantations cover about 90% of the cultivated land. Sugar remains a major export: tea and tobacco are also grown. Agriculture and fishing employ 14% of the workforce, compared with 36% in construction and industry. Textiles and clothing are the leading exports. Mauritius has growing industrial and financial sectors and is now a major tourist center.

PORT LOUIS

Capital of Mauritius, a seaport in the northwest of the island. It was founded by the French in 1735. Taken by the British during the Napoleonic Wars, it grew in importance as a trading port after the opening of the Suez Canal. Industries include sugar, electrical equipment, and textiles.

MEXICO

Mexico's flag dates from 1821. The stripes were inspired by the French tricolor. The emblem in the center contains an eagle, a snake, and a cactus. It is based on an ancient Aztec legend about the founding of their capital, Tenochtitlán (now Mexico City).

The United Mexican States is the world's largest Spanish-speaking country. It is largely mountainous. The Sierra Madre Occidental begins in the northwest state of Chihuahua, and runs parallel to Mexico's west coast and the Sierra Madre Oriental. Monterrey lies in the foothills of the latter. Between the two ranges lies the Mexican Plateau. The southern part of the plateau contains a series of extinct volcanoes, rising to Orizaba, at 5,700 m [18,701 ft]. The southern highlands of the Sierra Madre del Sur include the archaeological sites in Oaxaca. Mexico contains two large peninsulas: Baja California in the northwest; and the Yucatán peninsula in the southeast.

Area 756,061 sq mi [1,958,201 sq km]
Population 104,960,000
Capital (population) Mexico City (22,000,000)
Government Federal republic
Ethnic groups Mestizo 60%, Amerindian 30%, White 9%
Languages Spanish (official)
Religions Roman Catholic 90%, Protestant 6%
Currency Mexican peso = 100 centavos
Website www.presidencia.gob.mx

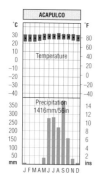

CLIMATE

Mexico's climate is hugely varied according to altitude. Most rain occurs between June and September. More than 70% of Mexico experiences desert or semidesert conditions.

HISTORY

Many Native American civilizations flourished in Mexico. The Olmec (800–400 BC), the Maya (AD 300–900), the Toltec Empire (900–1200). But it was the Aztec who dominated the central plateau from their capital at Tenochtitlán.

In 1519, Spanish conquistadors captured the capital and the Aztec emperor Montezuma. In 1535, the territory became the Viceroyalty of New Spain. Christianity was introduced. Spanish rule was harsh, divisive, and unpopular. Hidalgo y Costillo's revolt (1810) failed to win the support of creoles.

Mexico became independent in 1821 and a republic in 1824. War with Texas escalated into the Mexican War (1846–8) with the United States. Under the terms of the Treaty of Guadalupe-Hidalgo (1848), Mexico lost 50% of its territory. Liberal forces, led by Benito Juárez, triumphed in the War of Reform (1858–61), but conservatives with support from France installed Maximilian of Austria as Emperor in 1864. In 1867, republican rule was restored and Juárez became president. In 1876, an armed revolt gave Porfirio Díaz the presidency: his dictatorship lasted until 1910. Huerta's took power in 1913 and his dictatorship prolonged the Mexican Revolution (1910–40) and led to US intervention. During the 1920s and 1930s, Mexico introduced land and social reforms. After World War II, Mexico's economy developed with the introduction of liberal reforms. Relations with the US improved but problems remain over Mexican economic migration and drug trafficking.

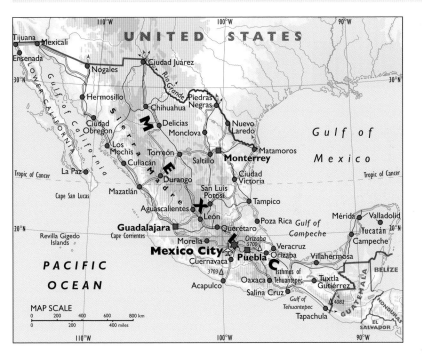

POLITICS

The Institutional Revolutionary Party (PRI) ruled Mexico continuously from 1929 to 2000. In 1994 the Zapatista National Liberation Army (ZNLA) staged an armed revolt in the southern state of Chiapas, calling for land reforms and recognition of Native American rights. Vicente Fox became the first non-PRI leader of Mexico in 2000. In 2001, after a nationwide march by the Zapatistas, the Mexican parliament passed a new rights bill for indigenous peoples.

ECONOMY

Mexico is classified as an "upper-middle-income" developing country. Agriculture is important. Petroleum and petroleum products are the chief exports, while manufacturing is the most valuable activity. Many factories, known as *maquiladoras*, near the northern border assemble goods such as automobile parts and electrical products for US companies.

MEXICO CITY

Capital of Mexico, in a volcanic basin at 7,800 ft [2,380 m], in the center of the country. It is the nation's political, economic, and cultural center and suffers from overcrowding and high levels of pollution. Hernán Cortés destroyed the former Aztec capital (Tenochtitlán), in 1521. A new city was constructed, which acted as the capital of Spain's New World colonies for 300 years. US troops occupied the city in 1847, during the Mexican War. In 1863, French troops conquered the city and established Maximilian as Emperor. Benito Juárez's forces recaptured it in 1867. Between 1914 and 1915, the revolutionary forces of Emiliano Zapata and Francisco Villa captured and lost the city three times.

MOLDOVA

Adopted in 1990, using the same colors as the Romanian flag to emphasize their ties. Blue stands for Transylvania, yellow for Wallachia, and red for Moldavia. The eagle symbolizes the Byzantine Empire; and the bison's head, star, rose, and crescent, the medieval principality of Moldavia.

The Republic of Moldova is a small country sandwiched between Ukraine and Romania. It was formerly one of the 15 republics that made up the Soviet Union. Much of the land is hilly and the highest areas are near the center of the country. The main river is the Dniester, which flows through eastern Moldova.

Forests of hornbeam and oak grow in northern and central Moldova. In the drier south, most of the region is now used for farming, with rich pasture along the rivers.

Area 13,070 sq mi [33,851 sq km]
Population 4,446,000
Capital (population) Chisinau (658,000)
Government Multiparty republic
Ethnic groups Moldovan/Romanian 65%, Ukrainian 14%, Russian 13%, others
Languages Moldovan/Romanian, Russian (official)
Religions Eastern Orthodox 98%
Currency Moldovan leu = 100 bani
Website www.moldova.org/index/eng/

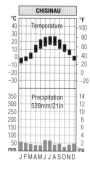

CLIMATE

Moldova has a moderately continental climate, with warm summers and fairly cold winters when temperatures dip below freezing point. Most of the rain falls in the warmer months.

HISTORY

Moldavia was a historic Balkan region, between the Carpathian Mountains in Romania and the Dniester River.

Under Roman rule, it formed the major part of the province of Dacia, and today's population is Romanian-speaking. In the 14th century, it became an independent principality ruled by the Vlachs; its lands included Bessarabia and Bukovina. In 1504, the Turks conquered Moldavia, and it remained part of the Ottoman Empire until the 19th century.

In 1775, the Austrians gained Bukovina, and in 1815 Russia conquered Bessarabia. After the Russo-Turkish War (1828–9), Russia became the dominant power. In 1856, the twin principalities of Moldavia and Wallachia gained considerable autonomy. Three years later, they united under one crown to form Romania, but Russia reoccupied southern Bessarabia in 1878.

In 1920, Bessarabia and Bukovina were incorporated into the Romanian state. In 1924, the Soviet republic of Moldavia was formed, which in 1947 was enlarged to include Bessarabia and northern Bukovina. In 1989, the Moldavians asserted their independence by making Romanian the official language, and in 1991, following the dissolution of the Soviet Union, Moldavia became the independent republic of Moldova.

POLITICS

Following independence in 1991, the majority Moldovan population wished to rejoin Romania, but this alienated the Ukrainian and Russian populations east of the Dniester, who declared their independence from Moldova as the Transdniester Republic. War raged between the

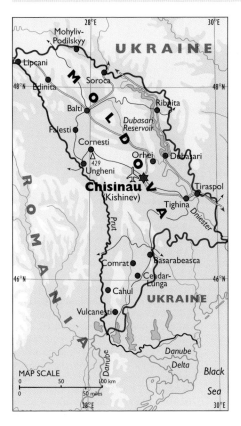

the Party of the Moldovan Communists (PCRM) won the highest share of the votes. The constitution was changed in 2000, turning Moldova from a semipresidential republic to a parliamentary republic. In 2001, the Communist leader Vladimir Voronin was elected president. The Communist party was reelected in 2005, though it now advocates close ties with the West, a matter of some concern to Russia.

ECONOMY

According to the World Bank, Moldova is a lower-middle income developing country, and in terms of GNP per capita, Europe's poorest country. It is fertile and agriculture remains central to the economy. Major products include fruits, corn, tobacco, and grapes for winemaking. Farmers also raise livestock, including dairy cattle and pigs.

There are few natural resources within Moldova, and the government must import materials and fuels for its industries. Light industries, such as food processing and the manufacturing of household appliances, are expanding.

two, with Transdniester supported by the Russian 14th Army. In August 1992, a ceasefire was declared. The former Communists of the Agrarian Democratic Party won multiparty elections in 1994. A referendum rejected reunification with Romania. Parliament voted to join the Commonwealth of Independent States (CIS).

In 1994 a new constitution established a presidential parliamentary republic. In 1995, Transdniester voted in favor of independence in a referendum. In 1996, Russian troops began their withdrawal and Petru Lucinschi was elected president. On 1 January 1997, a former Communist, Petru Lucinschi, became president. In 1998 and 2001,

CHISINAU (KISHINEV)

Capital of Moldova, in the center of the country, on the River Byk. Founded in the early 15th century, it came under Turkish then Russian rule. Romania held the city from 1918 to 1940 when it was annexed by the Soviet Union. In 1991 it became capital of independent Moldova. It has a 19th-century cathedral and a university (1945). Industries include plastics, rubber, textiles, and tobacco.

MONGOLIA

Mongolia's flag contains blue, the national color, together with red for Communism. The traditional Mongolian golden "soyonbo" symbol represents freedom. Within this, the flame is seen as a promise of prosperity and progress.

Mongolia, which is sandwiched between China and Russia, is the world's largest landlocked country. It consists mainly of high plateaus, the highest of which are in the west, between the Altai Mountains (or Aerhtai Shan) and the Hangayn Mountains (or Hangayn Nuruu).

The Altai Mountains contain the country's highest peaks (14,311 ft [4,362 m]). The land descends toward the east and south, where part of the Gobi Desert is situated.

Area 604,826 sq mi [1,566,500 sq km]
Population 2,751,000
Capital (population) Ulan Bator (760,000)
Government Multiparty republic
Ethnic groups Khalkha Mongol 85%, Kazakh 6%
Languages Khalkha Mongolian (official), Turkic, Russian
Religions Tibetan Buddhist Lamaism 96%
Currency Tugrik = 100 möngös
Website www.mongoliatourism.gov.mn

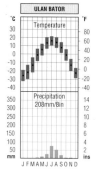

CLIMATE

Due to its remote position, Mongolia has an extreme continental climate, with long, bitterly cold winters and short, warm summers. Annual rainfall ranges from no more than 20 in [500 mm] in the highlands to 5 in [125 mm] in the lowlands.

HISTORY

In the 13th century, the great Mongol conqueror Genghis Khan united the Mongol people, created a ruthless army, and founded the largest land empire in history. Under his grandson, Kublai Khan, the Mongol empire stretched from Korea and China, across Asia into what is now Iraq. In the northwest, Mongol rule extended beyond the Black Sea into eastern Europe. Learning flourished under Kublai Khan, but, after his death in 1294, the empire broke up into several parts. It was not until the late 16th century that Mongol princes reunited Mongolia. During their rule, they introduced Lamaism (a form of Buddhism).

In the early 17th century, the Manchu leaders of Manchuria took over Inner Mongolia. They conquered China in 1644 and Outer Mongolia some 40 years later. Present-day Mongolia then became a remote Chinese province scarcely in contact with the outside world.

GOBI (SHA-MOH)

Desert area in central Asia, extending over much of southern Mongolia and northern China. One of the world's largest deserts, it is on a plateau, 900–1,500 m [3,000–5,000 ft] high. The fringes are grassy and inhabited by nomadic Mongolian tribes who rear sheep and goats. The Gobi has cold winters, hot summers, and fierce winds and sandstorms. It covers 1.3 million sq km [500,000 sq mi].

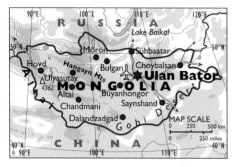

Outer Mongolia broke away from China following the collapse of the Qing Dynasty in 1911, and the Mongols appointed a priest, the Living Buddha, as their king. Legally, Outer Mongolia remained Chinese territory, but China and Russia agreed to grant it control over its own affairs in 1913. Russian influence increased and, in 1921, Mongolian and Russian Communists took control of Outer Mongolia, proclaiming the Mongolian People's Republic in 1924.

POLITICS

Mongolia became an ally of the Soviet Union, its support being particularly significant from the 1950s, when the Soviet Union was in dispute with Mongolia's neighbor, China. The Soviet Union helped develop Mongolia's mineral reserves so by the late 1980s, minerals had overtaken agriculture as the country's main source of revenue.

In 1990, the people, influenced by reforms taking place in the Soviet Union, held demonstrations, demanding more freedom. Free elections in June 1990 resulted in victory for the Communist Mongolian People's Revolutionary Party (MPRP). The new government began to move away from Communist policies, launching into privatization and developing a free-market economy. The "People's Democracy" was abolished in 1992 and democratic institutions were introduced.

The MPRP was defeated in elections in 1996 by the opposition Mongolian Democratic Union coalition. The Democratic Union ran into economic problems and, in the presidential elections of 1997, the MPRP candidate, Natasagiyn Babagandi, defeated the Democratic Union nominee. This achievement was followed by the parliamentary elections in July 2000, which resulted in a landslide victory for the MPRP, who gained 72 out of the 76 available seats in the Great Hural (parliament). The MPRP chairman, Nambaryn Enhbayar, became prime minister. Following disputed elections in 2004 a coalition government was set up.

ECONOMY

The World Bank classifies Mongolia as a "lower-middle-income" developing country. Many Mongolians were once nomads, moving around with their livestock. Under Communist rule, most were moved into permanent homes on government-owned farms. Livestock and animal products remain important.

The Communists developed mining and manufacturing and by 1996, mineral products accounted for nearly 60% of the country's exports. Minerals produced in Mongolia include coal, copper, fluorspar, gold, molybdenum, tin, and tungsten. The leading manufactures are textiles and metal products.

ULAN BATOR

Capital of Mongolia, on the River Tola. Ulan Bator dates back to the founding of the Lamaistic Temple of the Living Buddha in 1639. It grew as a stop for caravans between Russia and China and was later a focus for the Mongolian autonomy movement, becoming the capital in 1921. It is the political, cultural, and economic center of Mongolia. Industries include textiles, building materials, leather, paper, alcohol, and carpets.

MOROCCO

Morocco has flown a red flag since the 16th century. The green pentagram (five-pointed star), called the Seal of Solomon, was added in 1915. This design was retained when Morocco gained its independence from French and Spanish rule in 1956.

The Kingdom of Morocco lies in northwestern Africa. Its name comes from the Arabic Maghreb-el-Aksa (the furthest west). Behind the western coastal plain the land rises to a broad plateau and the Atlas Mountains. The High (Haut) Atlas contains the highest peak, Djebel Toubkal, at 13,665 ft [4,165 m]. Other ranges include the Anti Atlas, the Middle (Moyen) Atlas and the Rif Atlas (or Er Rif). East of the mountains the land lies the arid Sahara.

Area 172,413 sq mi [446,550 sq km]
Population 32,209,000
Capital (population) Rabat (1,220,000)
Government Constitutional monarchy
Ethnic groups Arab-Berber 99%
Languages Arabic (official), Berber dialects, French
Religions Islam 99%
Currency Moroccan dirham = 100 centimes
Website www.mincom.gov.ma

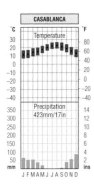

CLIMATE

The Atlantic coast is cooled by the Canaries Current. Inland, summers are very hot and dry while winters are mild. From October to April the southwesterly Atlantic winds bring rain; there is frequent snowfall in the High Atlas.

HISTORY

The original people of Morocco were the Berbers. In the 680s, Arab invaders introduced Islam and the Arabic language. By the early 20th century, France and Spain controlled Morocco. It finally became an independent kingdom in 1956.

POLITICS

King Hassan II ruled the country in an authoritarian way until his death in 1999. His successor, King Mohamed VI, faced a number of problems, including finding a solution to the future of Western Sahara. Relations with Spain became strained in 2002 over the disputed island of Leila (Perejil in Spanish), in the Strait of Gibraltar. Diplomatic relations were restored in 2003. Another problem faced by Morocco is activity by Islamic extremists. Its opposition to extremism led the United States to designate Morocco as a major non-NATO ally in 2004.

ECONOMY

Morocco is classified as a "lower-middle-income" developing country. It is the world's third largest producer of phosphate rock. Farming employs 38% of Moroccans. Fishing and tourism are important.

RABAT

Capital of Morocco, on the Atlantic coast, northern Morocco. Rabat dates from Phoenician times, but the fortified city was founded in the 12th century by the Almohad ruler, Abd al-Mumin. Under French rule, it was made the capital of the protectorate of Morocco.

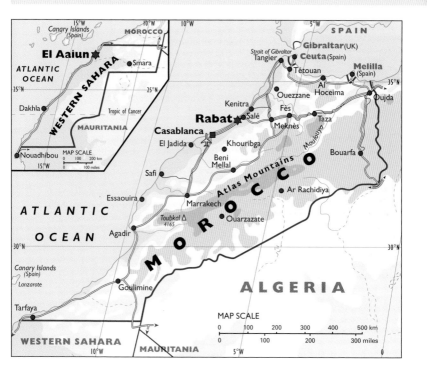

WESTERN SAHARA

Desert territory on the Atlantic coast of northwest Africa; the capital is El Aaiun. It comprises two districts: Saguia el Hamra in the north, and Río de Oro in the south. The population consists of Arabs, Berbers, and pastoral nomads, most of whom are Sunni Muslims. Livestock-rearing dominates agriculture. The first European discovery was in 1434, but it remained unexploited until Spain took control of the coastal area in the 19th century. In 1957, a nationalist movement temporarily overthrew the Spanish but they regained control of the region in 1958, and merged the two districts to form the province of Spanish Sahara. Large phosphate deposits were discovered in 1963. The Polisario Front began a guerrilla war in 1973, eventually forcing a Spanish withdrawal in 1976. Within a month, Morocco and Mauritania partitioned the country. Polisario (backed by Algeria) continued to fight for independence, renaming the country the Saharawi Arab Democratic Republic. In 1979, Mauritania withdrew and Morocco assumed full control. In 1982, the Saharawi Republic became a member of the Organization of African Unity. By 1988 it controlled most of the desert up to the Moroccan defensive line. Fragile ceasefires were agreed in 1988 and 1991. About 200,000 Saharawis continue to live in refugee camps, mostly in Algeria. Talks between Morocco and Western Sahara began in 1997 and have been inconclusive. It covers 102,680 sq mi [266,769 sq km].

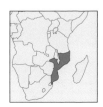

Mozambique's flag was adopted when the country became independent from Portugal in 1975. The green stripe represents fertile land, the black stands for Africa, and the yellow for mineral wealth. The badge on the red triangle contains a rifle, a hoe, a cogwheel, and a book.

The Republic of Mozambique borders the Indian Ocean in southeastern Africa. The coastal plains are narrow in the north but broaden to the south making up nearly half of the country. Inland lie plateaus and hills, which make up another two-fifths of the country, with highlands along the borders with Zimbabwe, Zambia, Malawi, and Tanzania.

Area 309,494 sq mi [801,590 sq km]
Population 18,812,000
Capital (population) Maputo (1,015,000)
Government Multiparty republic
Ethnic groups Indigenous tribal groups (Shangaan, Chokwe, Manyika, Sena, Makua, others) 99%
Languages Portuguese (official), many others
Religions Traditional beliefs 50%, Christianity 30%, Islam 20%
Currency Metical = 100 centavos
Website www.mozambique.mz

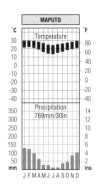

CLIMATE

Most of Mozambique has a tropical maritime climate, with two main seasons. The hot, wet season runs from November to March, with a dry, milder season between April and October. Rainfall varies, being greatest on the northwestern highlands and lowest on the southeastern lowlands.

Temperatures in the lowlands vary from between 79–86°F [20–30°C] in January, and between 52–59°F [11–15°C] in January. The interior highlands are much cooler and generally less humid.

HISTORY

Arab traders began to operate in the area in the 9th century AD, with Portuguese explorers arriving in 1497. The Portuguese set up trading stations in the early 16th century and the area became a source of slaves. When the European powers divided Africa in 1885, Mozambique was recognized as a Portuguese colony. Black African opposition to European rule gradually increased and in 1961, the Front for the Liberation of Mozambique (FRELIMO) was founded to oppose Portuguese rule. FRELIMO launched a guerrilla war in 1964, which continued for ten years. Mozambique achieved independence in 1975, when the Marxist-Leninist FRELIMO, took over the government.

MAPUTO (LOURENÇO MARQUES)

Capital and chief port of Mozambique, on Maputo Bay, southern Mozambique. It was visited by the Portuguese in 1502, and was made the capital of Portuguese East Africa in 1907, being known as Lourenço Marques until 1976. It is linked by rail to South Africa, Swaziland, and Zimbabwe, and is a popular resort area. Industries include footwear, textiles, and rubber.

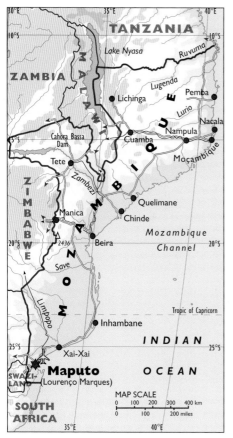

ended in 1992 and multiparty elections in 1994 were won by FRELIMO, whose leader, Joaquim A. Chissano, became president. RENAMO's leader, Afonso Dhlakama, accepted the election results and stated that the civil war would not be resumed. This led to a period of relative stability. In 1995, Mozambique became the 53rd member of the Commonwealth, joining its English-speaking allies in southern Africa.

ECONOMY

By the early 1990s, Mozambique was one of the world's poorest countries. Battered by a civil war, which had killed around a million people and had driven 5 million from their homes, and combined with devastating droughts and floods, the economy collapsed.

By the end of the twentieth century, economists were praising Mozambique for its economic recovery. Although 80% of the people are poor, support from the World Bank and other international institutions, privatization, and rescheduling of the country's foreign debts led to an expansion of the economy and the bringing down of inflation to less than 10% by 1999.

Massive floods at the start of 2000 affected about a quarter of the population making thousands homeless and devastating the economy for many years to come.

Agriculture is important. Crops include cassava, cotton, cashew nuts, fruits, corn, rice, sugarcane, and tea. Fishing is important and shrimp, cashew nuts, sugar, and copra are exported. Despite its large hydroelectric plant at the Cahora Bassa Dam on the River Zambezi, manufacturing is on a small scale. Electricity is exported to South Africa.

POLITICS

After independence, Mozambique became a one-party state. Its government aided African nationalists in Rhodesia (now Zimbabwe) and South Africa. However, the white governments of these countries helped an opposition group, the Mozambique National Resistance Movement (RENAMO) to lead an armed struggle against Mozambique's government. This civil war, combined with severe droughts, caused much human suffering in the 1980s.

In 1989, FRELIMO declared that it had dropped its Communist policies and ended one-party rule. The war officially

NAMIBIA

Namibia adopted this flag in 1990 when it gained its independence from South Africa. The red diagonal stripe and white borders are symbols of Namibia's human resources. The green and blue triangles and the gold sun represent the country's resources.

The Republic of Namibia lies on the Atlantic coast to the south of Angola and to the north of South Africa. The coastal region contains the arid Namib Desert, mostly between 2,950 and 6,560 ft [900–2,000 m] above sea level, which is virtually uninhabited. Inland is a central plateau, bordered by a rugged spine of mountains stretching north–south.

Eastern Namibia contains part of the Kalahari, a semidesert area which extends into Botswana. The Orange River forms Namibia's southern border, while the Cunene and Cubango rivers form parts of the northern borders.

Area 318,259 sq mi [824,292 sq km]
Population 1,954,000
Capital (population) Windhoek (147,000)
Government Multiparty republic
Ethnic groups Ovambo 50%, Kavango 9%, Herero 7%, Damara 7%, White 6%, Nama 5%
Languages English (official), Afrikaans, German, indigenous dialects
Religions Christianity 90% (Lutheran 51%)
Currency Namibian dollar = 100 cents
Website www.grnnet.gov.na

CLIMATE

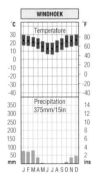

Namibia has a warm and largely arid climate. Daily temperatures range from about 75°F [24°C] in January to 68°F [20°C] in July. Annual rainfall ranges from about 20 in [500 mm] in northern areas to between 1 and 6 in [25–150 mm] in the south. Most of the rain falls in summer.

HISTORY

The earliest people in Namibia were the San (also called Bushmen) and the Damara (Hottentots). Later arrivals were people who spoke Bantu languages. They migrated into Namibia from the north and included the Ovambo, Kavango, and Herero. From 1868,

Germans began to operate along the coast and, in 1884, Germany annexed the entire territory which they called German South West Africa. In the 1890s, the Germans forcibly removed the Damara and Herero from the Windhoek area. About 65,000 Herero were killed when they revolted against their eviction.

South African troops took over the territory in 1915 and five years later the League of Nations gave South Africa a mandate to govern the country, but South Africa chose to rule it as though it were a South African province.

After World War II, many people challenged South Africa's right to govern the territory. A civil war began during the 1960s between African guerrillas and South African troops, with a ceasefire as finally being agreed in 1989. The country became independent in 1990.

POLITICS

After achieving independence, the government pursued a policy of "national reconciliation." An enclave on Namibia's coast, called Walvis Bay (Walvisbaai), remained part of South Africa until

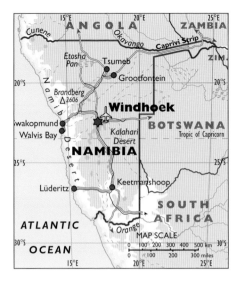

Farming employs around two out of every five Namibians, although many farmers live at subsistence level, contributing little to the economy. Because most of the land in Namibia has too little rainfall for arable farming, the principal agricultural activities are cattle and sheep raising. However, livestock raising has been hit in the last 20 years by extended droughts that have depleted the number of farm animals. The chief crops are corn, millet, and vegetables.

Fishing in the Atlantic Ocean is also important, though overfishing has reduced the yields of Namibia's fishing fleet. The country has few manufacturing industries apart from jewelry-making, some metal smelting, the processing of farm products such as karakul pelts (sheepskins that are used to make fur coats) and textiles. Tourism is developing, especially in the Etosha National Park in northern Namibia, which is rich in wildlife.

1994, when South Africa transferred it to Namibia. In 2004, Sam Nujoma of the South West African People's Organization (SWAPO), who had been president since independence, retired. His successor was Hifikepunye Pohama.

Namibia's Caprivi Strip is a geographical oddity. The strip was given to Germany by European powers in the late 19th century in order that Germany would have access to the River Zambezi. It became the scene of a rebellion in 1999 when a small band of rebels tried, unsuccessfully, to seize the regional capital, Kutima Mulilo, as part of an attempt to make the Caprivi Strip independent. The strip is populated mainly by Lozi people, who resent SWAPO rule. Lozi separatists also live in Botswana and Zambia.

ECONOMY

Namibia has important mineral reserves, including diamonds, zinc, uranium, copper, lead, and tin. Mining is the most valuable economic activity and, by the mid-1990s, minerals accounted for as much as 90% of the exports, with diamonds making up over half the total revenue from minerals.

WINDHOEK

Capital and largest city of Namibia, situated some 190 mi [300 km] inland from the Atlantic at a height of 5,410 ft [1,650 m]. Originally serving as the headquarters of a Nama chief, in 1892 it was made the capital of the new German colony of South West Africa. It was taken by South African troops in World War I. In 1990, it became capital of independent Namibia. An important world trade market for karakul sheepskins, its industries include diamonds, copper, and meat-packing. The German heritage is still very much in evidence with German restaurants selling traditional food and the German language still prevalent everywhere.

NEPAL

This Himalayan kingdom's uniquely shaped flag was adopted in 1962. It came about in the 19th century when two triangular pennants—the royal family's crescent moon symbol and the powerful Rana family's sun symbol—were joined together.

The Kingdom of Nepal in central Asia lies between India to the south and China to the north. More than three-quarters of the country is in the Himalayan mountain heartland, culminating in the world's highest peak Mount Everest (or Chomolongma in Nepali), at 29,035 ft [8,850 m].

Nepal comprises three distinct regions. A southern lowland area (*terai*) of grassland and forests is the main location of Nepal's agriculture and timber industry. The central Siwalik mountains and valleys are divided between the basins of the Ghaghara, Gandak, and Kosi rivers. Between the Gandak and Kosi lies Katmandu valley, Nepal's most populous area. The last region is the main section of the Himalayas. Vegetation varies widely according to altitude.

Area 56,827 sq miles [147,181 sq km]
Population 27,071,000
Capital (population) Katmandu (695,000)
Government Constitutional monarchy
Ethnic groups Brahman, Chetri, Newar, Gurung, Magar, Tamang, Sherpa and others
Languages Nepali (official), local languages
Religions Hinduism 86%, Buddhism 8%, Islam 4%
Currency Nepalese rupee = 100 paisa
Website www.welcomenepal.com

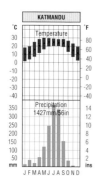

CLIMATE

The huge differences in altitude give Nepal a wide variety of climatic regions.

HISTORY

In 1482 the kingdom of Nepal was divided into three, Bhadgaon, Kathmandu, and Patan. It was nearly four hundred years later in 1768 when the three kingdoms finally resolved their differences and unified to form what is now known as Nepal. Between 1815 and 1816 the Anglo-Nepalese War took place as a result of rivalry between Nepal and the British East India Company over the annexation of minor states bordering Nepal. In exchange for autonomy the Nepalese signed The Treaty of Sugauli ceding parts of the Terrai and Sikkim to the British East India Company.

From 1846 to 1951, hereditary prime ministers from the Rana family ruled Nepal. In 1923, Britain recognized Nepal as a sovereign state. Gurkha soldiers fought in the British army during both World Wars. In 1951, the Rana government was overthrown and the monarchy reestablished. The first national constitution was adopted in 1959, and free elections were held. In 1960, King Mahendra

---

HIMALAYAS

System of mountains in southern Asia, extending 1,500 mi [2,400 km] north–south in an arc between Tibet and India-Pakistan. The mountains are divided into three ranges: the Greater Himalayas (north), which include Mount Everest and K2; the Lesser Himalayas; and the Outer Himalayas (south)

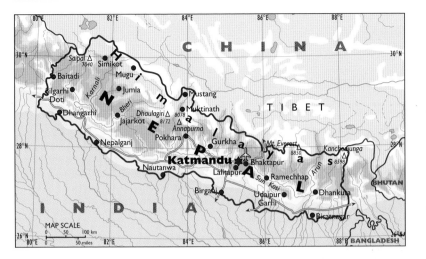

dissolved parliament and introduced a political system based on village councils (*panchayat*). In 1972, Birendra succeeded his father as king.

POLITICS

In 1990, after mass protests, a new constitution limited the power of the monarchy. In 1991 the Nepali Congress Party (NCP), led by G.P. Koirala, won multiparty elections. The NCP dominated the unstable politics of the 1990s, and Koirala led nine governments in ten years. Since the 1990s, a Maoist revolt has claimed more than 3,500 lives. A brief ceasefire was agreed in 2003, but fighting continued.

In 2001, King Birendra, his queen and six other members of his family were shot dead by his heir, Crown Prince Dipendra, who then took his own life. Gyanendra, Birendra's brother, became king.

Increasing Maoist activity led the king to take direct control of the government and appoint a new cabinet early in 2005.

ECONOMY

Nepal is one of the world's poorest countries, with a per capita gross national product of US$220 in 1999. Agriculture employs over 80% of the workforce, accounting for two-fifths of the gross domestic product. Export crops include herbs, jute, rice, spices, and wheat. Tourism, which is centered around the high Himalayas, has grown in importance since 1951, when the country first opened to foreigners. The government is highly dependent on aid to develop the infrastructure. There are also plans to exploit the hydroelectric potential offered by the Himalayan rivers.

KATMANDU (KATHMANDU)

Capital of Nepal, situated 4,500 ft [1,370 m] above sea level in a Himalayan valley. Founded in AD 723, it was independent from the 15th century to 1768, when Gurkhas captured it. Katmandu is Nepal's administrative, commercial, and religious center. Sights of interest include many beautiful temples including Kasthamandap (from which the city derives its name), the Royal Palace (Narayanhity Durbar) and the neoclassical Singha Durbar, former private residence of Rana prime ministers.

NETHERLANDS, THE

The flag of the Netherlands, one of Europe's oldest, dates from 1630, during the long struggle for independence from Spain which began in 1568. The tricolor became a symbol of liberty which inspired many other revolutionary flags around the world.

The Kingdom of the Netherlands lies at the western end of the North European Plain, which extends to the Ural Mountains in Russia. The country is largely flat, about 40% being below sea level at high tide. To prevent flooding, dikes have been built to hold back the waves. There are large areas called polders made up of land reclaimed from the sea.

Area 16,033 sq mi [41,526 sq km]
Population 16,318,000
Capital (population) Amsterdam (729,000); The Hague (seat of government, 440,000)
Government Constitutional monarchy
Ethnic groups Dutch 83%, Indonesian, Turkish, Moroccan and others
Languages Dutch (official), Frisian
Religions Roman Catholic 31%, Protestant 21%, Islam 4%, others
Currency Euro = 100 cents
Website www.holland.com

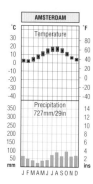

CLIMATE

Because of its position on the North Sea, the Netherlands has a temperate climate. Winters are mild, with rain coming from the Atlantic depressions which pass over the country. North Sea storms often batter the coasts. Storm waves have periodically breached the dikes, causing flooding and sometimes loss of life.

Netherlands, then containing both Belgium and Luxembourg, became an independent kingdom. Belgium broke away in 1830 and Luxembourg followed in 1890.

The Netherlands was neutral in World War I, but occupied by German troops in World War II. Much of the Dutch fleet escaped and served with the Allies, but around three-quarters of the country's Jews were murdered, while many other people were forced to work in German factories. By the end of the war, about 270,000 Netherlanders had been killed or had died of starvation.

HISTORY

Before the 16th century, the area that is now the Netherlands was under a succession of foreign rulers, including the Romans, the Germanic Franks, the French, and the Spanish. The Dutch declared their independence from Spain in 1581, this status finally being recognized by Spain in 1648. In the 17th century the Dutch built up a great overseas empire, especially in Southeast Asia.

France controlled the Netherlands from 1795 to 1813 and in 1815 the

POLITICS

In 1948, the Netherlands formed an economic union called Benelux with Belgium and Luxembourg and, in 1949, it became a member of NATO. Economic recovery was rapid and in 1957 it became a founder member of the EEC.

In 1949, after much fighting, the Dutch recognized the independence of its largest overseas possession, Indonesia. In 1954, Suriname and the Netherlands Antilles were granted self-government. In 1962, the Dutch handed over Nether-

lands New Guinea to the United Nations, which handed it over, as Irian Jaya, to Indonesia in 1963. Suriname became fully independent in 1975.

In 1953, waves penetrated the coastal defenses in the southwestern delta region, flooding about 4.3% of the country, destroying or damaging more than 30,000 houses and killing 1,800 people. Within three weeks, a commission of enquiry had recommended the Delta Plan, a huge project to protect the delta region. Completed in 1986, it involved the construction of massive dams and floodgates, which are closed during severe storms.

The Maastricht Treaty, which transformed the EEC into the European Union, was signed in the Dutch city of Maastricht in 1991. Since 1 January 2002, the euro has been its sole currency.

ECONOMY

The Netherlands has the world's 14th largest economy and it is a highly industrialized country. Manufacturing and commerce are the most valuable activities. Mineral resources include china clay, natural gas, petroleum, and salt. It imports most materials needed by its industries. The products are wideranging, including airplanes, chemical products, electronic equipment, machinery, textiles, and vehicles. South of Rotterdam, the Dutch have constructed a vast port and industrial area, Europoort. Together with Rotterdam's own facilities, the complex is the largest and busiest in the world. Agriculture employs 5% of the workforce, but use of scientific techniques gives high yields. The Dutch cut and sell more than 3 billion flowers a year. Dairy farming is the leading farming activity. In the areas above sea level, farming includes both cattle and crops. Major food crops include barley, potatoes, sugarbeet and wheat.

AMSTERDAM

Capital and largest city in the Netherlands, on the River Amstel, linked to the North Sea by the North Sea Canal. Amsterdam was chartered in 1300 and joined the Hanseatic League in 1369. The Dutch East India Company (1602) brought great prosperity to the city. It became a notable center of learning and book printing during the 17th century. It was captured by the French in 1795 and blockaded by the British during the Napoleonic Wars. Amsterdam was badly damaged during the German occupation during World War II. A major port and one of Europe's leading financial and cultural centers, it has an important stock exchange and diamond-cutting industry.

NEW ZEALAND

New Zealand's flag was designed in 1869 and adopted as the national flag in 1907 when New Zealand became an independent dominion. The flag includes the British Blue Ensign and four of the five stars in the Southern Cross constellation.

New Zealand lies about 994 mi [1,600 km] southeast of Australia. It consists of two main islands and several other small ones. New Zealand is mountainous and partly volcanic. The Southern Alps contain the country's highest peak, Aoraki Mount Cook, at 12,313 ft [3,753 m]. Minor earthquakes are common and there are several areas of volcanic and geothermal activity, especially on North Island.

About 75% of New Zealand lies above the 650 ft [200 m] contour. In the southeast, broad, fertile valleys have been cut by rivers between the low ranges. The only extensive lowland area of New Zealand is the Canterbury Plains. As a result of its isolation, almost 90% of the indigenous plants are unique to the country.

Much of the original vegetation has been destroyed and only small areas of the kauri forests have survived. Mixed evergreen forest grows on the western side of South Island. Beech forests grow in the highlands and large plantations are grown for timber.

Area 270,534 sq km [104,453 sq miles]
Population 3,994,000
Capital (population) Wellington (167,000)
Government Constitutional monarchy
Ethnic groups New Zealand European 74%, New Zealand Maori 10%, Polynesian 4%
Languages English and Maori (both official)
Religions Anglican 24%, Presbyterian 18%, Roman Catholic 15%, others
Currency New Zealnd dollar = 100 cents
Website www.govt.nz

CLIMATE

Auckland in the north has a warm, humid climate throughout the year. Wellington has cooler summers, while in Dunedin, to the southeast, temperatures sometimes dip below freezing in winter. The rainfall is heaviest on the western highlands.

HISTORY

Early Maori settlers arrived in New Zealand more than 1,000 years ago. The Dutch navigator Abel Janszoon Tasman reached the area in 1642, but after several of his men were killed by Maoris, he

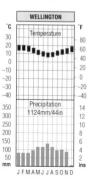

WELLINGTON

Capital and region of New Zealand, in the extreme south of North Island, on Port Nicholson, an inlet of Cook Strait. First visited by Europeans in 1826, it was founded in 1840. In 1865 it replaced Auckland as capital. Wellington's excellent harbor furthered its development as a transportation and trading center.

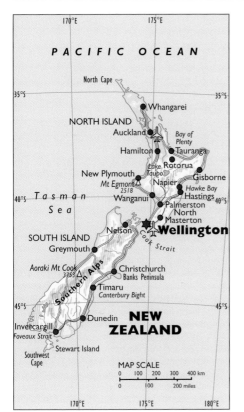

170°E   175°E

PACIFIC OCEAN

North Cape

35°S   35°S

Whangarei

NORTH ISLAND
Auckland   Bay of
    Plenty
Hamilton   Tauranga
    Lake Rotorua
New Plymouth   Taupo
Mt Egmont   Gisborne
2518   Napier
    Hawke Bay
T a s m a n   Wanganui   Hastings   40°S
S e a   Palmerston
    North
    Masterton
Nelson   **Wellington**
SOUTH ISLAND   *Cook Strait*
Greymouth

Aoraki Mt Cook   Christchurch
3753   Banks Peninsula
Timaru
    Canterbury Bight

45°S   45°S

Dunedin   **NEW**
Invercargill   **ZEALAND**
Foveaux Strait
    Stewart Island
Southwest
Cape   MAP SCALE
    0   100   200   300   400 km
    0   100   200 miles

170°E   175°E   180°E

mutual defense pact with Australia and the United States. Troops from New Zealand served in the Korean War (1950–3) and a few units later served in the war in Vietnam.

POLITICS
After Britain joined the EEC (now the EU) in 1973, New Zealand's exports to Britain shrank from 70% to 10%. Along with its reevaluation of its defense position through ANZUS, it also had to reassess its economic strategy. This has involved seeking new markets in Asia, cutting subsidies to farmers, privatization, and cutting back on its extensive welfare programs in the 1990s. The rights of Maoris and the preservation of their culture are other major political issues in New Zealand. In 1998, New Zealand completed a NZ$170 million settlement on the Ngai Tahu group on South Island in compensation for forced land purchases in the 19th century. The government expressed its profound regret for past suffering and for injustices that had impaired the development of the Ngai Tahu. Ties with Britain have been gradually reduced and in 2005 the prime minister, Helen Clark, stated that the country would eventually abolish the monarchy and become a republic.

made no further attempt to land. His discovery was not followed up until 1769, when the British Captain James Cook rediscovered the islands.

British settlers arrived in the early 19th century and in 1840, under the Treaty of Waitangi, Britain took possession of the islands. Clashes occurred with the Maoris in the 1860s, but from the 1870s the Maoris were gradually integrated into society. In 1893, New Zealand became the first country in the world to give women the vote and in 1907, it became a self-governing dominion in the British Empire.

New Zealanders fought alongside the Allies in both World Wars. In 1952, New Zealand signed the ANZUS treaty, a

ECONOMY
Manufacturing now employs twice as many people as agriculture. Meat and dairy products are the most valuable agricultural products. The country has more than 45 million sheep, 4.3 million dairy cattle, and 4.6 million beef cattle. Major crops include barley, fruits, potatoes and other vegetables, and wheat. Fishing is also important. The chief manufactures are processed food products, including butter, cheese, and frozen meat as well as woollen products.

NICARAGUA

The colors of the flag are in homage to Argentina. Central America i s represented by the white stripe; the Atlantic and Pacific Oceans are the blue stripes. The coat of arms features five volcanoes, a rainbow for hope, and red Liberty cap for freedom.

The Republic of Nicaragua is the largest country in Central America. The Central Highlands rise in the northwest Cordillera Isabella to more than 8,000 ft [2,400 m] and are the source for many of the rivers that drain the eastern plain. The Caribbean coast forms part of the Mosquito Coast. Lakes Managua and Nicaragua lie on the edge of a narrow volcanic region, which contains Nicaragua's major urban areas, including the capital, Managua, and the second-largest city, León. This region is highly unstable, with many active volcanoes, and is prone to earthquakes.

Rainforests cover large areas in the east, with trees such as cedar, mahogany, and walnut. Tropical savanna is common in the drier west.

Area 50,193 sq mi [130,000 sq km]
Population 5,360,000
Capital (population) Managua (1,109,000)
Government Multiparty republic
Ethnic groups Mestizo 69%, White 17%, Black 9%, Amerindian 5%
Languages Spanish (official)
Religions Roman Catholic 85%, Protestant
Currency Córdoba oro (gold córdoba) = 100 centavos
Website www.intur.gob.ni/index_eng.html

In 1821, Nicaragua gained independence, later forming part of the Central American Federation from 1825 to 1838. In the mid-19th century, civil war and US and British interference ravaged Nicaragua. The USA sought the construction of a transisthmian canal through Nicaragua. In 1855, William

CLIMATE

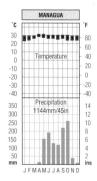

Nicaragua has a tropical climate, with a rainy season from June to October. Cooler weather is found in the Central Highlands. The wettest part is the Mosquito Coast, with 165 in [4,200 mm] of rain.

HISTORY

Spanish explorer Christopher Columbus reached Nicaragua in 1502, and claimed the land for Spain. Colonization claimed the lives of c. 100,000 Native Americans. By 1518 Nicaragua had become part of the Spanish Captaincy-General of Guatemala.

SANDINISTA

(Sandinista National Liberation Front) Revolutionary group in Nicaragua. They took their name from Augusto César Sandino, who opposed the dominant Somoza family and was killed in 1934. The Sandinistas overthrew the Somoza regime in 1979, and formed a government led by Daniel Ortega. They were opposed by right-wing guerrillas, the Contra, supported by the USA. The conflict ended when the Sandinista agreed to free elections. They lost, but the Contra were disbanded and the Sandinista retain political influence.

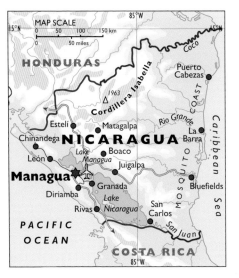

Walker invaded and briefly established himself as president. José Santos Zemalya's dictatorship from 1893 to 1909 gained control of Mosquito Coast and formed close links with the British. Following his downfall, civil war raged once more. In 1912, US marines landed to protect the pro-US regime, and in 1916 the USA gained exclusive rights to the canal. Opposition to US occupation resulted in guerrilla war, led by Augusto César Sandino. In 1933, the US marines withdrew but set up a National Guard to help defeat the rebels.

In 1934 Anastasio Somoza, director of the National Guard, assassinated Sandino. Somoza became president in 1937. His dictatorial regime led to political isolation. Somoza was succeeded by his sons Luis in 1956 and Anastasio in 1967. Anastasio's diversion of international relief aid following the devastating 1972 Managua earthquake cemented opposition.

POLITICS

In 1979, the Sandinista National Liberation Front (FSLN) overthrew the Somoza regime. The Sandinista government, led by Daniel Ortega, instigated wide-ranging socialist reforms. The USA, concerned about the Sandinista's ties with communist regimes, sought to destabilize the government by supporting the Contra rebels. A ten-year civil war devastated the economy and led to political dissatisfaction.

In 1990 elections, the National Opposition Union coalition, led by Violeta Chamorro, defeated the Sandinistas. Chamorro's coalition partners and the Sandinista-controlled trade unions blocked many of her reforms. In 1996 elections, Liberal leader Arnoldo Aleman defeated Chamorro. In 1998, Hurricane Mitch killed c. 4,000 people and caused extensive damage. Enrique Bolanos became president at elections in 2001. In 2003, former president Arnoldo Aleman was sentenced to 20 years in prison for corruption.

ECONOMY

Nicaragua faces problems in rebuilding its economy and introducing free-market reforms. Agriculture is the main activity, employing 50% of the workforce and accounting for 70% of exports. Major cash crops include coffee, cotton, sugar, and bananas. Rice is the main food crop.

There is some copper, gold, and silver, but mining is underdeveloped. Most of the country's manufacturing is based around Managua.

MANAGUA

Capital of Nicaragua, on the southern shore of Lake Managua, west-central Nicaragua. It became the capital in 1855. Managua suffered damage from earthquakes in 1931 and 1962. It is the economic, industrial, and commercial hub of Nicaragua. Industries include textiles, tobacco, and cement.

NIGER

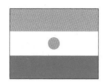

This flag was adopted shortly before Niger became independent from France in 1960. The orange stripe represents the Sahara in the north and the green represents the grasslands in the south. Between them, the white stripe represents the River Niger, with a circle for the sun.

The Republic of Niger is a land-locked nation in north-central Africa. The northern plateaux lie in the Sahara, while north-central Niger contains the rugged Aïr Mountains, which reach a height of 6,632 ft [2,022 m] above sea level near Agadez. The rainfall in the mountains—averaging around 7 in [175 mm] per year—is sufficient in places to permit the growth of thorny shrub. Severe droughts since the 1970s have crippled the traditional lifestyle of the nomads in northern and central Niger as the Sahara has slowly advanced south. The southern region has also been hit by droughts.

The south consists of broad plains. The Lake Chad Basin lies in southeastern Niger on the borders with Chad and Nigeria. The only permanent rivers are the Niger and its tributaries in the southwest. The narrow Niger Valley is the country's most fertile and densely populated region and includes the capital, Niamey. Yet Niger, a title which comes from a Tuareg word meaning "flowing water," seems scarcely appropriate for a country which consists mainly of hot, arid, sandy, and stony basins.

Buffaloes, elephants, giraffes, and lions are found in the "W" National Park, which Niger shares with Benin and Burkina Faso. Most of southern Niger lies in the Sahel region of dry grassland. The Aïr Mountains support grass and scrub. The northern deserts are generally barren.

CLIMATE

Niger is one of the world's hottest countries. The warmest months are March to May, when the harmattan wind blows from the Sahara. Niamey has a tropical

Area 489,189 sq mi [1,267,000 sq km]
Population 11,361,000
Capital (population) Niamcy (732,000)
Government Multiparty republic
Ethnic groups Hausa 56%, Djerma 22%, Tuareg 8%, Fula 8%, others
Languages French (official), Hausa, Djerma
Religions Islam 80%, indigenous beliefs, Christianity
Currency CFA franc = 100 centimes
Website www.un.int/niger

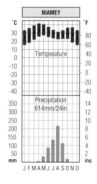

climate, with a rainy season from June to September. Rainfall decreases from south to north. Northern Niger is practically rainless. The far south consists of tropical savanna.

HISTORY

Neolithic remains have been found in the northern desert. Nomadic Tuareg settled in the Aïr Mountains in the 11th century AD, and by the 13th century established a state centered around Agadez and the trans-Saharan trade. In the 14th century, the Hausa settled in southern Niger. In the early 16th century, the Songhai Empire controlled much of Niger, but the Moroccans supplanted the Songhai at the turn of the century.

Later on, the Hausa and then the Fulani set up kingdoms in the region. In the early 19th century, the Fulani gained control of much of southern Niger. The first French expedition arrived in 1891,

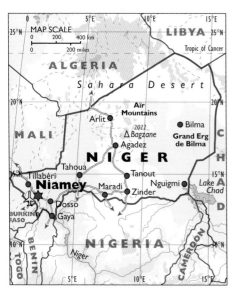

by the National Movement for a Development Society (MNSD), but a military coup, led by Colonel Ibrahim Bare Mainassara, seized power. In 1995, the government and the Tuaregs signed a peace accord. Elections in 1996 confirmed Mainassara as president. In 1999, bodyguards assassinated Mainassara and he was replaced briefly by Major Daouda Malam Wanke. Parliamentary rule was restored and, later that year, Tandjou Mamadou was elected president. He was reelected in 2004.

ECONOMY
Droughts have caused great hardship and food shortages in Niger, and have destroyed much of the traditional nomadic lifestyle. Niger's chief resource is uranium, and it is the world's second-largest producer. Uranium accounts for more than 80% of exports, most of which goes to France. Some tin and tungsten are also mined. Other mineral resources are largely unexploited.

Niger is one of the world's poorest countries, despite its resources. Farming employs 85% of the workforce, although only 3% of the land is arable and 7% is used for grazing. Food crops include beans, cassava, millet, rice, and sorghum. Cotton and peanuts are leading cash crops.

but Tuareg resistance prevented full occupation until 1914.

POLITICS
In 1922, Niger became a colony within French West Africa. In 1958, Niger voted to remain an autonomous republic within the French Community. It gained full independence in 1960, and Hamani Diori became Niger's first president. He maintained close ties with France.

Drought in the Sahel began in 1968, and killed many livestock and destroyed crops. In 1974, a group of army officers, led by Lieutenant Colonel Seyni Kountché, overthrew Hamani Diori and suspended the constitution. Kountché died in 1987, and was succeeded by his cousin General Ali Saibou. In 1991, the Tuareg in northern Niger began an armed campaign for greater autonomy. A national conference removed Saibou and established a transitional government. In 1993 multiparty elections, Mahamane Ousmane of the Alliance of Forces for Change (AFC) coalition became president. The collapse of the coalition led to fresh elections in 1995, which were won

NIAMEY
Capital of Niger, West Africa, in the southwestern part of the country, on the River Niger. It became capital of the French colony of Niger in 1926. It grew rapidly after World War II and is now the country's largest city and its commercial and administrative center. Manufactures include textiles, ceramics, plastics, and chemicals.

273

NIGERIA

Nigeria's flag was adopted in 1960 when Nigeria became independent from Britain. It was selected after a competition to find a suitable design. The green represents Nigeria's forests. The white in the center stands for peace.

The Federal Republic of Nigeria is the most populous nation in Africa. The country's main rivers are the Niger and Benue, which meet in central Nigeria. North of the two river valleys are high plains and plateaus. The Lake Chad Basin is in the northeast, with the Sokoto plains in the northwest. Southern Nigeria contains hilly uplands and broad coastal plains, including the swampy Niger Delta. Highlands form the border with Cameroon. Mangrove swamps line the coast, behind which are rainforests. The north contains large areas of savanna with forests along the rivers. Open grassland and semidesert occur in drier areas.

Area 356,667 sq mi [923,768 sq km]
Population 137,253,000
Capital (population) Abuja (339,000)
Government Federal multiparty republic
Ethnic groups Hausa and Fulani 29%, Yoruba 21%, Ibo (or Igbo) 18%, Ijaw 10%, Kanuri 4%
Languages English (official), Hausa, Yoruba, Ibo
Religions Islam 50%, Christianity 40%, traditional beliefs
Currency Naira = 100 kobo
Website www.nigeria.gov.ng

CLIMATE
The south of the country has high temperatures and rain all year. Parts of the coast have an average annual rainfall of 150 in [3,800 mm]. The north has a marked dry season and higher temperatures than the south.

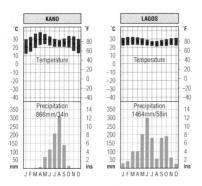

HISTORY
Nigeria has a long artistic tradition. Major cultures include the Nok (500 BC to AD 200), Ife, which developed about 1,000 years ago, and Benin, which flourished between the 15th and 17th centuries.

Britain outlawed slavery in 1807 and soon afterward the British began to trade in agricultural products. In 1851, Britain made Lagos a base from which they could continue their efforts to stop the slave trade. During the second half of the 19th century, Britain gradually extended its influence over Nigeria. By 1914 it ruled the entire country.

POLITICS
Nigeria became independent in 1960 and a federal republic in 1963. A federal constitution dividing the country into regions was necessary because Nigeria contains more than 250 ethnic and linguistic groups, as well as several religious ones. Local rivalries have long been a threat to national unity. In 1967, in an attempt to meet the demands of more ethnic groups, the country's four regions were replaced by 12 states. The division of the Eastern Region provoked an uprising. In 1967, the governor of the Eastern Region, Colonel Odumegwu Ojukwu,

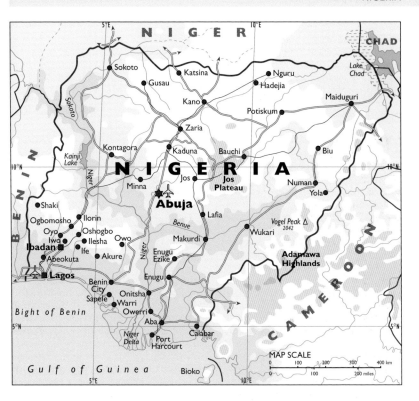

proclaimed it an independent republic called Biafra. Civil war continued until Biafra's surrender in January 1970.

The country had only nine years of civilian government between independence in 1960 and 1998. In 1998-9 civilian rule was restored. A former military leader, Olusegun Obasanjo, was elected president and reelected in 2003. Ethnic and religious differences are a threat to national unity. In the late 1990s and early 2000s, ethnic riots broke out between Yorubas and Hausas in the southwest, while the introduction of *sharia* (Islamic law) in northern states has caused friction between Muslims and Christians. The government declared in 2004 that it had put down an uprising in the northeast aimed at creating a Muslim state, while ethnic and religious conflict continued in other parts of the country.

ECONOMY

Despite its many natural resources, including petroleum reserves, metals, forests, and fertile farmland, Nigeria is a "low-income" developing economy. Agriculture employs 43% of the workforce and Nigeria is one of the world's leading producers of cocoa beans, peanuts, palm oil and kernels, and natural rubber. Leading food crops include beans, cassava, corn, millet, plantains, rice, sorghum, and tropical yams.

ABUJA

Nigeria's administrative capital since 1991. The new city was designed by the Japanese architect Kenzo Tange, and work began in 1976.

NORWAY

This flag became the national flag of Norway in 1898, although merchant ships had used it since 1821. The design is based on the Dannebrog, the flag of Denmark, the country which ruled Norway from the 14th century until the early 19th century.

The Kingdom of Norway forms the western part of the mountainous Scandinavian Peninsula. The landscape is dominated by rolling plateaus, the *vidda*, which are generally between 1,000 and 3,000 ft [300–900 m] high, but some peaks rise from 5,000 to 8,000 ft [1,500–2,500 m] in the area between Oslo, Bergen, and Trondheim. The highest areas retain permanent ice fields, as in the Jotunheimen Mountains above Sognefjord.

Norway's jagged coastline is the longest in Europe. The *vidda* are cut by long, narrow, steep-sided fjords on the west coast. The largest of the fjords, is Sognefjord, which is 127 mi [203 km] long and less than 3 mi [5 km] wide.

Area 125,049 sq mi [323,877 sq km]
Population 4,575,000
Capital (population) Oslo (513,000)
Government Constitutional monarchy
Ethnic groups Norwegian 97%
Languages Norwegian (official)
Religions Evangelical Lutheran 86%
Currency Norwegian krone = 100 øre
Website www.norge.no

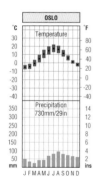

CLIMATE

The warm North Atlantic Drift flows off the coast and moderates the country's climate, with milder winters and cooler summers. Most of Norway's ports remain ice-free all year. Inland, away from the moderating effects of the sea, the climate becomes more severe. Winters are bitterly cold with snow cover for at least three months of the year.

HISTORY

From about AD 800, Vikings from Norway roamed the northern seas, raiding and founding colonies around the coasts of Britain, Iceland, and even North America. In about 900, Norway was united under Harold I, the country's first king. Viking power ended in the late 11th century. In 1380, Norway was united with Denmark and in 1397 Sweden joined the union. Sweden broke away in 1523 and, in 1526, Denmark, which had become increasingly powerful, made Norway a Danish province.

In 1814, Denmark ceded Norway to Sweden, but retained Norway's colonies of Greenland, Iceland, and the Faeroe Islands. Norway finally ended its union with Sweden in 1905. The Norwegians chose as their king a Danish prince, who took the title Haakon VII.

Despite Norway being neutral in World War I, and seeking to remain so in World War II, German troops invaded in 1940.

OSLO

Capital of Norway, southern Norway. Founded in the mid-11th century and largely destroyed by fire in 1624, Christian IV rebuilt the city, naming it Christiania. In 1905, it became the capital of independent Norway. It acquired the name Oslo in 1925.

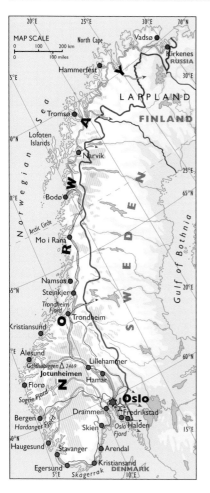

Soviet Union. In 1960, Norway formed the European Free Trade Association with six other countries while continuing to work with its Scandinavian neighbors through the Nordic Council. In 1994, Norwegians again voted against membership of the EU. The 1990s–2000s saw Norwegian diplomats seeking to broker peace deals in Sri Lanka and Palestine.

ECONOMY

Norway's chief resources and exports are petroleum and natural gas. Dairy farming and meat production are the chief farming activities, though Norway has to import food. Industries include petroleum products, chemicals, aluminum, wood products, machinery, and clothing.

SVALBARD

Archipelago in the Arctic Ocean, 400 mi [640 km] north of Norway. Spitsbergen is the largest of nine islands. The administrative center and largest settlement is Long-yearbyen on Spitsbergen. Ice fields and glaciers cover more than half the land, although the western edge of the islands is ice free most of the year. The area abounds in Arctic flora and fauna. The islands are an important wildlife refuge with polar bears, walrus, and whales. Protective measures have saved certain mammals from extinction. In the 17th century, Svalbard was an important whaling center, and in the 18th century, fur traders hunted the lands. Large coal deposits were found on Spitsbergen at the end of the 19th century, and Norway, Russia, and Sweden mined the area. The islands became a sovereign territory of Norway in 1925, in return for allowing mining concessions to other nations. It covers an area of 24,000sq mi [62,000sq km], with a population of 2,400.

POLITICS

After World War II, Norwegians worked to rebuild their economy and their merchant fleet. The economy was boosted in the 1970s, when Norway began producing petroleum and natural gas from wells in the North Sea. Rapid economic growth has ensured that Norwegians are among the most prosperous in Europe.

In 1949, it became a member of NATO, though neither NATO bases nor nuclear weapons were permitted on its soil for fear of provoking its neighbor, the

OMAN

White symbolizes peace. Green is traditional Islamic color, also standing for the fertility of the land. Red represents the blood shed in the struggle for liberation. The Sultanate's white coat of arms consists of two crossed swords, a *khnajar* (dagger), and belt.

The Sultanate of Oman is the oldest independent nation in the Arab world. It occupies the southeastern corner of the Arabian peninsula and includes the tip of the Musandam Peninsula, which is separated from the rest of Oman by UAE territory. This peninsula overlooks the strategic Strait of Hormuz.

The Al Halar al Gharbi range, rising to 9,904 ft [3,019 m] above sea level, borders the narrow coastal plain in the north. This fertile plain along the Gulf of Oman is called Al Battinah. Inland are deserts, including part of the Rub' al Khali (Empty Quarter). Much of the land along the Arabian Sea is barren, but the province of Zufar (or Dhofar) in the southeast is a hilly, fertile region.

Area 119,498 sq mi [309,500 sq km]
Population 2,903,000
Capital (population) Muscat (41,000)
Government Monarchy with consultative council
Ethnic groups Arab, Baluchi, Indian, Pakistani
Languages Arabic (official), Baluchi, English
Religions Islam (mainly Ibadhi), Hinduism
Currency Omani rial = 100 baizas
Website www.omanet.om/english/home.asp

CLIMATE

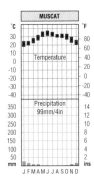

Temperatures in Oman can reach 129°F [54°C] in summer, but winters are mild to warm. Rainfall in the northern mountains can exceed 16 in [400 mm] per year, while in the southeast it can be up to 25 in [630 mm], but for most of Oman the desert climate means less than 6 in [150 mm] per year. Sandstorms, duststorms and droughts feature and occasionally, tropical cyclones bring stormy weather.

HISTORY

Oman first became a major trading region 5,000 years ago. Islam was introduced into the area in the 7th century, and today, 75% of the population follow the strict Ibadi Islamic sect.

The Portuguese conquered its ports in the early 16th century, but local Arabs forced them out in 1650. The Al Bu Said family came to power in the 1740s and has ruled the country ever since. British influence dates back to the end of the 18th century, when the two countries entered into the first of several treaties.

In 1920, Britain brokered an agreement whereby the interior was ruled by imams, with coastal areas under the control of the Sultan. Clashes between the two groups continued into the 1950s, but Sultan Said bin Taimur regained control of the whole country in 1959.

POLITICS

Under Sultan Said bin Taimur, Oman had been an isolated, feudal country. Its economy was backward compared to its petroleum-rich Gulf neighbors. However, after Sultan Said bin Taimur was deposed by his son, Sultan Qaboos ibn Said, in 1970, Oman made substantial strides. With the help of soldiers from Iran and Jordan, he saw an end to war against Yemen-backed separatist guerrillas in the province of Zufar (1965–1975).

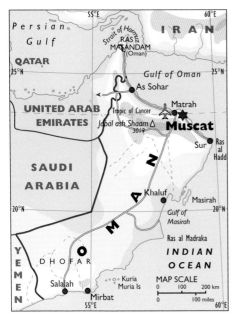

elections were held to the Consultative Council. For the first time, all citizens over 21 were allowed to vote although no parties are allowed. In 2004, the Sultan appointed the first woman minister with portfolio. In 2005, nearly 100 suspected Islamists were arrested and 31 were convicted of trying to overthrow the government, but they were later pardoned.

ECONOMY

The World Bank classifies Oman as an "upper-middle-income" developing country. It has sizeable petroleum and natural gas deposits, a large trade surplus and low inflation. Petroleum accounts for more than 90% of Oman's export revenues. Huge natural gas deposits, equal to all the finds of the previous 20 years, were discovered in 1991. Although only about 0.3% of the land is cultivated, agriculture and fishing are the traditional economic activities. Major crops include alfalfa, bananas, coconuts, dates, tobacco, and wheat. Water supply is a major problem. Oman depends on water from underground aquifers, which will eventually run dry, and from desalination plants. Industries include copper smelting, cement, and chemicals, as well as food processing and import substitution. Tourism is a growing activity.

He also led the way in developing an expanding economy based on petroleum reserves far larger than expected when production began in 1967. Qaboos opened up Oman to the outside world, ending the isolation it had long endured. At home, he avoided the prestigious projects often favored by Arab leaders to concentrate on social programs, including the education of girls. His leadership proved popular despite the lack of a democratic government.

In 1991, Oman took part in the military campaign to liberate Kuwait. In 1997, Oman held its first direct elections to a Consultative Council. Unusually for the Gulf region, two women were elected. In 1999, Oman and the United Arab Emirates signed an agreement confirming most of the borders between them. In 2001, while a military campaign was being launched in Afghanistan, Britain held military exercises in the Omani desert. This was an example of the long-standing political and military relationship between the two countries. In 2003,

MUSCAT (MASQAT, MASKAT)

Capital of Oman, on the Gulf of Oman, in the south-east Arabian Peninsula. The Portuguese held Muscat from 1508 to 1650, when it passed to Persia. After 1741 it became capital of Oman. In the 20th century, Muscat's rulers developed treaty relations with Britain.

PACIFIC OCEAN

AMERICAN SAMOA

Area 77 sq mi [199 sq km]
Population 58,000
Capital (population) Pago Pago (4,000)
Government Territory of the USA
Ethnic groups Native Pacific Islander 93%, Asian, White
Languages Samoan 91%, English 3%, Tongan 2%, other Pacific islander 2%
Religions Christian Congregationalist 50%, Protestant 30%, Roman Catholic 20%
Currency US dollar = 100 cents
Website www.asg-gov.net

The Territory of American Samoa is an unincorporated territory of the United States and lies in the south-central Pacific Ocean. Two of its islands are coral islands, the other five are extinct volcanoes. The US took control of the islands between 1900 and 1904. The main industry is the canning of tuna; fish products dominate the economy.

COOK ISLANDS

Area 113 sq mi [293 sq km]
Population 21,388
Capital (population) Avarua (12,188)
Government Self-governing parliamentary democracy
Ethnic groups Cook Island Maori (Polynesian) 88%
Languages English (official), Maori
Religions Cook Islands Christian Church 57%, Roman Catholic 17%, Seventh Day Saint 8%, Church of Latter Day Saints 4%, other Protestant 6%, others
Currency New Zealand dollar = 100 cents
Website www.ck

A group of 15 islands in the south Pacific Ocean to the northeast of New Zealand consisting of the Northern (Manihiki) Cook Islands and the Southern (Lower) Cook Islands. A self-governing territory in free association with New Zealand. Discovered in 1773 by Captain James Cook, the islands became a British protectorate in 1888 and were annexed to New Zealand in 1901. They became self-governing in 1965. Products include copra and citrus fruits.

FRENCH POLYNESIA

Area 1,544 sq mi [4,000 sq km]
Population 266,000
Capital (population) Papeete (24,000)
Government Overseas territory of France
Ethnic groups Polynesian 78%, Chinese 12%, local French 6%, metropolitan French 4%
Languages French 61%, Polynesian 31% (both official), Asian languages
Religions Protestant 54%, Roman Catholic 30%
Currency Comptoirs Francais du Pacifique franc = 100 centimes
Website www.presidence.pf

French Polynesia consists of 130 islands, scattered over 1 million sq mi [2.5 million sq km] of the Pacific Ocean. Tribal chiefs agreed to a French protectorate in 1843. They gained increased autonomy in 1984. Links with France ensure a high standard of living. Some favor independence. Following a struggle for power in 2004, the pro-independence Union for Democracy party, ousted the pro-French ruling party.

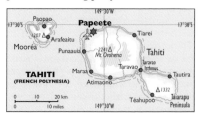

GUAM

Area 212 sq mi [549 sq km]
Population 166,000
Capital (population) Hagatna (Agana) (1,000)
Government Territory of the USA
Ethnic groups Chamorro 37%, Filipino 26%, other Pacific islander 11%, white 7%, other Asian 6%
Languages English 38%, Chamorro 22%, Philippine languages 22%, other Pacific island languages 7%, Asian languages 7%, others
Religions Roman Catholic 85%, others 15%
Currency US dollar = 100 cents
Website www.visitguam.org

The Territory of Guam is a strategically important "unincorporated territory" of the USA and the largest of the Mariana Islands in the Pacific Ocean. It is composed of a coralline limestone plateau. Guam was ruled by Spain from 1668 until it was ceded to the United States in 1898 after the Spanish-American War.

KIRIBATI

Area 280 sq mi [726 sq km]
Population 101,000
Capital (population) Bairiki (on Tarawa) (32,000)
Government Multiparty republic
Ethnic groups Micronesian 99%
Languages I-Kiribati and English (both official)
Religions Roman Catholic 52%, Protestant (Congregational) 40%, others
Currency Australian dollar = 100 cents
Website www.janeresture.com/kirihome/index.htm

The Republic of Kiribati is an independent nation in the west Pacific Ocean. It comprises about 33 islands, including the Gilbert, Phoenix, and Line Islands, and straddles the Equator over an area of 2 million sq mi [5 million sq km]. The islands are threatened by global warming and consequent rising sea levels. Rainfall is abundant.

British navigators first visited the Gilbert and Ellice Islands during the late 18th century. They became a British protectorate in 1892 and a colony in 1915. In 1975 the Ellice Islands, following a referendum, officially severed links with the Gilbert Islands and became a separate territory called Tuvalu in 1978. The Gilbert Islands became fully independent within the Commonwealth of Nations in 1979 as the Republic of Kiribati.

Agriculture is now the major economic activity.

MARSHALL ISLANDS

Area 70 sq mi [181 sq km]
Population 58,000
Capital (population) Majuro (20,000)
Government Constitutional government in free association with the US
Ethnic groups Micronesian
Languages Marshallese 98% and English (both official)
Religions Protestant 55%, Assembly of God 26%, Roman Catholic 9%, Bukot nan Jesus 3%, Mormon 2%
Currency US dollar = 100 cents
Website www.yokwe.net

The Republic of the Marshall Islands consists of 31 coral atolls, five single islands and more than 1,000 islets. It lies north of Kiribati in the region of Micronesia.

The islands came under German rule in 1885 and became a Japanese mandate after World War I. US forces took the main islands in 1944 and they became a US Trust Territory in 1947. Independence was achieved in 1991, but the islands remain heavily dependent on US aid. The main activities are agriculture and tourism.

PACIFIC OCEAN

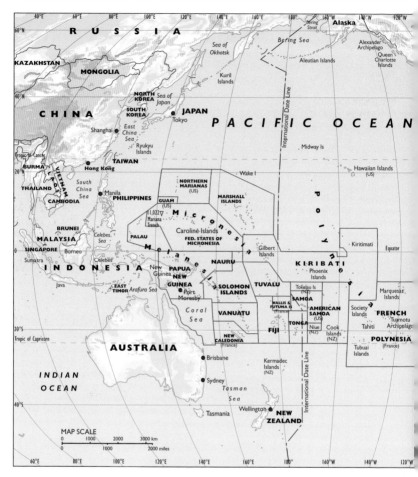

MICRONESIA

Area 271 sq mi [702 sq km]
Population 108,000
Capital (population) Palikir (on Pohnpei) (5,000)
Government Constitutional government in free association with the US
Ethnic groups Micronesian and Polynesian
Languages English (official), others
Religions Roman Catholic 50%, Protestant 47%
Currency US dollar = 100 at
Website www.visit-fsm.org

Federated States of Micronesia is a republic in the western Pacific Ocean, consisting of all the Caroline Islands except Belau. The 607 islands of the republic divide into four states: Chuuk, Pohnpei, Yap, and Kosrae. After 1874 the islands were under a succession of rulers from Spain to Germany in 1899, to Japan in 1920. They came under US administration in 1947. In 1979, the Federated States of Micronesia came into being, with Belau remaining a US trust territory. In 1986, a compact of free association with the USA was signed. In 1991

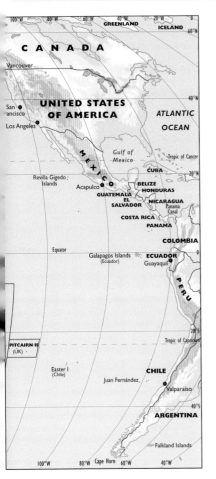

Micronesia became a full member of the UN. The economy depends heavily on US aid. Land use is limited to subsistence agriculture.

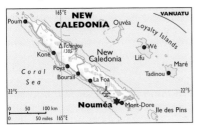

NAURU

Area 8 sq mi [21 sq km]
Population 13,000
Capital (population) Yaren (4,500)
Government Multiparty republic
Ethnic groups Nauruan 58%, other Pacific Islander 26%, Chinese 8%, European 8%
Languages Nauruan (official), English
Religions Protestant 66%, Roman Catholic 33%
Currency Australian dollar = 100 cents
Website www.un.int/nauru/

A former UN Trust Territory ruled by Australia, Nauru became independent in 1968. Located in the western Pacific close to the Equator, it is the world's smallest republic. Nauru's prosperity is based on phosphate mining, but reserves are running out.

NEW CALEDONIA

Area 7,172 sq mi [18,575 sq km]
Population 214,000
Capital (population) Nouméa (76,000)
Government Overseas territory of France
Ethnic groups Melanesian 43%, European 37%, Wallisian 8%, Polynesian 4%, Indonesian 4%
Languages French (official), 33 Melanesian-Polynesian dialects
Religions Roman Catholic 60%, Protestant 30%
Currency Comptoirs Français du Pacifique franc = 100 centimes
Website www.newcaledoniatourism-south.com

New Caledonia is the most southerly of the Melanesian countries in the Pacific. A French possession since 1853 and an Overseas Territory since 1958. In 1998, France announced an agreement with local Melanesians that a vote on independence would be postponed until 2014. The country is rich in mineral resources. Experts claim that it has about a quarter of the world's nickel reserves.

PACIFIC OCEAN

NIUE

Area 100 sq mi [260 sq km]
Population 2,000
Capital (population) Alofi (404)
Government Self-governing parliamentary democracy
Ethnic groups Niuen 78%, Pacific islander 10%, European 4%, others
Languages Niuean, English
Religions Ekalesia Niue 61%, Latter-Day Saints 9%, Roman Catholic 7%, others
Currency New Zealand dollar = 100 cents
Website www.niueisland.nu

Niue is an island territory in the southern Pacific Ocean, 2,160 km [1,340 mi] northeast of New Zealand. The largest coral island in the world, Niue was first visited by Europeans in 1774. In 1901 it was annexed to New Zealand. In 1974 it achieved self-government in free association with New Zealand. Its economy is mainly agricultural; the major export is coconut.

NORFOLK ISLAND

Area 13 sq mi [34 sq km]
Population 1,828
Capital (population) Kingston
Government Territory of Australia
Ethnic groups Descendants of the Bounty mutineers, Australian, New Zealander, Polynesianss
Languages English, Norfolk
Religions Anglican 35%, Roman Catholic 12%, Uniting Church in Australia 11%, others
Currency Australian dollar = 100 cents
Website www.nf

The Territory of Norfolk Island lies in the southwest Pacific Ocean 900 mi [1,450 km] east of Australia. Visited in 1774 by Captain James Cook, it was a British penal colony. Many Pitcairn Islanders, decendents of the *Bounty* mutineers, resettled here in 1856. The chief economic activities are agriculture and tourism.

NORTHERN MARIANA ISLANDS

Area 179 sq mi [464 sq km]
Population 78,000
Capital (population) Saipan (39,000)
Government Commonwealth in political union with the US
Ethnic groups Asian 56%, Pacific islander 36%
Languages Philippine languages 24%, Chinese 23%, Chamorro 22%, English 10%, others
Religions Christianity
Currency US dollar = 100 cents
Website www.cnmi.net

The Commonwealth of the Northern Mariana Islands contains 16 mountainous islands north of Guam in the western Pacific Ocean. In a 1975 plebescite, the islanders voted for Commonwealth status in union with the USA and in 1986 they were granted US citizenship.

PALAU

Area 177 sq mi [459 sq km]
Population 20,000
Capital (population) Koror (11,000)
Government Multiparty republic
Ethnic groups Palauan 70%, Filipino 15%, Chinese 5%, others
Languages Palauan (official except in Sonsoral (Sonsoralese and English official), Tobi (Tobi and English official), and Angaur (Angaur, Japanese, and English official), Filipino, English Chinese
Religions Roman Catholic 42%, Protestant 23%, Modekngei 9% (indigenous to Palau),
Currency US dollar = 100 cents
Website www.visit-palau.com

The Republic of Palau became fully independent in 1994 after the US refused to accede to a 1979 referendum that declared this island nation a nuclear-free zone. The economy relies on US aid, tourism, fishing, and subsistence agriculture. The main crops include cassava, coconuts, and copra.

PITCAIRN ISLAND

Area 21 sq mi [55 sq km]
Population 46
Capital (population) Adamstown
Government Overseas territory of the UK
Ethnic groups Descendants of the *Bounty* mutineers and their Tahitian wives
Languages English (official), Pitcairnese
Religions Seventh-Day Adventist
Currency New Zealand dollar = 100 cents
Website www.government.pn/homepage.htm

Pitcairn Island is a British overseas territory in the Pacific Ocean. Its inhabitants are descendents of the original settlers—nine mutineers from HMS *Bounty* and 18 Tahitians who arrived on this formerly uninhabited island in 1790.

SAMOA

Area 1,093 sq mi [2,831 sq km]
Population 178,000
Capital (population) Apia (32,000)
Government Parliamentary democracy and constitutional monarchy
Ethnic groups Samoan 93%, Euronesians
Languages Samoan, English
Religions Congregationalist 35%, Roman Catholic 20%, Methodist 15%, Latter-Day Saints 13%
Currency Tala = 100 sene
Website www.visitsamoa.ws

The Independent State of Samoa comprises two islands in the South Pacific Ocean. The ownership of these Polynesian islands was disputed by European powers, but Germany took control in 1900. Following Germany's defeat in World War I, New Zealand governed from 1920 until 1961. The country became independent in 1962. The economy is based on agriculture, plus coconut products, copra, and fishing.

SOLOMON ISLANDS

Area 11,157 sq mi [28,896 sq km]
Population 524,000
Capital (population) Honiara (49,000)
Government Parliamentary democracy
Ethnic groups Melanesian 95%, others
Languages Melanesian pidgin
Religions Church of Melanesia 33%, Roman Catholic 19%, South Seas Evangelical 17%, Seventh-Day Adventist 11%, United Church 10%
Currency Solomon Islands dollar = 100 cents
Website www.solomons.com

The Solomon Islands, a chain of mainly volcanic islands in the Pacific Ocean, extending 1,400 mi [2,250 km]. A British territory between 1893 and 1978. In 2003 an Australian peacekeeping force was sent in the belief that the islands were threatened with anarchy. Fish, coconuts, and cocoa are important.

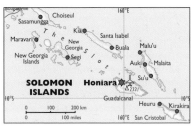

PACIFIC OCEAN

TOKELAU

Area 3.86 sq mi [10 sq km]
Population 1,405
Government Self-administering territory of New Zealand
Ethnic groups Polynesian
Languages Tokelauan, English
Religions Congregational Christian Church 70%, Roman Catholic 28%
Currency New Zealand dollar = 100 cents
Website www.tokelau.org.nz

Tokelau was originally settled by Polynesian emigrants from surrounding islands. It was made a British protectorate in 1889 and transferred to New Zealand administration in 1925. It is made up of three villages, there is little economic development and agriculture is at subsistence level. The people produce copra, postage stamps, souvenir coins, and handicrafts, but rely heavily on aid from New Zealand.

TONGA

Area 251 sq mi [650 sq km]
Population 110,000
Capital (population) Nuku'alofa 22,000)
Government Constitutional monarchy
Ethnic groups Polynesian
Languages Tongan, English
Religions Christian (Free Wesleyan Church)
Currency Pa'anga = 100 seniti
Website http://pmo.gov.to

Originally called the Friendly Islands, the Kingdom of Tonga is an island kingdom in the South Pacific, 1,370 mi [2,200 km] northeast of New Zealand. The archipelago consists of c. 170 islands in five administrative groups. Only 36 of the islands are inhabited. They are mainly coral atolls, but the western group are volcanic, with some active craters. The largest island is Tongatapu, the seat of the capital, Nukualofa, and home to 66% of the population.

The northern islands were discovered by Europeans in 1616, and the rest by Abel Tasman in 1643. During the 19th century, British missionaries converted the indigenous population to Christianity. In 1900, Tonga became a British Protectorate.

In 1970, the country achieved independence. The economy is dominated by agriculture, the chief crops are yams, tapioca and fish.

TUVALU

Area 10 sq mi [26 sq km]
Population 11,000
Capital (population) Fongafale (3,000)
Government Constitutional monarchy with a parliamentary democracy
Ethnic groups Polynesian 96%, Micronesian 4%
Languages Tuvaluan, English, Samoan, Kiribati
Religions Church of Tuvalu (Congregationalist) 97%, others
Currency Tuvaluan dollar and Australian dollar = 100 cents
Website www.tuvaluislands.com

Tuvalu, formerly the Ellice Islands (see Kiribati), is an independent republic in western Pacific Ocean. None of the cluster of nine low-lying coral islands rises more than 15 ft [4.6 m] out of the Pacific, making them vulnerable to rising sea levels.

The first European to discover the islands was the Spanish navigator Alvaro de Mendaña in 1568. Between 1850 and 1880, the population was reduced from around 20,000 to just 3,000 by Europeans abducting workers for other Pacific plantations. In 1892, the British assumed control, and Tuvalu was subsequently administered with the nearby Gilbert Islands (now Kiribati). In

1978, Tuvalu became a separate self-governing colony within the Commonwealth.

Poor soil restricts vegetation to coconut palms, breadfruit, and bush. The population survives by subsistence farming, raising pigs and poultry, and by fishing. Copra is the only significant export crop, but more foreign exchange is derived from the sale of elaborate postage stamps.

The Republic of Vanuatu, formerly the Anglo-French Condominium of the New Hebrides, became independent in 1980. The word Vanuatu means "Our Land Forever." The republic consists of a chain of 80 islands in the South Pacific Ocean. Its economy is based on agriculture and it exports copra, beef and veal, timber, and cocoa. Fishing, offshore financial services, and tourism are also important.

VANUATU

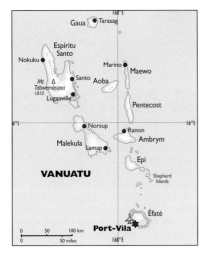

Area 4,706 sq mi [12,189 sq km]
Population 203,000
Capital (population) Port Vila (19,000)
Government Multiparty republic
Ethnic groups Ni-Vanuatu 99%
Languages Local languages (more than 100) 73%, pidgin (known as Bislama or Bichelama) 23%, others
Religions Presbyterian 31%, Anglican 13%, Roman Catholic 13%, Seventh-Day Adventist 11%, indigenous beliefs 6% (including Jon Frum cargo cult), others
Currency Vatu
Website www.vanuatugovernment.gov.vu

WALLIS & FUTUNA

Area 77 sq mi [200 sq km]
Population 16,000
Capital (population) Mata-Utu (1,000)
Government Overseas territory of France
Ethnic groups Polynesian
Languages Wallisian 59%, Futunian 30%, French 11%
Religions Roman Catholic
Currency Comptoirs Français du Pacifique franc = 100 centimes
Website www.wallis.co.nc

The Territory of the Wallis and Futuna Islands in the South Pacific Ocean form the smallest and the poorest of France's overseas territories, although they were in fact discovered by the Dutch and the British in the 17th and 18th centuries.

In 1959, the inhabitants of the islands voted to become a French overseas territory. A French dependency since 1842, the territory comprises two groups of islands. The Isles de Hoorn, which includes Futuna, are situated to the northeast of the Fiji Islands. The other is the Wallis Archipelago.

The economy is based on subsistence agriculture, and 80% of the workforce makes their living from either coconuts and vegetables, livestock, or fishing.

Other revenue comes from the licensing of fishing rights to Japan and South Korea.

287

PAKISTAN

Pakistan's flag was adopted in 1947, when the country gained independence. The color green, the crescent Moon and the five-pointed star are all traditional symbols of Islam. The white stripe represents the other religions in Pakistan.

The Islamic Republic of Pakistan contains high mountains, fertile plains and rocky deserts. The Karakoram range contains K2, the world's second highest peak at 28,251 ft [8,611 m], and lies in the northern part of Jammu and Kashmir. The Thar (or Great Indian) Desert straddles the border with India in the southeast. The arid Baluchistan Plateau lies in the south.

Area 307,372 sq mi [796,095 sq km]
Population 159,196,000
Capital (population) Islamabad (529,000)
Government Federal republic
Ethnic groups Punjabi, Sindhi, Pashtun (Pathan), Baluchi, Muhajir
Languages Urdu (official), many others
Religions Islam 97%, Christianity, Hinduism
Currency Pakistani rupee = 100 paisa
Website www.infopak.gov.pk

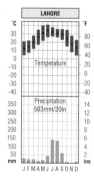

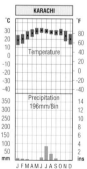

CLIMATE

Most of Pakistan has hot summers and cool winters. Winters in the mountains are cold and snowy. Rainfall is sparse over much of the country.

HISTORY

Pakistan was the site of the Indus Valley civilization which developed about 4,500 years ago. It broke into smaller cultures around 1700 BC. Historians believe that its breakup may have been caused by changes in the courses of the rivers.

Islam was introduced in AD 711. In 1526, it became part of the Mogul Empire, under which Sikhism emerged, combining elements of Hinduism and Islam. In the 1840s, the British East India Company gained areas in Punjab and Sind. The region became known as British India in 1858.

POLITICS

British India was divided into Pakistan and India in 1947. Slaughter ensued as Hindus and Sikhs fled Pakistan and Muslims fled India. Pakistan consisted of West and East Pakistan. Following a bitter civil war, East Pakistan broke away in 1971 to become Bangladesh. In 1948–9, 1965 and 1971, Pakistan and India clashed over the disputed territory of Kashmir. In 1998 Pakistan responded in kind to a series of Indian nuclear weapon tests, provoking global controversy. 2003–2005 saw a series of peace moves

ISLAMABAD

Capital of Pakistan, in the north of the country. In 1967 Islamabad became the official capital, replacing Karachi. It lies at the heart of an agricultural region. Administrative and governmental activities predominate.

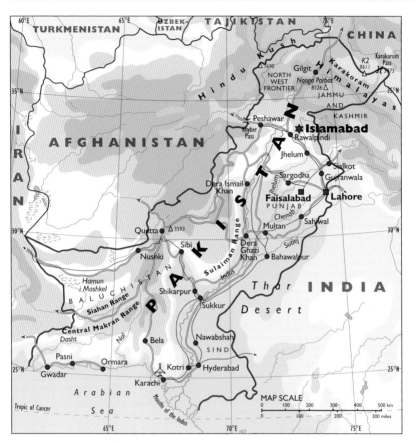

by Pakistan, raising hopes of a settlement, though militant activity continued.

Elections in 1988 led to Benazir Bhutto becoming prime minister. She was removed from office in 1990 but returned as prime minister from 1993 to 1996. In 1997 Narwaz Sharif was elected prime minister but a military coup brought General Pervez Musharraf to power. In 2001 Pakistan supported the Western assault on Taliban forces in Afghanistan. In 2002, voters agreed to extend Musharraf's term in office by five years; he then changed the constitution to increase his own powers. Pakistan's declaration of support for the coalition against terrorism provoked a backlash by Islamic fundamentalists. In 2004, Musharraf announced that he would remain army chief as well as head of state. In October 2005 nearly 75,000 people were killed and 3 million left homeless in an earthquake centered on Pakistan-administered Kashmir.

ECONOMY

Pakistan is a "low-income" developing country. Agriculture employs 44% of the workforce. Major crops are cotton, fruits, rice, sugarcane, vegetables, and wheat. Livestock include goats and sheep. Manufactures include bicycles, automobile tires, cement, industrial chemicals, and jute.

PANAMA

The blue quarter stands for the Conservative Party. The red quarter represents the Liberal Party. The white quarters symbolize peace between the parties The blue star stands for purity and honesty. The red star denotes government and law.

The Republic of Panama forms an isthmus linking Central America to South America. The narrowest part of Panama is less than 37 mi [60 km] wide. The Panama Canal, which is 50.7 mi [81.6 km] long and cuts straight across the isthmus, has made the country a major transportation center. and most Panamanians live within 12 mi [20 km] of it. Most of the land between the Pacific and Caribbean coastal plains is mountainous, rising to 11,400 ft [3,475 m] at the volcano Barú.

Tropical forests cover approximately 50% of Panama. Mangrove swamps line the coast, though in recent years more than 150 sq mi [400 sq km] have been lost to agriculture, ranching, and shrimp mariculture. Subtropical woodland grows on the mountains, while tropical savanna occurs along the Pacific coast.

Area 29,157 sq mi [75,517 sq km]
Population 3,000,000
Capital (population) Panamá (484,000)
Government Multiparty republic
Ethnic groups Mestizo 70%, Black and Mulatto 14%, White 10%, Amerindian 6%
Languages Spanish (official), English
Religions Roman Catholic 85%, Protestant 15%
Currency US dollar; Balboa = 100 centésimos
Website www.visitpanama.com

cross Panama and see the Pacific Ocean. The indigenous population was soon wiped out and Spain established control. In 1821, Panama became a province of Colombia. The USA exerted great influence from the mid-19th century.

CLIMATE

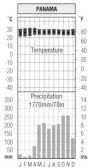

Panama has a tropical climate, though the mountains are much cooler than the coastal plains. The rainy season is between May and December. The Caribbean side of has about twice as much rain as the Pacific side.

HISTORY

Christopher Columbus landed in Panama in 1502. In 1510, Vasco Núñez de Balboa became the first European to

PANAMA CANAL

Waterway connecting the Atlantic and Pacific oceans across the Isthmus of Panama. A canal, begun in 1882 by Ferdinand de Lesseps, was subsequently abandoned because of bankruptcy. The US government decided to finance the project to provide a convenient route for its warships. The main construction took about ten years to complete, and the first ship passed through in 1914. The 51 mi [82 km] waterway reduces the sea voyage between San Francisco and New York by about 7,800 mi [12,500 km]. Control of the canal passed from the USA to Panama at the end of 1999.

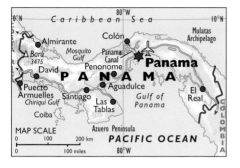

captured, and taken to the USA for trial.

Pérez Balladares became president in 1994 elections. In 1999, Mireya Moscoso, Panama's first woman president, succeeded Balladares. She was succeeded in 2004 by Martin Torrijos, son of a former military dictator.

Revenues from the canal rose in the early 21st century, but overall the economy slowed, causing social discontent and problems for the government.

After a revolt in 1903, Panama declared independence from Colombia. In 1904, the USA began construction of the Panama Canal, and established the Panama Canal Zone. Since it opened in 1914, the status of the Canal has dominated Panamanian politics. The Panama Canal Zone, a strip of land along the canal, was then administered by the United States. US forces intervened in 1908, 1912 and 1918 to protect US interests.

POLITICS

Panama has been politically unstable throughout the 20th century, with a series of dictatorial regimes and military coups.

Civil strife during the 1950s and 1960s led to negotiations with the USA for the transfer of the Canal Zone. In 1977, a treaty confirmed Panama's sovereignty over the canal, while providing for US bases in the Canal Zone. The USA agreed to hand over control of the canal on 31 December 1999. In 1979, the Canal Zone disestablished.

In 1983, General Noriega took control of the National Guard and ruled Panama through a succession of puppet governments. In 1987, the USA withdrew its support for Noriega after he was accused of murder, electoral fraud, and aiding drug smuggling. In 1988, the USA imposed sanctions and in 1989, Noriega annulled elections, made himself president, and declared war on the USA. On 20 December 1989, 25,000 US troops invaded Panama. Noriega was quickly

ECONOMY

The World Bank classifies Panama as a "lower-middle-income" developing country. The Panama Canal is a major source of revenue, generating jobs in commerce, trade, manufacturing, and transportation. The other main activity is agriculture, employing 27% of the workforce. Rice is the main food crop. Bananas, shrimp, sugar, and coffee are exported. Tourism is also important. Many ships are registered under Panama's flag, due to its low taxes.

PANAMA CITY

Capital of Panama, on the shore of the Gulf of Panama, near the Pacific end of the Panama Canal. It was founded by Pedro Arias de Avila in 1519, and was destroyed and rebuilt in the 17th century. It includes an area known as Casco Antiguo (Colonial Panama), which was constructed inland after the destruction of the first city. Casco Antiguo has been declared "Patrimony of Humanity" by UNESCO. The city became the capital of Panama in 1903 and it developed rapidly after the construction of the Panama Canal in 1914. Industries include brewing, shoes, textiles, petroleum refining, and plastics.

PAPUA NEW GUINEA

Papua New Guinea's flag was first adopted in 1971, four years before the country became independent from Australia. It includes a local bird of paradise, the "kumul," in flight, together with the stars of the Southern Cross. The colors are those often used by local artists.

The Independent State of Papua New Guinea is part of a southwest Pacific island region called Melanesia 100 mi [160 km] northeast of Australia that includes the eastern part of New Guinea, Bismarck Archipelago, Solomon Islands, New Hebrides, the Trobriand and D'Entrecasteaux Islands, the Louisiade Archipelago, and the Tonga group.

The land is largely mountainous, rising to Mount Wilhelm, at 14,790 ft [4,508 m], eastern New Guinea. In 1995, two volcanoes erupted in Eastern New Britain. East New Guinea also has extensive coastal lowlands.

Forests cover more than 70% of the land. The dominant vegetation is rainforest. Mangrove swamps line the coast. "Cloud" forest and tussock grass are found on the higher peaks.

Area 178,703 sq mi [462,840 sq km]
Population 5,420,000
Capital (population) Port Moresby (193,000)
Government Constitutional monarchy
Ethnic groups Papuan, Melanesian, Micronesian
Languages English (official), Melanesian Pidgin, more than 700 other indigenous languages
Religions Traditional beliefs 34%, Roman Catholic 22%, Lutheran 16%, others
Currency Kina = 100 toea
Website www.pngonline.gov.pg

ern New Guinea (now the Papua part of Indonesia) in 1828, but it was not until 1884 that Germany took northeastern New Guinea as German New Guinea and Britain formed the protectorate of British New Guinea in southeast New

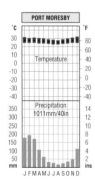

CLIMATE

The climate is tropical. It is hot all year with most rain occurring during the monsoon from December to April, when winds blow from the northeast. Winds blow from the southwest during the dry season.

HISTORY

The Portuguese made the first European sighting of the island in 1526, although no settlements were established until the late 19th century. The Dutch took west-

---

BOUGAINVILLE

Volcanic island in the southwest Pacific Ocean, east of New Guinea; a territory of Papua New Guinea. It was discovered in 1768 by Louis de Bougainville. The island was under German control from 1884, and then under Australian administration after 1914 and again in 1945 (after the Japanese wartime occupation). It has been the scene of guerrilla warfare since the late 1980s. Kieta is the chief port. Industries include copper mining, copra, cocoa, timber. It covers an area of 10,049 sq km [3,880 sq mi], with a population of 204,800.

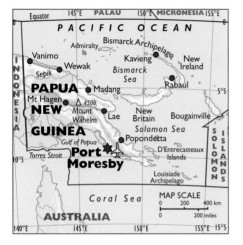

mercenaries created unrest in the army. In 1997, troops and civilians surrounded Parliament, forcing the resignation of Prime Minister Sir Julius Chan. He was succeeded by Bill Skate. In April 1998, a cease-fire was declared on Bougainville.

In July 1998, a tidal wave hit northern Papua New Guinea, killing more than 1,600 people. Local autonomy was granted to Bougainville in 2000. In 2004 Australia sent police to the country to help fight crime after a report had stated that the country was heading for social and economic collapse. In 2005 global warming lead to the proposed evacuation of all the residents of the Cartaret atolls due to rising sea levels.

Guinea. In 1906, Britain handed the southeast over to Australia. It then became known as the Territory of Papua. When World War I broke out in 1914, Australia took German New Guinea. In 1921, the League of Nations gave Australia a mandate to rule the area, which was named the Territory of New Guinea.

Japan invaded New Guinea in 1942, but the Allies reconquered it in 1944. In 1949, Papua and New Guinea were combined into the Territory of Papua and New Guinea. In 1973, the Territory achieved self-government as a prelude to full independence as Papua New Guinea in 1975.

POLITICS

Since independence, the government has worked to develop mineral reserves. One of the most valuable reserves was a copper mine at Panguna on Bougainville. Conflict developed when the people of Bougainville demanded a larger share in mining profits.

Following an insurrection, the Bougainville Revolutionary Army proclaimed independence in 1990. Bougainville's secession was not recognized internationally. In 1992 and 1996, Papua New Guinea launched offensives against the rebels. The use of highly paid

ECONOMY

The World Bank classifies Papua New Guinea as a "lower-middle-income" developing country. Agriculture employs 75% of the workforce, many at subsistence level. Minerals, notably copper and gold, are the most valuable exports. Papua New Guinea is the world's ninth-largest producer of gold.

PORT MORESBY

Capital of Papua New Guinea, on the southeast coast of New Guinea, built around Fairfax Harbor, the island's largest harbor. Settled by the British in the 1880s and named after British explorer John Moresby, its sheltered harbor was the site of an important Allied base in World War II. It developed rapidly in the post-war period. In recent years it has experienced problems with a growing disparity in income which has lead to an increase in crime. Exports include gold, copper, and rubber.

PARAGUAY

The front (obverse) side of Paraguay's tricolor flag, which evolved in the early 19th century, contains the state emblem, which displays the May Star, commemorating liberation from Spain in 1811. The reverse side shows the treasury seal—a lion and staff.

The Republic of Paraguay is a land-locked country in South America. Rivers form most of its borders. They include the Paraná in the south and the east, the Pilcomayo (Brazo Sur) in the southwest, and the Paraguay in the northeast. West of the River Paraguay is a region known as the Gran Chaco, which extends into Bolivia and Argentina. The Gran Chaco is mostly flat, but the land rises to the northwest. East of the Paraguay is a region of plains, hills and, in the east, the Paraná Plateau region.

Area 406,752 sq km [157,047 sq mi]
Population 6,191,000
Capital (population) Asunción (547,000)
Government Multiparty republic
Ethnic groups Mestizo 95%
Languages Spanish and Guarani (both official)
Religions Roman Catholic 90%, Protestant
Currency Guarani = 100 céntimos
Website www.paraguay.com

CLIMATE

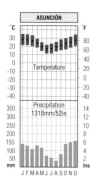

The northern half of Paraguay lies in the tropics, while the southern half is subtropical. Most of the country has a warm, humid climate. The Gran Chaco is the driest and hottest part of the country. Rainfall increases to the Paraná Plateau in the southeast.

HISTORY

The Guarani, an Amerindian people, were the indigenous people of what is now Paraguay. Spanish and Portuguese explorers reached the area in the early 16th century and, in 1537, a Spanish expedition built a fort at Asunción, which later became the capital of Spain's colonies in southeastern South America. The Spaniards were attracted by the potential labor supply of the Guarani and the chance to find a shortcut to the silver mines of Peru. From the late 16th century, Jesuit missionaries arrived to convert the Guarani to Christianity and to protect them against those who wanted to exploit them as cheap labor. Complaints against the Jesuits' power led to their expulsion in 1767.

From 1776, Paraguay formed part of the Rio de la Plata Viceroyalty, with its capital at Buenos Aires. However, this proved unpopular and Paraguay broke free in 1811, achieving its independence from Buenos Aires in 1813.

Between 1865 and 1870, war against Brazil, Argentina and Uruguay cost the country more than half of its 600,000 population, and much of its territory. Some territory was regained after the Chaco Wars against Bolivia between 1920 and 1935, and, in 1947, a period of civil war was followed by a spell of political and economic stability. While most other South American countries were attracting European settlers and foreign capital, Paraguay remained isolated and forbidding.

POLITICS

In 1954, General Alfredo Stroessner seized power and assumed the presiden-

mate, General Lino Oviedo, who had been imprisoned for attempting a coup against the previous president, Juan Carlos Wasmosy. In March 1999, Paraguay's vice-president, an opponent of Cubas, was assassinated and the Congress impeached Cubas, who resigned and fled to Argentina. In 2003, Nicanor Duarte Frutos was elected president.

ECONOMY

The World Bank classifies Paraguay as a "lower-middle-income" developing country. Agriculture and forestry are the leading activities, employing 48% of the workforce. The country has very large cattle ranches, while crops are grown in the fertile soils of eastern Paraguay. Major exports include cotton, soybeans, timber, vegetable oils, coffee, tannin, and meat products. The country has abundant hydroelectricity and exports power to Argentina and Brazil. Its factories produce cement, processed food, leather goods, and textiles.

cy. During his dictatorship, there was considerable economic growth, with an emphasis on developing hydroelectricity. By 1976, Paraguay was self-sufficient in electrical energy due to the completion of the Aracay complex. A second hydroelectric project, the world's largest, started production in 1984, at Itaipu. This was a joint US$20 billion venture with Brazil to harness the Paraná. Paraguay was then generating 99.9% of its electricity from water power. However, demand slackened and income declined, making it difficult for Paraguay to repay foreign debts incurred on the projects. High inflation and balance of payments problems followed.

Stroessner's regime was an unpleasant variety of nepotism. He ruled with an increasing disregard for human rights during nearly 35 years of fear and fraud until his supporters deposed him in 1989.

Three elections were held in the 1990s. The fragility of democracy was demonstrated in 1998, when the newly elected president, Raul Cubas Grau, was threatened with impeachment after issuing a decree freeing his former running

ASUNCIÓN

Capital, chief port, and largest city of Paraguay, located on the east bank of the Paraguay River near its junction with the River Pilcomayo. Founded by the Spanish 1536 as a trading post, Asunción was the scene of the Communeros rebellion against Spanish rule in 1721 and was later occupied by Brazil (1868–76). City sights include the Pantéon Nacional, Encarnación Church, National University (1889), and the Catholic University (1960). It is an administrative, industrial, and cultural center. Industries include vegetable oil and textiles.

PERU

Peru's flag was adopted in 1825. The colors are said to have been inspired by a flock of red and white flamingos which the Argentine patriot General José de San Martín saw flying over his marching army when he arrived in 1820 to liberate Peru from Spain.

The Republic of Peru lies in the tropics in western South America. A narrow coastal plain borders the Pacific Ocean in the west. Inland are ranges of the Andes Mountains, which rise to 22,205 ft [6,768 m] at Mount Huascarán, an extinct volcano. The Andes also contain active volcanoes, windswept plateaus, broad valleys, and, in the far south, part of Lake Titicaca, the world's highest navigable lake. To the east, the Andes descend to a hilly region and a huge plain. Eastern Peru is part of the Amazon basin.

Area 496,222 sq mi [1,285,216 sq km]
Population 27,544,000
Capital (population) Lima (5,681,000)
Government Constitutional republic
Ethnic groups Mestizo (Spanish-Indian) 44%, Creole (mainly African American) 30%, Mayan Indian 11%, Garifuna (Black-Carib Indian) 7%, others 8%
Languages English (official), Creole, Spanish
Religions Roman Catholic 62%, Protestant 30%
Currency Belize dollar = 100 cents
Website www.peru.info/perueng.asp

CLIMATE

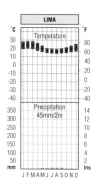

Lima, on the coastal plain, has an arid climate. The coastal region is chilled by the cold offshore Humboldt Current. In the Andes, temperatures are moderated by the altitude and many mountains are snow-capped. The eastern lowlands are hot and humid.

HISTORY

Amerindian people reached the area about 12,000 years ago. Several civilizations developed in the Andes region. By about AD 1200, the Inca were established in southern Peru. In 1500, their empire extended from Ecuador to Chile. The Spanish adventurer Francisco Pizarro visited Peru in the 1520s. Hearing of Inca riches, he returned in 1532. By 1533, he had conquered most of Peru.

In 1820, the Argentinian José de San Martín led an army into Peru and declared the country to be independent. However, Spain still held large areas. In 1823, the Venezuelan Simón Bolívar led another army into Peru and, in 1824,

LIMA

Capital and largest city of Peru, on the River Rímac at the foot of the Cerro San Cristóbal. Founded in 1535, it functioned as the capital of Spain's New World colonies until the 19th century. Chilean forces occupied Lima during the War of the Pacific (1881–3). It is the commercial and cultural center of Peru. With the petroleum-refining port of Callao, the Lima metropolitan area forms the third-largest city of South America, and handles more than 75% of all Peru's manufacturing. Lima has the oldest university (1551) in the Western Hemisphere.

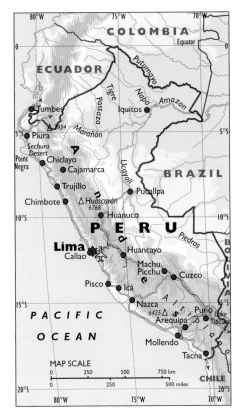

A new constitution was introduced in 1993, giving increased power to President Albert Fujimori. In 1996, Tupac Amaru (MRTA) rebels seized the Japanese ambassador's residence, taking hostages and demanding the release of guerrilla prisoners. The stalemate ended in April 1997, when Peruvian troops attacked and freed the remaining 72 hostages.

Peru faced many problems in the 1990s, including a cholera outbreak, the worst El Niño in the 20th century, and a border dispute with Ecuador which was finally settled in 1998. Fujimori began his third term as president in 2000, but, in November, the Congress declared him "morally unfit" to govern. He resigned and sought sanctuary in Japan. In his absence he was banned from holding office until 2011. In 2005 the government began its attempt to extradite him and try for financial corruption and sanctioning death squads. In 2001, Alejandro Toledo became the first Peruvian of Amerindian descent to hold the office of president. Toledo faced many problems including in 2003–4 a resurgence of activity by the "Shining Path" guerillas.

one of his generals defeated the Spaniards at Ayacucho. The Spaniards surrendered in 1826. Peru suffered much instability throughout the 19th century.

POLITICS

Instability continued into the 20th century. When civilian rule was restored in 1980, a left-wing group called the Sendero Luminoso (Shining Path), began guerrilla warfare against the government. In 1990, Alberto Fujimori, son of Japanese immigrants, became president. In 1992, he suspended the constitution and dismissed the legislature. The guerrilla leader, Abimael Guzmán, was arrested in 1992, but instability continued.

ECONOMY

The World Bank classifies Peru as a "lower-middle-income" developing country. Agriculture employs 35% of the workforce and major food crops include beans, corn, potatoes, and rice. Coffee, cotton, and sugar are the chief cash crops. Many farmers live at subsistence level. Other farms are cooperatives. Fishing is important.

Peru is one of the world's main producers of copper, silver, and zinc. Iron ore, lead, and petroleum are also produced, while gold is mined in the highlands. Most manufacturing is small-scale.

297

PHILIPPINES

This flag was adopted in 1946, when the country won its independence from the United States. The eight rays of the large sun represent the eight provinces which led the revolt against Spanish rule in 1898. The three smaller stars stand for the three main island groups.

The Republic of the Philippines is an island country in southeastern Asia. It includes about 7,100 islands, of which 2,770 are named and about 1,000 are inhabited. Luzon and Mindanao, the two largest islands, make up more than two-thirds of the country.

The land is mainly mountainous, it is also unstable and prone to earthquakes. The islands also have several active volcanoes, one of which is the highest peak, Mount Apo, at 9,692 ft [2,954 m].

Area 115,830 sq mi [300,000 sq km]
Population 86,242,000
Capital (population) Manila (1,581,000)
Government Multiparty republic
Ethnic groups Christian Malay 92%, Muslim Malay 4%, Chinese and others
Languages Filipino (Tagalog) and English (both official), Spanish and many others
Religions Roman Catholic 83%, Protestant 9%, Islam 5%
Currency Philippine peso = 100 centavos
Website www.gov.ph

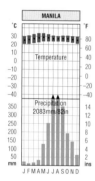

CLIMATE

The climate is tropical with high temperatures all year. The dry season runs from December to April. The rest of the year is wet. Typhoons periodically strike the east coast bringing high rainfall.

HISTORY

The first European to reach the Philippines was Ferdinand Magellan in 1521. Spanish explorers claimed the region in 1565 when they established their first permanent settlement on Cebu. Manila was founded in 1571. The Spaniards regarded their new territory as a stepping stone to the Spice Islands to the south. But they also converted most people (except the Muslims on Mindanao and Sulu) to Roman Catholicism.

The economy grew from the late 18th century when the islands were opened up to foreign trade. In 1896 a secret revolu-

tionary society called Katipunan launched a revolt against Spanish rule. The revolt was put down and the rebel leader Emilio Aguinaldo left the country. In 1898, the United States declared war

MANILA

Capital of the Philippines, on Manila Bay, southwest Luzon island. It is the industrial, commercial, and administrative heart of the Philippines. The River Pasig bisects the city. On the south bank stands the old walled city (Intramuros), built by the Spanish in the 16th century on the site of a Muslim settlement. It became a trading center for the Pacific area. On the north bank lies Ermita, the administrative and tourist center. Japan occupied the city in 1942 and in 1945 a battle between Japanese and Allied forces destroyed the old city.

the islands but the Philippines finally achieved independence on July 4, 1946.

POLITICS

From 1946 until 1971, the country was governed under a constitution similar to that of the United States. In 1971, constitutional changes were proposed, but before ratification, President Ferdinand Marcos declared martial law. In 1977, the main opposition leader, Benigno Aquino, Jr., was sentenced to death. He was allowed a stay of execution and went to the United States for medical treatment. Martial law was lifted in 1981, but Aquino was shot dead on his return to the Philippines in 1983.

Following presidential elections in 1986, Marcos was proclaimed president, but the elections proved to be fraudulent and his opponent, Corazon Aquino, the widow of Benigno Aquino, became president. In 2001 Gloria Macapagal-Arroyo, became president and set out to try to find peace in the southern Philippines. In 2003, the government put down military rebellion. Gloria Arroyo was reelected president in 2004 and a ceasefire was agreed in the south with the Moro Islamic Liberation Front. This ceasefire was broken in 2005.

ECONOMY

The Philippines is a developing country with a "lower-middle-income" economy. Agriculture employs 40% of the workforce. Rice and corn are the main food crops, along with bananas, cassava, coconuts, coffee, cocoa, fruits, sugarcane, sweet potatoes, and tobacco. Farm animals include water buffalo, goats, and pigs.

Forests cover nearly half the land and forestry is a valuable industry. Sea fishing is also important and shellfish are obtained from inshore waters.

on Spain and the first major engagement was the destruction of all the Spanish ships in Manila Bay. Aguinaldo returned to the Philippines and formed an army which fought alongside the Americans. He proclaimed the Philippines an independent nation and a peace treaty between Spain and the United States was signed with the US taking over the government of the Philippines. However, Aguinaldo still wanted independence and fighting continued between 1899 and 1901.

The Philippines became a self-governing US Commonwealth in 1935 and was guaranteed full independence after a ten-year transitional period. During World War II Japanese troops occupied

POLAND

Poland's flag was adopted when the country became a republic in 1919. Its colors were taken from the 13th-century coat of arms of a white eagle on a red field. This coat of arms still appears on Poland's merchant flag.

The Republic of Poland faces the Baltic Sea in north-central Europe. Behind the lagoon-fringed coast is a broad plain. The land rises to a plateau region in the southeast of the country. The Sudeten Highlands straddle the border with the Czech Republic. Part of the Carpathian Range lies on the southeastern border with the Slovak Republic.

Area 124,807 sq mi [323,250 sq km]
Population 38,626,000
Capital (population) Warsaw (1,615,000)
Government Multiparty republic
Ethnic groups Polish 97%, Belarusian, Ukranian, German
Languages Polish (official)
Religions Roman Catholic 95%, Eastern Orthodox
Currency Zloty = 100 groszy
Website www.poland.pl

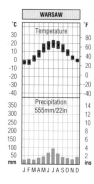

CLIMATE

Poland's climate is influenced by its geographical position. Warm, moist air masses come from the west, while cold air masses come from the north and east. Summers are warm and winters cold and snowy.

Silesia and Breslau (now Wroclaw), in the northwest the Baltic port of Stettin (now Szczecin), and in the north the port of Danzig (now Gdańsk). Acquisition of a length of Baltic coastline gave Poland an opportunity to develop maritime interests.

HISTORY

Poland's boundaries have changed several times in the last 200 years. It disappeared from the map in the late 18th century, when a Polish state called the Grand Duchy of Warsaw was set up. But in 1815, the country was partitioned, between Austria, Prussia and Russia. Poland became independent in 1918, but in 1939 it was divided between Germany and the Soviet Union. The country again became independent in 1945, when it lost land (poor agricultural land), and around 6 million people, to the Soviet Union. In compensation, it gained parts of Germany as far as the River Oder, an important industrial region in the west. Other gains were, in the southwest,

WARSAW

Capital and largest city of Poland, on the River Vistula. It dates from the 11th century and became Poland's capital in 1596. Controlled by Russia from 1813 to 1915, German troops occupied it during World War I. The 1939 German invasion and occupation of Warsaw marked the beginning of World War II. In 1940, the Germans isolated the Jewish ghetto (500,000 people). In January 1945, the Red Army liberated Warsaw and found only 200 surviving Jews. The old town was painstakingly reconstructed and Warsaw became a major transportation and industrial center.

POLITICS

Communists took power in 1948, but opposition mounted and became focused through an organization called Solidarity, led by a trade unionist, Lech Walesa. A coalition government was formed between Solidarity and the Communists in 1989. In 1990, the Communist Party was dissolved and Walesa became president. He faced many problems in turning Poland toward a market economy. Solidarity divided in 1990 over personality and the speed of reform. The adoption of its reforms was interrupted in 1993, when the former Communists won the parliamentary elections.

In 1995, the ex-Communist Aleksander Krasniewski defeated Walesa in presidential elections, but he continued to follow westward-looking policies. Poland became a member of NATO in 1999 and of the EU in 2004. Having lost elections in 1997, Krasniewski was reelected president in 2000. But, in 2005, a swing to the right resulted in Lech Kaczynski of the Law and Justice party becoming president.

ECONOMY

Poland has large reserves of coal and deposits of various minerals which are used in its factories. Manufactures include chemicals, processed food, machinery, ships, steel, and textiles. Major crops include barley, potatoes, rye, sugarbeet, and wheat.

301

PORTUGAL

Portugal's colors, which were adopted in 1910 when the country became a republic, represent the soldiers who died in the war (red), and hope (green). The armillary sphere—an early navigational instrument—reflects Portugal's leading role in world exploration.

The Portuguese Republic shares the Iberian Peninsula with Spain. It is the most westerly of Europe's mainland countries. The land rises from the coastal plains on the Atlantic Ocean to the western edge of the huge plateau, or Meseta, which occupies most of the Iberian Peninsula. In central Portugal, the Sera da Estrela contains Portugal's highest point, at 6,537 ft [1,993 m]. Portugal also contains two autonomous regions, the Azores and Madeira island groups.

Area 34,285 sq mi [88,797 sq km]
Population 10,524,000
Capital (population) Lisbon (663,000)
Government Multiparty republic
Ethnic groups Portuguese 99%
Languages Portuguese (official)
Religions Roman Catholic 94%, Protestant
Currency Euro = 100 cents
Website www.portugal.org

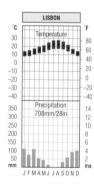

CLIMATE

Winds blowing in from the Atlantic Ocean moderate the climate. Portugal experiences cooler summers and milder winters than in other lands on the Mediterranean.

HISTORY

The Romans completed their conquest of the Iberian Peninsula around 2,000 years ago and Christianity was introduced in the 4th century AD. The Romans called Portugal Lusitania. Following the collapse of the Roman Empire in the 5th century, Portugal was conquered by the Christian Visigoths, but in the early 8th century, the Iberian Peninsula was conquered by Muslim Moors. The Christians strove to drive out the Muslims and, by the mid-13th century, they had retaken Portugal and most of Spain.

In 1143, Portugal became a separate country, independent from Spain. In the 15th century, the Portuguese, who were skilled navigators, led the "Age of Exploration," pioneering routes around Africa onward to Asia.

Portugal set up colonies in Africa and Asia, though the most valuable was Brazil. Portugal became wealthy through trade and the exploitation of its colonies, but its power began to decline in the 16th century, when it could no longer defend its far-flung empire. Spain ruled Portugal from 1580 until 1640, when Portugal's independence was restored by John, Duke of Braganza, who took the title of John IV. England supported Portuguese independence and several times defended it from invasion or threats by Spain and its allies. However, in 1822, Portugal lost Brazil.

POLITICS

Portugal became a republic in 1910, but its first attempts at democracy led to great instability. Portugal fought alongside the Allies in World War I. A coup in 1926 brought an army group to power. They abolished the parliament and set up a dictatorial regime. In 1928, they selected António de Oliviera Salazar, an economist, as minister of finance. He became prime minister in 1932 and

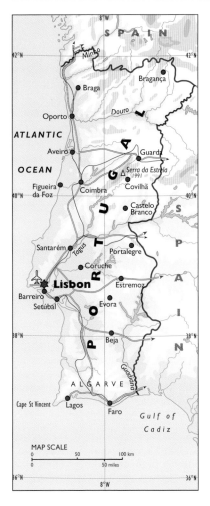

colonies. Free elections were held in 1978 and full democracy was restored in 1982, when a new constitution abolished the military Council of the Revolution and reduced the powers of the president.

Portugal joined the European Community (now the EU) in 1986, and in 1999 became one of the 12 EU countries to adopt the euro, the single currency of the EU. In 2005 the Socialists led by a moderate, José Socrates, won a decisive victory in parliamentary elections.

ECONOMY

Although its economy was growing strongly in the late 1990s, Portugal remains one of the EU's poorer members. Agriculture and fishing were the mainstays of the economy until the mid-20th century. But manufacturing is now the most valuable sector. Textiles, processed food, paper products, and machinery are important manufactures. Major crops include grapes for winemaking, olives, potatoes, rice, corn, and wheat. Cattle and other livestock are raised and fishing catches include cod, sardines, and tuna.

Forest products including timber and cork are important, though forest fires often cause much damage.

LISBON (LISBOA)

Capital, largest city, and chief port of Portugal, at the mouth of the River Tagus, on the Atlantic Ocean. An ancient Phoenician settlement, conquered by the Romans in 205 BC. After Teutonic invasions in the 5th century AD, it fell to the Moors in 716. In 1147, the Portuguese reclaimed Lisbon, and in 1260 it became the capital. It declined under Spanish occupation (1580–1640). In 1755, an earthquake devastated the city and Marques de Pombal oversaw its reconstruction. It is an international port and tourist center.

ruled as a dictator from 1933. After World War II, when other European powers began to grant independence to their colonies, Salazar was determined to maintain his country's empire. Colonial wars flared up and weakened Portugal's economy. Salazar suffered a stroke in 1968 and died two years later. His successor, Marcello Caetano, was overthrown by another military coup in 1974 and the new military leaders set about granting independence to Portugal's

QATAR

The vertical serrated white stripe, added at the request of the British, denotes friendly Arab states. The nine-point serrated line indicates that Qatar is the ninth member of the reconciled Emirates of the Arabian Gulf. The maroon area represents blood shed in the 19th-century wars.

The State of Qatar occupies a long, narrow peninsula jutting into the Persian Gulf. The peninsula is about 124 mi [200 km] long, with a greatest width of 56 mi [90 km]. The land is mostly flat desert covered by gravel and loose, wind-blown sand. Sand dunes occur in the south-east. There are also some barren salt flats. The highest point, on a central limestone plateau, is only 321 ft [98 m] above sea level. Qatar also includes several offshore islands and coral reefs. Freshwater is scarce and much of the water supply comes from desalination plants.

Area 4,415 sq mi [11,437 sq km]
Population 522,000
Capital (population) Doha (264,000)
Government Absolute monarchy
Ethnic groups Arab 40%, Pakistani 18%, Indian 18%, Iranian 10%
Languages Arabic (official), English
Religions Islam 95% (all native Qataris are Wahhabi Sunni)
Currency Rial = 100 dirham
Website http://english.mofa.gov.qa

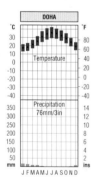

CLIMATE

The weather from May to September is extremely hot and dry, with the temperature soaring to 120°F [49°C]. Sand and dust storms are common. Winters are mild to warm, with the weather generally sunny and pleasant. The total annual rainfall seldom exceeds 4 in [100 mm]. Most of the rain occurs in winter.

HISTORY

In the 18th century, migrants established trading settlements along the coast of the peninsula. Since the mid-19th century, members of the Al-Thani family have been the leaders of Qatar. Between 1871 and 1913, the Ottoman Turks, with Qatar's consent, occupied a garrison on the peninsula. In 1916, Qatar agreed that Britain would take responsibility for the country's foreign affairs. Petroleum was struck in 1939 but exploitation was delayed by World War II (1939–45). Commercial exploitation began in 1949, leading to the rapid development and modernization of the country's infrastructure.

In 1968, Britain announced that it would withdraw its forces from the Gulf. Qatar negotiated with Bahrain and the United Arab Emirates concerning the formation of a federation, but this proposal was finally rejected.

POLITICS

Qatar became fully independent on September 3, 1971. In 1972, because of rivalries in the ruling family, the deputy ruler Khalifa bin Hamad Al Thani seized power, from his cousin, Emir Ahmad in Al-Thani, in a coup. In 1982, Qatar together with Bahrain, Kuwait, Oman, Saudi Arabia, and the United Arab Emirates united to form the Gulf Cooperation Council, which is concerned with such matters as defense and economic development.

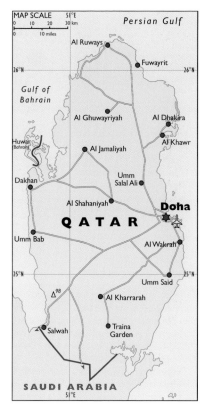

In 1990, following Iraq's invasion of Kuwait, Qatar agreed to allow foreign troops on its soil and, in 1991, Qatari troops were involved in the military campaign to free Kuwait. In 1995, Qatar signed a security pact with the United States. A bloodless coup occurred in 1995, when the heir apparent, Sheikh Hamad bin Khalifa Al-Thani, deposed his father, while Khalifa was abroad. An attempted countercoup failed in 1996.

In 1996, the Al-Jazeera satellite television was launched in Qatar. It soon won a worldwide reputation for tackling controversial issues, especially those connected with the Arab world. In 2001, it became famous when it became the first station to air recorded statements by the al Qaida leader Osama bin Laden and,

from 2003, it covered the conflict in Iraq graphically. Qatar is an emirate, ruled by the Emir and his appointed Council of Ministers. Municipal elections in 1999 heralded moves towards democracy.

A new constitution introduced in 2004 provided for a 45-member Consultative Council. This Council consisted of 45 members, two-thirds of whom would be elected by the public and one-third appointed by the Emir. The new constitution came into force in 2005 with elections expected by 2007. Qatar resolved longstanding boundary disputes with Bahrain and Saudi Arabia in 2001. In 2003, the US Central Command forward base on Qatar became the main center for the US-led invasion of Iraq.

ECONOMY

The people of Qatar enjoy a high standard of living, which derives from petroleum revenues. The country has a comprehensive welfare system, and many of its services are free or highly subsidized. Petroleum production has given Qatar one of the world's highest per capita incomes and accounts for more than 80% of the country's export revenues. Qatar has about 5% of the world's proven reserves of petroleum and more than 15% of the world's proven natural gas reserves.

Besides petroleum refining, they produce ammonia, cement, fertilizers, petrochemicals, and steel bars. Wells have been dug to develop agriculture and produce includes beef, dairy products, fruits, poultry, and vegetables.

DOHA

Capital of Qatar, on the east coast of the Qatar peninsula, in the Persian (Arabian) Gulf. Doha was a fishing village until petroleum production began in 1949. It is now a modern city and trade center. Industries include petroleum refining, shipping, engineering.

ROMANIA

Romania's flag, adopted in 1948, uses colors from the arms of the provinces, which united in 1861 to form Romania. A central coat of arms, added in 1965, was deleted in 1990 after the fall of the Communist regime under the dictator Nicolae Ceaucescu.

Romania is on the Black Sea in eastern Europe. Eastern and southern Romania form part of the Danube River Basin. The delta region, where the river flows into the Black Sea, is one of Europe's finest wetlands. The southern part of the coast contains several resorts.

The country is dominated by the Carpathian mountains which curve around the plateaus of Transylvania in central Romania. The southern arm of the mountains, including Mount Moldoveanu (8,341 ft [2,543 m]), is known as the Transylvanian Alps. On the border with Serbia and Montenegro, the River Danube (Dunav/Dunărea) has cut a gorge, the Iron Gate (Portile de Fier), whose rapids have been tamed by a huge dam. Forests cover large areas in Transylvania and the Carpathians, while farmland dominates in the Danubian lowlands and the plateaus.

Area 92,043 sq mi [238,391 sq km]
Population 22,356,000
Capital (population) Bucharest (2,001,000)
Government Multiparty republic
Ethnic groups Romanian 89%, Hungarian 7%, Roma 2%, Ukranian
Languages Romanian (official), Hungarian, German
Religions Eastern Orthodox 87%, Protestant 7%, Roman Catholic 5%
Currency Leu = 100 bani
Website www.gov.ro/engleza/

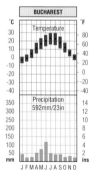

CLIMATE

Romania has hot summers and cold winters. Rainfall is heaviest in spring and early summer, when thundery showers are common.

HISTORY

Around 2,300 years ago, Romania was called Dacia. After the Romans conquered the area in AD 106, the Dacians embraced Roman culture and language so completely that the region became known as Romania. The first step toward the creation of the modern state occurred in the 14th century when two principalities were formed: Walachia (or Valachi) in the south and Moldavia in the east. But they were conquered by the Ottoman Turks around 1500.

Walachia and Moldavia united in 1861 to form modern Romania. After World War I, Romania, which had fought with the Allies, gained much land, including Transylvania, almost doubling the country's size and population. In 1939 Romania lost territory to Bulgaria, Hungary, and the Soviet Union. Romania fought with Germany in World War II, and was occupied by the Soviet Union in 1944. Hungary returned northern Transylvania to Romania in 1945, but Bulgaria and the Soviet Union kept former Romanian territory when King Michael was forcibly removed from the throne.

In the 1960s, Romania's Communist Party, led by Gheorghe Gheorghiu-Dej, began to oppose Soviet control, a policy continued by Nicolae Ceaucescu, who became Communist Party chief in 1965.

Under Ceaucescu Romania developed industries based on its petroleum and natural gas reserves. His rule was

but not fraudulent. A new constitution enshrining pluralist democracy, human rights, and a market economy was passed by parliament in 1991. There were strikes and protests against the new authorities and also against the effects of the switch to a market economy, which caused food shortages, rampant inflation, and increased unemployment. Foreign investment was sluggish, deterred by the political instability. Presidential elections in 1996 led to defeat for Iliescu and victory for the center-right Emil Constantinescu. In 2000, Iliescu was reelected president, though the government continued its privatization policies. He stood down in 2004. Romania became a member of NATO in 2004 and is expected to join the EU in 2007.

corrupt and selfseeking, but he won plaudits from the West for his independent stance against Soviet control, including a knighthood from Queen Elizabeth II. However, he pursued a strict Stalinist approach and the remorseless industrialization and urbanization programs of the 1970s caused severe debt. In the 1980s, he cut imports and diverted output to exports. Self-sufficiency turned to subsistence and shortages, with savage rationing of food and energy.

Ceaucescu's building schemes desecrated some of the country's finest architecture and demolished thousands of villages. In December 1989, mass antigovernment demonstrations were held in Timisoara with protests across Romania. Security forces fired on crowds, causing many deaths. But after army units joined the protests, Nicolae Ceaucescu and his wife Elena fled from Bucharest on December 22 . Both were executed on Christmas Day on charges of genocide and corruption. A provisional government of the National Salvation Front (NSF), took control, much of the old administrative apparatus was dismantled, and the Communist Party was dissolved.

POLITICS

In May 1990, under Ion Iliescu, the NSF won Romania's first free elections since World War II, a result judged to be flawed

ECONOMY

According to the World Bank, Romania is a "lower-middle-income" economy. Natural gas and petroleum are the chief mineral resources and the aluminum, copper, lead and zinc industries use domestic supplies. Manufactures include cement, processed food, petroleum products, textiles, and wood. Agriculture employs nearly a third of the workforce. Crops include fruits, corn, potatoes, sugarbeet, and wheat. Sheep are the chief livestock.

BUCHAREST (BUCURESTI)

Capital and largest city of Romania, on the River Dimbovita, southern Romania. Founded in the 14th century on an important trade route, it became capital in 1862 and was occupied by Germany in both World Wars. It is an industrial, commercial and cultural center. The seat of the patriarch of the Romanian Orthodox Church, it has notable churches, museums, and galleries.

RUSSIA

In August 1991, Russia's traditional flag, which had first been used in 1699, was restored as Russia's national flag. It uses colors from the flag of the Netherlands. This flag was suppressed when Russia was part of the Soviet Union.

The Russian Federation is the world's largest country. About 25% lies west of the Ural Mountains (Uralskie Gory) in European Russia, where 80% of the population lives. It is mostly flat or undulating, but the land rises to the Caucasus Mountains in the south, with Russia's highest peak, Elbrus (18,481 ft [5,633 m]). Siberia, contains vast plains and plateaus, with mountains in the east and south. The Kamchatka peninsula in the far east has many active volcanoes. Russia contains many of the world's longest rivers, including Yenisey-Angara and the Ob-Irtysh. It also includes part of the world's largest inland body of water, the Caspian Sea, and Lake Baikal, the world's deepest lake.

Area 6,592,812 sq mi [17,075,400 sq km]
Population 143,782,000
Capital (population) Moscow (8,297,000)
Government Federal multiparty republic
Ethnic groups Russian 82%, Tatar 4%, Ukrainian 3%, Chuvash 1%, more than 100 others
Languages Russian (official), many others
Religions Russian Orthodox, Islam Judaism
Currency Russian ruble = 100 kopeks
Website www.kremlin.ru/eng

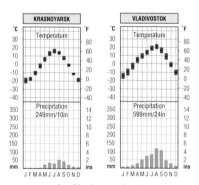

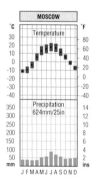

CLIMATE

The Moscow climate is continental with cold and snowy winters and warm summers. While Krasnoyarsk in south-central Siberia has a harsher, drier climate, it is not as severe as parts of northern Siberia.

HISTORY

In the 9th century AD, a state called Kievan Rus was formed at the junction of the forest and the steppe in what is now Ukraine. Other states then formed further to the north and all were eventually united under the principality of Muscovy. In the 13th century, Mongol armies from the east penetrated the forests and held sway over the Slavic peoples there. It was only in the 16th century that the Mongol yoke was thrown off as the Slavs, under Ivan the Terrible (1530–84), began to advance across the steppes.

There began a period of expansion from the core area of Slavic settlement to the south, east, and west. Expansion across Siberia was rapid and the first Russian settlement on the Pacific, Okhotsk, was established in 1649. By 1696, Azov, the key to the Black Sea, was secured. A series of struggles in the 17th and 18th centuries against the Swedes and the Poles resulted in the addition of the Gulf of Finland, the Baltic coast, and part of Poland to the growing Russian

Empire, while, in the 19th century, the Caucasus, Central Asia, and new territories in the Far East were added.

The Russian Revolution took place in 1917; Tsar Nicholas II was forced to abdicate and a Bolshevik (Communist) government was established under Vladimir Ilyich Lenin (1870–1924). The Union of Soviet Socialist Republics (the USSR or the Soviet Union) was established in 1922.

POLITICS

From 1924 Joseph Stalin introduced a socialist economic program, suppressing all opposition both within the Party and among the population. His authority was consolidated with the Great Purge, a period of widespread arrests and executions which reached its peak in 1937. His introduction of five-year plans and collective farming transformed the USSR from a largely peasant society to a major world industrial power by the end of the 1930s. Collectivization was violently resisted by many peasants, resulting in millions of casualties from famine and mass repression of peasants (*kulaks*) by the authorities. After Stalin's death, the Soviet leaders modified some policies, but remained true to the principles of Communism until Mikhail Gorbachev changed the face of Russia in the 1980s.

The USSR joined World War II (the Great Patriotic War) in June 1941. The armed forces of the USSR inflicted about 80% of losses suffered by German land forces in World War II (about 3 million soldiers). The USSR suffered enormous losses with 25 million dead.

Under Soviet rule, changes took place in the distribution of the population so that the former pattern of a small highly populated core and "empty" periphery began to break down. As a result, a far higher proportion of the Russian population lives east of the Ural Mountains than before the Revolution. The redistribution was actively encouraged by a regime committed to developing the east. Migration to the towns and cities has also been marked and by 1997 73% of the population lived in cities and towns.

In the 1980s, Mikhail Gorbachev sought to introduce economic and political reforms necessitated by the failures of Communist economic policies. This was a time of *glasnost* (openness) and *perestroika* (restructuring). The Soviet Union broke up in December 1991. Russia maintained contact with 11 of the 15 former Soviet republics through a loose confederation called the Commonwealth of Independent States (CIS).

Despite Gorbachev's brave efforts at reform, his successor Boris Yeltsin inherited an economy in crisis. The abolition of price controls sent the cost of basic commodities rocketing, there were food

MOSCOW (MOSKVA)

Capital of Russia and largest city in Europe, on the River Moskva. Inhabited since Neolithic times, but not recorded until 1147. It was a principality by the end of the 13th century, and in 1367 the first stone walls of the Kremlin were constructed. By the end of the 14th century, Moscow emerged as the focus of Russian opposition to the Monguls. Moscow was the capital of the Grand Duchy of Russia from 1547 to 1712, when the capital moved to St. Petersburg. Occupied by Napoleon in 1812, who was forced to flee when the city burned to the ground. In 1918, following the Russian Revolution, it became the capital of the Soviet Union. The failure of the German army to seize the city in 1941 was the Nazis' first major setback in World War II. The Kremlin is the center of the city, and the administrative heart of the country. Adjoining it are Red Square, the Lenin Mausoleum, and the 16th-century cathedral of St. Basil the Blessed.

shortages and rising unemployment. Despite these difficulties, including rising corruption and crime, the government's program of reforms was supported in a 1993 referendum and Yeltsin returned as president in July 1996.

Yeltsin resigned on 31 December 1999 due to poor health and appointed the prime minister Vladimir Putin as the acting president. Putin, was elected president by a landslide in March 2000.

Fighting began in the secessionist Chechnya during the 1990s and flared up into full-scale war in 1999. The conflict slowed in 2000, but Russia faced a new threat, namely bombings of its cities by Chechen terrorists. After the attacks on the United States on September 11, 2001, Putin and President George W. Bush found common cause in the campaign against international terrorism and the assault on the Taliban in Afghanistan, though relations soured when Russia opposed the attack on Iraq in 2003. The

secessionist conflict in Chechnya mounted, and the occupation of a school by Muslim extremists in 2004 which led to more than 300 deaths caused international outrage.

ECONOMY

In the early 1990s the World Bank described Russia as a "lower-middle-income" economy. In 1997 Russia was admitted to the Council of Europe, the same year Russia attended the G7 summit, suggesting that it was now counted among the world's leading economies. Industry is the chief activity and light industries producing consumer goods are becoming important. Russia's resources include petroleum and natural gas, coal, timber, metal ores, and hydroelectric power. Russia is a major producer of farm products. Major crops include barley, flax, fruit, oats, rye, potatoes, sugarbeet, sunflower seeds, vegetables, and wheat.

RWANDA

Rwanda's new flag was adopted in 2002. The color blue is used to symbolize peace and tranquility. Yellow represents wealth as the country works to achieve sustainable economic growth, while green denotes prosperity, work, and productivity. The 24-ray golden sun symbolizes new hope.

The Republic of Rwanda is Africa's most densely populated country. It is a small state in the heart of Africa. The western border is formed by Lake Kivu and the River Ruzizi. Rwanda has a rugged landscape, dominated by high, volcanic mountains, rising to Mount Karisimbi, at 14,787 ft [4,507 m]. The capital, Kigali, stands on the central plateau. East Burundi consists of stepped plateaus, which descend to the lakes and marshland of the Kagera National Park on the Tanzania border.

The lush rainforests in the west are one of the last refuges for the mountain gorilla. Many of Rwanda's forests have been cleared and 35% of the land is now arable. The steep mountain slopes are intensively cultivated. Despite contour ploughing, heavy rains cause severe soil erosion.

Area 10,169 sq mi [26,338 sq km]
Population 7,954,000
Capital (population) Kigali (234,000)
Government Republic
Ethnic groups Hutu 84%, Tutsi 15%, Twa 1%
Languages French, English and Kinyarwanda (all official)
Religions Roman Catholic 57%, Protestant 26%, Adventist 11%, Islam 5%
Currency Rwandan franc = 100 centimes
Website www.gov.rw

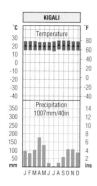

CLIMATE

Temperatures in Kigali are moderated by the altitude. The rainfall is abundant, but much heavier rain falls on the western mountains. The dry season is June–August. The floor of the Great Rift Valley is warmer and drier than the rest of the country.

HISTORY

The Twa, a pygmy people, were the first known people to live in Rwanda. About 1,000 years ago, a farming people, the Hutu, settled in the area, gradually displacing the Twa.

From the 15th century, a cattle-owning people from the north, the Tutsi, began to dominate the Hutu, who had to serve the Tutsi overlords.

By the late 18th century, Rwanda and Burundi formed a single Tutsi-dominated state, ruled by a King (Mwami). In 1890, Germany conquered the area and subsumed it into German East Africa. During World War I, Belgian forces occupied (1916) both Rwanda and Burundi. In 1919, it became part of the Belgian League of Nations mandate territory of Ruanda-Urundi (which in 1946 became a UN Trust Territory). The Hutu majority intensified their demands for political representation. In 1959, the Tutsi Mwami died. The ensuing civil war between Hutus and Tutsis claimed more than 150,000 lives. Hutu victory led to a mass exodus of Tutsis. The Hutu Emancipation Movement, led by Grégoire Kayibanda, won the 1960 elections. In 1961, Rwanda declared itself a republic. Belgium granted independence in 1962, and Kayibanda became president.

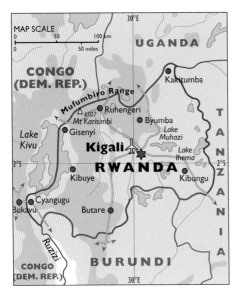

region. In 1997, Rwandan troops supported Laurent Kabila's successful overthrow of President Mobutu in Zaïre. Kabila failed to expel the Hutu militia from Congo, and Rwanda switched to supporting rebel forces. In 1998, the UN International Criminal Tribunal sentenced Rwanda's former prime minister Jean Kambanda to life imprisonment for genocide. Paul Kagame became president in 2000. He was reelected in 2003.

In the early 21st century, prosecutions began in both Belgium and Tanzania of people accused of genocide. Rwanda finally withdrew from DR Congo in late 2002 after signing a peace deal with Kinshasa.

ECONOMY

According to the World Bank, Rwanda is a "low-income" developing country. Agriculture employs 90% of the workforce, but many farmers live at subsistence level. Chief food crops include bananas, beans, cassava, plantains, potatoes, sorghum, and sweet potatoes. Some farmers raise cattle and other livestock. The chief cash crop is coffee, also the leading export, followed by tea, hides and skins. Rwanda also produces pyrethrum, which is used to make insecticide. The country produces some cassiterite (tin ore) and wolframite (tungsten ore). Manufacturing is small-scale and include beverages, cement, and sugar.

POLITICS

Rwanda was subject to continual Tutsi incursions from Burundi and Uganda. In 1973, Major General Habyarimana overthrew Kayibanda in a military coup. In 1978, Habyarimana became president. Drought devastated Rwanda in the 1980s. More than 50,000 refugees fled to Burundi. In 1990, the Tutsi-dominated Rwandan Patriotic Front (RPF) invaded Rwanda, forcing Habyarimana to adopt a multiparty constitution. In April 1994, Habyarimana and the president of Burundi died in a rocket attack. The Hutu army and militia launched an act of genocide against the Tutsi minority, massacring more than 800,000 of them. In July 1994, an RPF offensive toppled the government, creating 2 million Hutu refugees. A government of national unity, comprising Tutsis and Hutus, emerged. More than 50,000 people died in refugee camps in eastern Zaïre (now DR Congo). Hutu militia controlled the camps, their leaders facing prosecution for genocide. The sheer number of refugees (1995, 1,000,000 in Zaïre and 500,000 in Tanzania) destabilized the

KIGALI

Capital city lying in the center of Rwanda. It was a trade center under German and Belgian colonial administration, becoming the capital when Rwanda achieved independence in 1962. Industries include tin mining, cotton, and coffee.

SAUDI ARABIA

Saudi Arabia's flag was adopted in 1938. It is the only national flag with an inscription as its main feature. The Arabic inscription above the sword means "There is no God but Allah, and Muhammad is the Prophet of Allah."

The Kingdom of Saudi Arabia occupies about three-quarters of the Arabian Peninsula in southwest Asia. The land is mostly desert and includes the largest expanse of sand in the world, the Rub' al Khali (Empty Quarter), covering an area of 250,000 sq mi [647,500 sq km]. Mountains to the west border the Red Sea plains.

Area 829,995 sq mi [2,149,690 sq km]
Population 25,796,000
Capital (population) Riyadh (3,000,000)
Government Absolute monarchy
Ethnic groups Arab 90%, Afro-Asian 10%
Languages Arabic (official)
Religions Islam 100%
Currency Saudi riyal = 100 halalas
Website www.saudinf.com

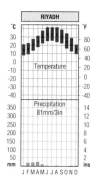

CLIMATE

Saudi Arabia has a hot, dry climate. In the summer, the temperatures are extremely high and often exceed 104°F [40°C], though the nights are cool.

HISTORY

Saudi Arabia contains the two holiest places in Islam—Mecca, the birthplace of the Prophet Muhammad in 570, and Medina where Muhammad and his followers went in 622.

In the mid-15th century, the Saud Dynasty established control over a small area near present-day Riyadh. In the mid-18th century an alliance was established with a religious leader, Muhammad Ibn Abd al-Wahhab, who wanted to restore strict observance of Islam. The Wahhabi movement swept across Arabia and the Saud family took over areas converted to the Wahhabi beliefs. By the early 19th century, they had taken Mecca and Medina. The Ottoman governor of Egypt attacked to halt their expansion and by the late 19th century, most of the

Arabian Peninsula was under the rule of Ottoman Turks.

In 1902 Abd al-Aziz Ibn Saud led a force from Kuwait, where he had been living in exile, and captured Riyadh. From 1906, the Saud family gradually won control over the territory held by their ancestors and extended their land following the defeat of the Ottoman Empire in World War I. After further conquests in the 1920s, Ibn Saud proclaimed the country the Kingdom of Saudi Arabia in 1932.

POLITICS

The first major petroleum discovery was made in 1938, and full-scale production began after World War II. Saudi Arabia eventually became the world's leading petroleum exporter and highly influential in the Arab world where it played a major role in supplying development aid.

Saudi Arabia supported Egypt, Jordan and Syria in the Six Day War against Israel, in 1967. It did not send troops, but gave aid to the Arab combatants.

King Fahd suffered a stroke in 1995 and appointed his half-brother, Crown Prince Abdullah Ibn Abdulaziz, to act on his behalf. Fahd died in 2005 and Abdullah succeeded him as king.

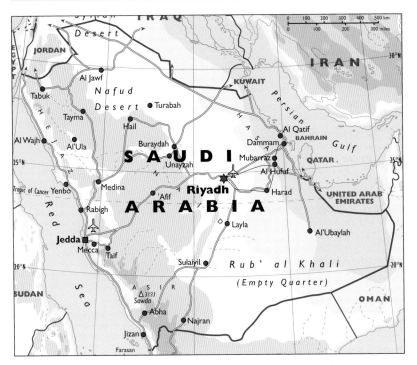

Although assisted by a Consultative Council, the monarch holds executive and legislative powers and is also the imam (supreme religious ruler). Saudi Arabia is an absolute monarchy with no formal constitution.

Despite its support of Iraq against Iran in the First Gulf War in the 1980s, Saudi Arabia asked for the protection of Western forces against possible Iraqi aggression following the invasion of Kuwait in 1990. In 1991, the country played a significant role in the quick victory over Iraq's Saddam Hussein. Relations between Saudi Arabia and the US became strained following the terrorist attacks on the US on September 11, 2001, in part because Osama bin Laden and many of his followers were Saudi-born. Saudi authorities denounced the attacks and severed relations with Afghanistan's Taliban regime. In 2003 and 2004, Saudi Arabia was hit by Islam-ic attacks. The government held nation-wide municipal elections in 2005, its first exercise in democracy. However, political parties are banned and activists who publicly broach the subject of reform risk jail.

ECONOMY

Saudi Arabia has about 25% of the world's known petroleum reserves, and petroleum and petroleum products make up 85% of its exports.

RIYADH

Capital of Saudi Arabia, in the east-central part of the country, 235 mi [380 km] inland from the Persian Gulf. In the early 19th century, it was the domain of the Saud dynasty, becoming capital of Saudi Arabia in 1932. The chief industry is petroleum refining.

SENEGAL

This flag was adopted in 1960 when Senegal became independent from France. It uses the three colors that symbolize African unity. It is identical to the flag of Mali, except for the five-pointed green star. This star symbolizes the Muslim faith of most of the people.

The Republic of Senegal is situated on the northwest coast of Africa. The volcanic Cape Verde (Cap Vert), on which Dakar stands, is the most westerly point in Africa. The country entirely surrounds Gambia. The Atlantic coastline from St. Louis to Dakar is sandy. Plains cover most of Senegal, though the land rises gently in the southeast. The north forms part of the Sahel. The main rivers are the Sénégal, which forms the north border, and the Casamance in the south. The River Gambia flows into the Gambia.

Desert and semidesert cover northeast Senegal. In central Senegal, dry grasslands and scrub predominate. Mangrove swamps border parts of the south coast. The far south is a region of tropical savanna, though large areas have been cleared for farming. Senegal has several protected parks, the largest is the Niokolo-Kobo Wildlife Park.

Area 75,954 sq mi [196,722 sq km]
Population 10,852,000
Capital (population) Dakar (880,000)
Government Multiparty republic
Ethnic groups Wolof 44%, Pular 24%, Serer 15%
Languages French (official), tribal languages
Religions Islam 94%, Christianity (mainly Roman Catholic) 5%, traditional beliefs 1%
Currency CFA franc = 100 centimes
Website www.senegalembassy.co.uk

CLIMATE

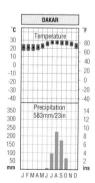

Dakar has a tropical climate, with a short rainy season between June and September when moist winds blow from the southwest. Temperatures are higher inland. Rainfall is greatest in the south.

HISTORY

From the 6th to the 10th century, Senegal formed part of the Empire of ancient Ghana. Between the 10th and 14th centuries, the Tukolor state of Tekrur dominated the Sénégal valley. The Almoravid dynasty of Zenega Berbers introduced Islam and it is from the Zenega that Senegal got its name. In the 14th century, the Wolof established the Jolof Empire. The Songhai Empire began to dominate the region.

In 1444, Portuguese sailors became the first Europeans to reach Cape Verde. Trading stations were rapidly established in the area. In the 17th century, France and the Netherlands replaced Portuguese influence. France gradually gained control of the valuable slave trade and founded St. Louis in 1658. By 1763 Britain had expelled the French from Senegal and in 1765, they set up

THE SAHEL

A band of semiarid scrub and savanna grassland in Africa to the south of the Sahara. Over the past 30 years the Sahara has encroached on the northern Sahel in the world's most notorious example of desertification.

Senegambia, the first British colony in Africa. In 1783 France regained control and in the mid-19th century, battled for control of the interior. The French founded Dakar in 1857. In 1895, Senegal became a French colony within the Federation of French West Africa. In 1902, the capital of this huge empire transferred from St. Louis to Dakar. Dakar became a major trading center. In 1946, Senegal joined the French Union.

POLITICS

In 1959, Senegal joined French Sudan (now Mali) to form the Federation of Mali. Senegal withdrew in 1960 to become the separate Republic of Senegal, within the French community. Its first post-colonial president, Léopold Sédar Senghor, was a noted African poet.

Following an unsuccessful coup in 1962, Senghor gradually assumed wider powers. During the 1960s Senegal's economy deteriorated and a succession of droughts caused starvation and widespread civil unrest.

During the 1970s southern Senegal was a base for guerrilla movements in Guinea and Portuguese Guinea (modern Guinea-Bissau). In 1974, Senegal was a founding member of the West African Economic Community.

Senghor continued in office until 1981, when he was succeeded by the prime minister, Abdou Diouf. In that same year, Senegalese troops suppressed a coup in the Gambia. In 1982 the two countries joined to form the Confederation of Senegambia, but the union collapsed in 1989. From 1989 to 1992, Senegal was at war with Mauritania.

In 2000, Diouf was surprisingly beaten in presidential elections by veteran opposition leader Abdoulaye Wade of the Senegalese Democratic Party, ending 40 years of Socialist Party rule. In 2001, the government signed a peace treaty with the separatist rebels in the southern Casamance province.

ECONOMY

According to the World Bank, Senegal is a "lower-middle-income" developing country. Agriculture still employs 65% of the population, though many farmers produce little more than they need to feed their families. Food crops include cassava, millet and rice. Senegal is the world's sixth largest producer of peanuts. Phosphates are the chief resource, and Senegal also refines petroleum imported from Gabon and Nigeria. Fishing is important.

DAKAR

Capital and largest city of Senegal. Founded in 1857 as a French fort, the city grew rapidly with the arrival of a railroad (1885). A major Atlantic port, it later became capital of French West Africa. There is a Roman Catholic cathedral and a presidential palace. Dakar has excellent educational and medical facilities, including the Pasteur Institute. Industries include textiles, petroleum refining, and brewing.

SERBIA AND MONTENEGRO

The tricolor flag uses the Pan-Slavic colors of blue, white and red. These colors derive from the 19th century flag of Russia.

Serbia and Montenegro were part of Yugoslavia, which broke up in the early 1990s. In 2003, the two republics became semi-independent and adopted the name of the Union of Serbia and Montenegro.

The country has a short coastline along the Adriatic Sea. Inland is a mountainous region including the Dinaric Alps and part of the Balkan Mountains. The Pannonian Plains, which are drained by the River Danube, are in the north.

Area 39,449 sq mi [102,173 sq km]
Population 10,826,000
Capital (population) Belgrade (1,594,000)
Government Federal republic
Ethnic groups Serb 62%, Albanian 17%, Montenegrin 5%, Hungarian 3%, others
Languages Serbian (official), Albanian
Religions Orthodox 65%, Islam 19%, others
Currency New dinar = 100 paras
Website www.info.gov.yu/start.php?jezik=e
www.srbija.sr.gov.yu/?change_lang=en
www.montenegro.yu/english/naslovna/index.htm

CLIMATE

The coast has a mild, Mediterranean climate, but the inland upland areas have a more continental climate. The highlands have cold, snowy winters, while the Pannonian Plains have hot, arid summers, with heavy rains in the spring and fall.

HISTORY

South Slavs began to move into the region around 1,500 years ago. Each group founded its own state, but by the 15th century Serbia and Montenegro were under the Turkish Ottoman Empire.

In 1914, Austria-Hungary declared war on Serbia, blaming it for the assassination of Archduke Franz Ferdinand of Austria-Hungary. This led to World War I and the defeat of Austria-Hungary.

In 1918, the South Slavs united in the Kingdom of the Serbs, Croats and Slovenes. In 1929, King Alexander abolished the constitution and renamed the country Yugoslavia. Ruling as a dictator, he sought to enforce the use of one language, Serbo-Croatian. His new political divisions failed to acknowledge the historic boundaries determined by the ethnic groups so the unity of the new state was under constant threat from nationalist and ethnic tensions. After the Germans invaded in 1941, Yugoslavs fought the Germans and themselves. The Communist-led partisans of Josip Broz Tito (a Croat) emerged victorious in 1945.

POLITICS

From 1945, the Communists controlled the country, then called the Federal People's Republic of Yugoslavia. But after Tito's death in 1980, the country faced many problems. Yugoslavia split in 1991–2 with Bosnia-Herzegovina, Croatia, Macedonia, and Slovenia each proclaiming their independence. Serbia and Montenegro became the new Yugoslavia.

Fighting broke out in Croatia and Bosnia-Herzegovina as rival groups struggled for power. In 1992, the United

stepped up attacks on Albanian-speaking villages, forcibly expelling the people, who fled into Albania and Macedonia. The NATO offensive ended when Serbian forces withdrew from Kosovo and the KLA was disbanded. In 2000, the Yugoslav leader Slobodan Milosevic was defeated in presidential elections and, in February 2002, he faced charges, including crimes against humanity, at the UN War Crimes Tribunal in The Hague.

Serbia and Montenegro signed an agreement in 2002 that renamed the country the Union of Serbia and Montenegro and made both states semi-independent.

ECONOMY

Resources include bauxite, coal, copper and other metals, together with petroleum and natural gas. Manufactures include aluminum, machinery, plastics, steel, textiles and vehicles. Agriculture remains important.

Nations withdrew recognition of Yugoslavia because of its failure to halt atrocities committed by Serbs living in Croatia and Bosnia-Herzegovina. In 1995, Yugoslavia took part in talks that led to the Dayton Peace Accord, but it had problems of its own as stringent international sanctions struck the war-ravaged economy.

In 1998, the fragility of the region was again highlighted in Kosovo, a former autonomous region in southern Serbia where most people are Albanian-speaking Muslims. Serbs forced Muslim Albanians to leave their homes, but they were opposed by the Kosovo Liberation Army (KLA). The Serbs hit back and thousands of civilians fled for their lives.

In March 1999, after attempts to find an agreement had failed, NATO forces intervened by launching aerial attacks on administrative and industrial targets in Kosovo and Serbia. Serbian forces

BELGRADE (BEOGRAD)

Capital of Serbia and Montenegro, at the confluence of the Sava and Danube rivers. Belgrade became capital of Serbia in the 12th century, but fell to the Ottoman Turks in 1521. Freed from Ottoman rule in 1867, it became capital of the newly created Yugoslavia in 1929. The city suffered much damage under German occupation in World War II. In 1999 Belgrade was further damaged by Allied air strikes after Milosevic sent federal troops into Kosovo. In October 2000, more than 300,000 people marched through the streets of Belgrade, forcing Milosevic to step down as president.

SIERRA LEONE

The green of the flag represents the nation's agriculture and its lush mountain slopes. Blue stands for the waters of the Atlantic that lap Sierra Leone's coast. White symbolizes the desire for peace, justice, and unity

The Republic of Sierra Leone on the west coast of Africa is about the same size as the Republic of Ireland. The coast contains several deep estuaries in the north, with lagoons in the south. The most prominent feature is the mountainous Freetown (or Sierra Leone) peninsula. North of the peninsula is the River Rokel estuary, west Africa's best natural harbor. Behind the coastal plain, the land rises to mountains, with the highest peak, Loma Mansa, reaching 6,391 ft [1,948 m].

Swamps cover large areas near the coast. Inland, much of the rainforest has been destroyed. The north is largely covered by tropical savanna.

Area 27,699 sq mi [71,740 sq km]
Population 5,884,000
Capital (population) Freetown (470,000)
Government Single-party republic
Ethnic groups Native African tribes 90%
Languages English (official), Mende, Temne, Krio
Religions Islam 60%, traditional beliefs 30%, Christianity 10%
Currency Leone = 100 cents
Website www.visitsierraleone.org

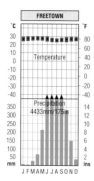

CLIMATE

The climate is tropical, with heavy rainfall. In the north, it is dry between December and March. In the south, it is dry in January and February.

HISTORY

Portuguese sailors reached the coast in 1460. In the 16th century, the area became a source of slaves. Freetown was founded in 1787 as a home for freed slaves. In 1808, the settlement became a British Crown Colony. The interior was made a Protectorate in 1896 and in 1951, the Protectorate and Colony united.

Sierra Leone gained independence in 1961 and in 1971 became a republic.

POLITICS

A 1991 referendum voted for the restoration of multiparty democracy, but the military seized power in 1992. A civil war raged between government forces and the Revolutionary United Front (RUF). The RUF fought to end foreign interference and to nationalize the diamond mines. After 1996 elections,

FREETOWN

Capital and chief port of Sierra Leone, west Africa. First explored by the Portuguese in the 15th century and visited by Sir John Hawkins in 1562. Freetown was founded by the British in 1787 as a settlement for freed slaves from England, Nova Scotia, and Jamaica. It was the capital of British West Africa (1808–74). West Africa's oldest university, Fourah Bay, was founded here in 1827. Freetown was made capital of independent Sierra Leone in 1961. Industries include platinum, gold, diamonds, petroleum refining, and palm oil.

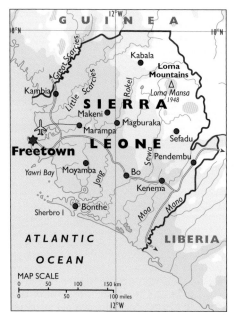

Ahmad Tejan Kabbah led a civilian government. In 1997, Major Johnny Paul Koroma seized power in a military coup. The Economic Community of West African States (ECOWAS) imposed sanctions, and Nigeria led an intervention force that restored Kabbah as president in 1998.

A 1999 peace treaty and the arrival of UN peacekeeping forces seemed to signal an end to the civil war, but in 2000 RUF rebels, led by Foday Sankoh and backed by Liberia, abducted UN troops and renewed the war. British soldiers arrived to bolster the UN peacekeeping effort. Disarmament continued throughout 2001 through a UN-brokered peace plan. Sankoh was captured and, in 2002, the war, which had left about 50,000 people dead, appeared to be over. Rebel raids from Liberia in 2003 failed to disturb the country's fragile peace. Stability was gradually restored and, in late 2005, the last contingent of UN soldiers left the country.

ECONOMY

The World Bank classifies Sierra Leone among the "low-income" economies. Agriculture provides a living for 70% of the workforce, though farming is mostly at subsistence level. Food crops include cassava, corn, and rice, the staple food. Export crops include cocoa and coffee. The most valuable exports include diamonds, bauxite, and rutile (titanium ore).

ROKEL RIVER

The Rokel River rises in the Guinea Highlands in northern Sierra Leone. It drains a 4,100 sq mi [10,620 sq km] basin and flows for 250 mi [400 km] in the direction of the Atlantic, emptying into the River Rokel estuary, the best natural harbor in West Africa.

DIAMOND

Crystalline form of carbon (C). The hardest natural substance known, it is found in kimberlite pipes and alluvial deposits. Appearance varies according to impurities. Bort (inferior in crystal and color), carborondo (an opaque gray to black variety), and other non-gem varieties are used in industry. Industrial diamonds are used as abrasives, bearings in precision instruments such as watches, and in the cutting heads of drills for mining. Synthetic diamonds, made by subjecting graphite, with a catalyst, to high pressure and temperatures of 5,400°F [3,000°C] are fit only for industry. Diamonds are weighed in carats (0.2gm) and points (¹⁄₁₀₀ carat).

SINGAPORE

Singapore's flag was adopted in 1959 and it was retained when Singapore became part of the Federation of Malaysia in 1963. The crescent stands for the nation's ascent. The stars stand for Singapore's aims of democracy, peace, progress, justice, and equality.

The Republic of Singapore is an island country at the southern tip of the Malay Peninsula. It consists of the large Singapore Island and 59 small islands, 20 of which are inhabited.

Singapore Island is 26 mi [42 km] wide and 14 mi [28 km] across. It is linked to the peninsula by a 3,465 ft [1,056 m] long causeway. The land is mostly low-lying; the highest point, Bukit Timah, is only 577 ft [176 m] above sea level. Its strategic position, at the convergence of some of the world's most vital shipping lanes, ensured its growth.

Rainforest once covered Singapore, but forests now grow on only 5% of the land. Today, about 50% of Singapore is built up. The distinction between island and city has all but disappeared. Most of the rest consists of open spaces, including parks, granite quarries, and inland waters. Farmland covers 4% of the land, and plantations of permanent crops make up 7%.

Area 264 sq mi [683 sq km]
Population 4,354,000
Capital (population) Singapore City (3,894,000)
Government Multiparty republic
Ethnic groups Chinese 77%, Malay 14%, Indian 8%
Languages Chinese, Malay, Tamil and English (all official)
Religions Buddhism, Islam, Hinduism, Christianity
Currency Singaporean dollar = 100 cents
Website www.gov.sg

Temasak ("sea town"), but was named Singapura ("city of the lion") when an Indian prince thought he saw a lion there. Singapore soon became a busy trading center within the Sumatran Srivijaya kingdom. Javanese raiders destroyed it in 1377. Subsumed into Johor, Singapore became part of the powerful Malacca sultanate.

In 1819, Sir Thomas Stamford Raffles, agent of the British East India Company, made a treaty with the Sultan of Johor which allowed the British to build a settlement on Singapore Island. In 1826,

CLIMATE

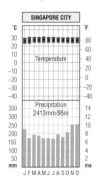

Singapore has a hot, humid equatorial climate, with temperatures averaging 86°F [30°C]. Total average annual rainfall is, 95 in [2,413 mm], with Rain occuring (on average) 180 days each year.

SIR THOMAS STAMFORD RAFFLES (1781–1826)

British colonial administrator, founder of Singapore City. When Java returned to Dutch rule (1816), Raffles bought the island of Singapore for the British East India Company (1819). Under his guidance, it developed rapidly into a prosperous free port.

HISTORY

According to legend, Singapore was founded in 1299. It was first called

Singapore, Penang (now Pinang), and Malacca formed the Straits Settlement. Singapore soon became the most important British trading center in Southeast Asia, and the Straits Settlement became a Crown Colony in 1867. Despite British defensive reinforcements in the early 20th century, Japanese forces seized the island in 1942.

POLITICS

British rule returned in 1945. In 1946, the Straits Settlement dissolved and Singapore became a separate colony. In 1959, Singapore achieved self-government. Following a referendum in 1963, Singapore merged with Malaya, Sarawak, and Sabah to form the Federation of Malaysia. In 1965, Singapore broke away from the Federation to become an independent republic within the Commonwealth of Nations.

The People's Action Party (PAP) has ruled Singapore since 1959. Its leader, Lee Kuan Yew, served as prime minister from 1959 until 1990, when he resigned and was succeeded by Goh Chok Tong. Under the PAP, the economy has expanded rapidly, although some people consider that the PAP's rule has been dictatorial and oversensitive to criticism. In 2004, Lee Hsien Loong, eldest son of Lee Kuan Yew, succeeded Goh Chok Tong as prime minister and called for a more open socie-ty. He also called for more people to marry and have babies, because of the country's falling birth rate.

ECONOMY

The World Bank classifies Singapore as a "high-income" economy. It is one of the world's fastest growing (tiger) economies. Historically, Singapore's economy has been based on transshipment, and this remains a vital component. It is one of the world's busiest ports, annually handling more than 290 million tonnes of cargo. Post-1945 the economy diversified. Singapore has a highly skilled and productive workforce. The service sector employs 65% of the workforce; banking and insurance provide many jobs.

Manufacturing is the largest export sector. Industries include computers and electronics, telecommunications, chemicals, machinery, scientific instruments, ships, and textiles. It has a large petroleum refinery. Agriculture is relatively unimportant. Most farming is highly intensive, and farmers use the latest technology and scientific methods.

SINGAPORE CITY

The capital of Singapore, on Singapore Island, the largest island in the Republic of Singapore, at the mouth of the Singapore River. The city is home to an ethnic mix of Chinese, Malaysians, and Indians with English the main language. It has a very high standard of living due to its very healthy export-based economy. Tourism is one of the largest industries with attractions including the Singapore Zoological Gardens and the Jurong Bird Park, not to mention the Orchard Road area which is the shopping and entertainment center.

323

SLOVAK REPUBLIC

The flag uses the typical red, white, and blue Slavic colors. The coat of arms is taken from part of the Hungarian arms and shows a double cross set on three hills to commemorate the arrival of Christianity in the Carpathian region in the 9th century.

The Slovak Republic (Slovakia), is a predominantly mountainous country; part of the Carpathian system that divides the Slovak Republic from Poland is found in the north. The highest peak (Gerlachovka 8,711 ft [2,655 m]) is in the scenic Tatra (Tatry) Mountains on the Polish border.

Forests cover much of the mountain slopes and there is also extensive pasture. The southwestern Danubian lowlands form a fertile lowland region. The Danube forms part of the southern border with Hungary.

Area 18,924 sq mi [49,012 sq km]
Population 5,424,000
Capital (population) Bratislava (449,000)
Government Multiparty republic
Ethnic groups Slovak 86%, Hungarian 11%
Languages Slovak (official), Hungarian
Religions Roman Catholic 60%, Protestant 8%, Orthodox 4%, others
Currency Slovak koruna = 100 halierov
Website www.government.gov.sk/english

CLIMATE

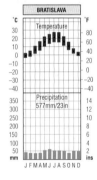

Slovakia has a transitional climate, in between the mild conditions of western Europe and the continental conditions of Russia to the east. The conditions in Kosice, eastern Slovakia are fairly typical. Temperatures can range from 27°F [–3°C] in January to 68°F [20°C] in July. Kosice has an average annual rainfall of 24 in [600 mm]. The mountains have a more extreme climate, with snow or rain throughout the year.

HISTORY

Slav peoples settled in the region in the 5th century AD. In the 9th century, the region, together with Bohemia and Moravia in what is now the Czech Republic, became part of the Greater Moravian Empire. Hungarians conquered this empire in 907 and ruled for nearly a thousand years. Religious wars in the 15th century led many Czech nobles to settle in what is now the Slovak Republic. Hungary was defeated by the Turkish Ottomans in 1526 and, soon afterward, the Ottomans occupied much of eastern and central Hungary. As a result, the center of Hungarian power shifted into Slovakia.

Slovak nationalism developed from the late 18th century, but it was kept in check by the Hungarians who enforced "Magyarization." In 1867, Hungary and Austria were united to form the dual monarchy of Austria-Hungary. At the end of World War I, Austria-Hungary collapsed and the Czechs and Slovaks united to form a new nation called Czechoslovakia. In 1938, Hungary forced Czechoslovakia to give up several areas with large Hungarian populations. These areas included Kosice in the east.

In 1939, fearing that it might be divided up between Germany, Poland, and Hungary, Slovakia declared itself independent, but the country was then conquered by Germany. At the end of World War II, Slovakia again became part of

Czechoslovakia. Communists seized control in 1948. In the late 1960s, many Czechs and Slovaks, led by Alexander Dubcek, tried to reform the Communist system. This movement, known as "the Prague Spring," was put down in 1968 by Soviet troops. Demands for democracy reemerged in the 1980s, when Soviet leader Mikhail Gorbachev launched a series of reforms in the USSR.

POLITICS

At the end of November 1989, Czechoslovakia's parliament abolished the Communist Party's sole right to govern. In December, the head of the Communists, Gustáv Hável, resigned. Non-Communists led by the playwright and dissident Václav Havel formed a new government, they then won a majority in the elections of June 1990.

In the elections of 1992, the Movement for Democratic Slovakia, led by Vladimir Meciar, campaigned for Slovak independence and won a majority in Slovakia's parliament. The Slovak National Council then approved a new constitution for the Slovak Republic, which came into existence on 1 January 1993.

The Slovak Republic became a member of the OECD in 1997 and maintained close contacts with its former partner. Slovak independence raised national aspirations among the Magyar-speaking community. Relations with Hungary were not helped when in 1996 the Slovak government initiated eight new administrative regions which the Hungarian minority claimed under-represented them politically. The government also made Slovak the only official language. The government's autocratic rule, human rights record, and apparent tolerance of organized crime led to mounting international criticism. In 1998, Meciar's party was defeated in a general election by a four-party coalition and Mikulas Dzurinda, leader of the center-right Slovak Democratic Coalition, became prime minister. Dzurinda narrowly won the parliamentary elections of 2002 and his government continued its policy of strengthening ties with the West. Slovakia became a member of both NATO and the EU in 2004.

ECONOMY

Communist governments developed manufacturing industries, producing chemicals, machinery, steel, and weapons. Since the late 1980s, many state-run businesses have been handed over to private owners. Manufacturing employs around 33% of workers. Bratislava and Kosice are the chief industrial cities. Products include ceramics, machinery, and steel. The armaments industry is based at Martin, in the northwest. Farming employs about 12% of the workforce. Major crops include barley, grapes for winemaking, corn, sugarbeet, and wheat.

BRATISLAVA

Capital of Slovakia, on the Danube, western Slovakia. It became part of Hungary after the 13th century, and was the Hungarian capital from 1526–1784. Incorporated into Czechoslovakia in 1918, it became the capital of Slovakia in 1992. Industries include petroleum refining, textiles, chemicals, electrical goods.

SLOVENIA

Slovenia's flag, which was based on the flag of Russia, was originally adopted in 1848. Under Communist rule, a red star appeared at the center. This flag, which was adopted in 1991 when Slovenia proclaimed its independence, has a new emblem, the national coat of arms.

The Republic of Slovenia was one of the six republics which made up Yugoslavia. Much of the land is mountainous and forested. The highest peak is Mount Triglav (9,393 ft [2,863 m]) in the Julian Alps (Julijske Alpe), an extension of the main Alpine ranges in the northwest. Much of central and eastern Slovenia is hilly. The River Sava which flows through central Slovenia is a tributary of the Danube, as is the Drava in the northeast.

Central Slovenia contains the limestone Karst region, with numerous underground streams and cave networks. The Postojna Caves, southwest of Ljubljana, are among the largest in Europe. The country has a short coastline on the Adriatic Sea.

Forests cover about half of Slovenia. Mountain pines grow on higher slopes, with beech, oak, and hornbeam at lower levels. The Karst region is largely bare of vegetation because of the lack of surface water. Farmland covers about a third of Slovenia.

Area 7,821 sq mi [20,256 sq km]
Population 2,011,000
Capital (population) Ljubljana (264,000)
Government Multiparty republic
Ethnic groups Slovene 92%, Croat 1%, Serb, Hungarian, Bosniak
Languages Slovenian (official), Serbo-Croatian
Religions Mainly Roman Catholic
Currency Tolar = 100 stotin
Website www.slovenia-tourism.si

HISTORY

The ancestors of the Slovenes, the western branch of a group of people called the South Slavs, settled in the area around 1,400 years ago. An independent Slovene state was formed in AD 623, but the area came under Bavarian-Frankish rule in 748. the Austrian royal family the Habsburgs took control of the region in 1278 and, apart from a short period of French rule between 1809 and 1815, it remained under Austrian control until 1918, when the dual monarchy of Austria-Hungary collapsed.

At the end of World War I, Slovenia became part of a new country called the Kingdom of the Serbs, Croats, and Slovenes, renamed Yugoslavia in 1929. Slovenia was invaded by Germany and Italy in 1941 and was partitioned between them and Hungary. At the end of the war, Slovenia again became one of the six republics of Yugoslavia.

In the late 1960s and early 1970s, some Slovenes called for the secession of their federal republic from Yugoslavia, but the dissidents were removed from the Communist Party by President Josip Broz Tito, whose strong rule maintained the unity of his country.

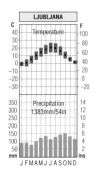

CLIMATE

The Slovenian coast has a mild Mediterranean climate. The climate inland is more continental, with snow capping the mountains in winter. Eastern Slovenia has cold winters and hot summers. Rain occurs in every month in Ljubljana, with late summer being the rainiest.

the Party of Democratic Reform became president, while Janez Drnovsek, of the center-left Liberal Democratic Party, became prime minister, heading a coalition government. The Liberal Democrat coalition government was returned again in 1996 and 2000. Slovenia became a member of NATO and the EU in 2004 and later that year, the center-right Slovenian Democratic Party topped the polls in parliamentary elections and a center-right coalition was formed.

ECONOMY

The reform of the formerly state-run economy, and the fighting in areas to the south, caused problems for Slovenia. It remains one of the fastest growing economies in Europe.

Manufacturing is the principal activity and manufactures include chemicals, machinery and transportation equipment, metal goods, and textiles. Slovenia mines some iron ore, lead, lignite and mercury. The leading crops are corn, potatoes, and wheat.

POLITICS

After Tito's death in 1980, the federal government in Belgrade found it increasingly difficult to maintain the unity of the disparate elements of the population. It was also weakened by the fact that Communism was increasingly seen to have failed in Eastern Europe and the Soviet Union. In 1990, Slovenia held multiparty elections and a non-Communist coalition was formed to rule the country.

Slovenia and neighboring Croatia proclaimed their independence in June 1991, but these acts were not accepted by the central government. After a few days of fighting between the Slovene militia and Yugoslav forces, Slovenia, the most ethnically homogenous of Yugoslavia's six component parts, found ready support from Italy and Austria (which had Slovene minorities of about 100,000 and 80,000, respectively), as well as Germany (an early supporter of Slovene independence). After a three-month moratorium, during which there was a negotiated, peaceful withdrawal, Slovenia became independent on October 8, 1991, thereby avoiding the conflict that was to plague other former Yugoslav states.

Slovenia's independence was recognized by the European Community in 1992. Multiparty elections were held and Milan Kucan (a former Communist) of

LJUBLJANA

Capital and largest city of Slovenia, at the confluence of the rivers Sava and Ljubljanica. In 34 BC Roman Emperor Augustus founded Ljubljana as Emona. From 1244 it was the capital of Carniola, an Austrian province of the Habsburg Empire. During the 19th century, it was the center of the Slovene nationalist movement. The city remained under Austrian rule until 1918, when it became part of the Kingdom of Serbs, Croats, and Slovenes (later Yugoslavia). When Slovenia achieved independence in 1991, Ljubljana became the capital.

SOMALIA

This flag was adopted in 1960, when Italian Somaliland in the south united with British Somaliland in the north to form Somalia. The colors are based on the United Nations flag and the points of the star represent the five regions of East Africa where Somalis live.

Somalia, is in a region known as the "Horn of Africa." A narrow, mostly barren, coastal plain borders the Indian Ocean and the Gulf of Aden. In the interior, the land rises to a plateau, of 3,300 ft [1,000 m]. In the north is a highland region. The south contains the only rivers, the Juba and the Scebeli.

Much of Somalia is dry grassland or semidesert. There are areas of wooded grassland, with trees such as acacia and baobab. Plants are most abundant in the the lower Juba valley.

Area 246,199 sq mi [637,657 sq km]
Population 8,305,000
Capital (population) Mogadishu (900,000)
Government Transitional, parliamentary federal government
Ethnic groups Somali 85%, Bantu, Arab, others
Languages Somali (official), Arabic, English, Italian
Religions Islam (Sunni Muslim)
Currency CFA franc = 100 centimes
Website www.unsomalia.net

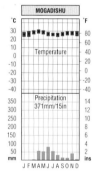

CLIMATE

Rainfall is light throughout, the wettest regions being in the south and the northern mountains. The country is prone to droughts, with temperatures on the plateaus and the plains often reaching 90°F [32°C].

HISTORY

In the 7th century, Arab traders established coastal settlements and introduced Islam. Around 900, Mogadishu was founded as a trading center. The interest of European imperial powers increased after the opening of the Suez Canal in 1869. In 1887, Britain established a Protectorate in what is now northern Somalia. In 1889, Italy formed a Protectorate in the central region, and extended its power to the south by 1905. The new boundaries divided the Somalis into five areas: the two Somalilands, Djibouti (taken by France in 1896), Ethiopia, and Kenya. In 1936 Italian

MOGADISHU

Capital and chief port of Somalia, on the Indian Ocean. It was founded by Arabs in the 10th century. In the 16th century, the Portuguese captured the city and it became a cornerstone of their trade with Africa. In 1871 the Sultan of Zanzibar took control. He first leased (1892) and then sold (1905) the port to the Italians. The city then became the capital of Italian Somaliland. It was occupied by the British during World War II. In 1960, Mogadishu became the capital of independent Somalia. The civil war of the 1980s/1990s devastated the city, its population was swelled by refugees escaping famine and drought in the outlying regions. In 1992, UN troops flew in to control aid distribution but withdrew in 1995 after little success.

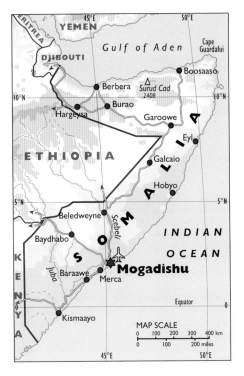

draw, but resistance continued, forcing one million refugees to flee to Somalia. In 1991, Barre was overthrown and the United Somali Congress (USC), led by Ali Mahdi Muhammad, gained power. Somalia disintegrated into civil war between rival clans. The Ethiopian-backed Somali National Movement (SNM) gained control of northwest Somalia, and seceded as the Somaliland Republic in 1991. An attack from the Somali National Alliance (SNA), led by General Muhammad Aideed, shattered Mogadishu. War and drought resulted in a devastating famine. The UN was slow to provide relief and unable to secure distribution. US marines led a taskforce to aid food distribution, but became embroiled in conflict with Somali warlords.

In 1994, 30 US marines died in the fighting and US forces withdrew. Civil strife continued, and in 1996 Aideed was killed. The Cairo Declaration (1997), signed by 26 of the 28 warring factions, held out hope of an end to factional feuding. In 2000, clan leaders elected Abdulkassim Salat Hassan as president, but factional fighting continued. An interim parliament was set up in Kenya (for safety) in 2004, but attempts to move it to Somalia in 2005 saw limited success.

Somaliland united with the Somali regions of Ethiopia to form Italian East Africa. During World War II, Italy invaded British Somaliland. But British forces conquered the region in 1941 and ruled both Somalilands until 1950, when Italian Somaliland returned to Italy as a UN Trust Territory. In 1960, both Somalilands gained independence and joined to form the United Republic of Somalia.

POLITICS

The new republic faced calls for the creation of a "Greater Somalia" to include the Somali-majority areas in Ethiopia, Kenya, and Djibouti. In 1969, the army, led by Siad Barre, seized power and formed a socialist, Islamic republic. During the 1970s, Somalia and Ethiopia fought for control of the Ogaden Desert, inhabited mainly by Somali nomads. In 1978 Ethiopia forced Somalia to with-

ECONOMY

Somalia is a developing country whose economy has been shattered by drought and war. Catastrophic flooding in late 1997 displaced tens of thousands of people, further damaging the country's infrastrucure, destroying hopes of economic recovery.

Many Somalis are nomads who raise livestock. Live animals, meat, and hides are major exports, plus bananas grown in the wetter south. Other crops include citrus fruits, cotton, corn, and sugarcane.

329

South Africa's flag was first flown in 1994 when the country adopted a new, nonracial constitution. It incorporates the red, white, and blue of former colonial powers, Britain and the Netherlands, together with the green, black, and gold of black organizations.

The Republic of South Africa is geologically very ancient, with few deposits less than 600 million years old. The country can be divided into two main regions, the interior plateau, the southern part of the huge plateau that makes up most of southern Africa, and the coastal fringes.

The interior consists of two main parts. Most of Northern Cape Province and Free State are drained by the Orange River and its right-bank tributaries that flow over level plateaus, varying in height from 4,000 to 6,000 ft [1,200–2,000 m]. The Northern Province is occupied by the Bushveld, an area of granites and igneous intrusions.

The Fringing Escarpment divides the interior from the coastal fringe. This escarpment makes communication within the country very difficult. In the east, the massive basalt-capped rock wall of the Drakensberg, at its most majestic near Mont-aux-Sources and rising to more than 10,000 ft [3,000 m], overlooks KwaZulu-Natal and Eastern Cape coastlands.

In the west there is a similar, though less well developed, divide between the interior plateau and the coastlands. The Fringing Escarpment also parallels the south coast, where it is fronted by a series of ranges, including the folded Cape Ranges.

CLIMATE

Most of South Africa has a mild, sunny climate. Much of the coastal strip, including the city of Cape Town, has warm, dry summers and mild, rainy winters, just like the Mediterranean lands in northern Africa. Inland, large areas are arid.

Area 471,442 sq mi [1,221,037 sq km]
Population 42,719,000
Capital (population) Cape Town (legislative, 855,000); Tshwane/Pretoria (administrative, 2,200,000); Bloemfontein (judiciary, 350,000)
Government Multiparty republic
Ethnic groups Black 76%, White 13%, Colored 9%, Asian 2%
Languages Afrikaans, English, Ndebele, Pedi, Sotho, Swazi, Tsonga, Tswana, Venda, Xhosa and Zulu (all official)
Religions Christianity 68%, Islam 2%, Hinduism 1%
Currency CFA franc = 100 centimes
Website www.gov.za

HISTORY

Early inhabitants were the Khoisan (also called Hottentots and Bushmen). However, the majority of the people today are Bantu-speakers from the north who entered the country, introducing a cattle-keeping, grain-growing culture. Arriving via the plateaus of the northeast, they continued southward into the well-watered zones below the Fringing Escarpment of KwaZulu-Natal and Eastern Cape. By the 18th century, these

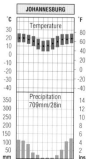

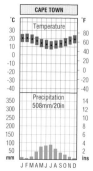

people had reached the southeast. They formed large groups, including the Zulu, Xhosa, Sotho, and Tswana.

Also at this time, a group of Europeans was establishing a supply base for the Dutch East India Company on the site of present-day Cape Town. The first group was led by Jan van Riebeeck who founded the base in 1652. In 1657, some company employees set up their own farms and became known as Boers (farmers). After Britain took over the Cape Town settlement in the early 19th century, many Boers, who resented British rule, began to move inland to develop their own Afrikaaner culture. Beginning in 1836, this migration was known as the Great Trek. Their advance

was channeled in the south by parallel coastal ranges, and eventually black and white met near the Kei River. To the north, once the Fringing Escarpment had been overcome, the level plateau surfaces allowed a rapid spread northward, with the Boers founding the Transvaal in 1852 and Orange Free State in 1854.

In 1870, diamonds were found near the site where Kimberley now stands. Both the British and the Boers claimed the area, but Britain annexed it in 1871. In 1880, the Boers rebelled and defeated the British in the First Boer War. In 1886, gold was discovered in the Witwatersrand in what is now Gauteng. Many immigrants, called *uitlanders* (foreigners), flooded to the area. Most of them were British and, to maintain their control, the Boers restricted their freedom. Tension developed, culminating in the Second Boer War (1899–1902). The Boer republics of Orange Free State and Transvaal then surrendered and became British colonies. Meanwhile, British forces had overcome Zulu resistance to European settlement. By 1898, all opposition had been suppressed and the black people had lost their independence.

POLITICS

In 1906, Transvaal was granted self-rule, followed by Orange Free State in 1907. The other two parts of the country, Cape Colony and Natal, already had self-rule. In 1910, the entire country was united as the Union of South Africa, a self-governing country within the British Empire. During World War I, two Boer generals led South African forces against Germany. In German South West Africa (now Namibia), General Louis Botha conquered the Germans, while General Jan Christiaan Smuts led Allied forces in German East Africa (now Tanzania). In 1920, the League of Nations gave South Africa control over South West Africa, under a trusteeship agreement. In 1931, Britain granted South Africa full independence as a member of the Commonwealth of Nations.

The development of mineral extraction and urban complexes in South Africa caused an even greater divergence between black and white. The African farmers gained little from the mineral boom. With taxes to pay, they had little alternative but to seek employment in the mines or on European-owned farms. Migrant labor became the normal way of life for many men, while agriculture in black areas stagnated. Groups of Africans took up urban life, living in communities set apart from the white settlements. These townships, with their rudimentary housing often supplemented by shanty dwellings and without any real services,

mushroomed during World War II and left South Africa with a major housing problem in the late 1940s. Nowhere was this problem greater than in Johannesburg, where a vast complex of brick boxes called SOWETO (South Western Townships) was built. The contrast between the living standards of blacks and whites increased rapidly.

At the start of World War II, opinion was divided as to whether South Africa should remain neutral or support Britain. The pro-British General Smuts triumphed. He became prime minister and South African forces served in Ethiopia, northern Africa, and Europe. During the war, Daniel Malan, a supporter of Afrikaner nationalism, reorganized the National Party. The Nationalists came to power in 1948, with Malan as prime minister, and introduced the policy of apartheid. The African National Congress, which had been founded in 1912, became the leading black opposition group. Opposition to South Africa's segregationist policies mounted around the world. Stung by criticism from Britain and other Commonwealth members, South Africa became a republic and withdrew from the Commonwealth in 1961. In 1966, the United Nations voted to end South Africa's control over South West Africa, though it was not until 1990 that the territory finally became independent as Namibia.

PRETORIA (TSHWANE)

Administrative capital of South Africa, Gauteng province. Founded in 1855 and named after Andries Pretorius. It became the capital of the Transvaal in 1860, and of the South African Republic in 1881. The Peace of Vereeniging, which ended the South African Wars, was signed here in 1902. In 1910, it became the capital of the Union of South Africa. Pretoria is an important communications center.

In response to continuing opposition, South Africa repealed some apartheid laws and, in 1984, under a new constitution, a new three-house parliament was set up. The three houses were for whites, Coloreds and Asians, but there was still no provision for the black majority. In 1986, the European Community (now the European Union), the Commonwealth, and the United States applied sanctions on South Africa, banning trade in certain areas. In 1989, F. W. de Klerk was elected president and in 1990 he released the banned ANC leader Nelson Mandela from prison.

In the early 1990s, more apartheid laws were repealed. The country began to prepare a new constitution giving all non-whites the right to vote, though progress toward majority rule was marred by fighting between the Zulu-dominated Inkatha Freedom Party and the ANC.

Elections held in 1994 resulted in victory for the ANC, and Nelson Mandela became president. Mandela advocated reconciliation between whites and non-whites, and his government sought to alleviate the poverty of Africans in the townships. The slow rate of progress disappointed many, as did other problems, including an increase in crime and the continuing massive gap in living standards between the whites and the blacks. However, in 1999, following the retirement of Nelson Mandela, his successor, Thabo Mbeki, led the African National Congress to an overwhelming electoral victory. Besides poverty, one of the biggest problems facing the country is the estimate given in a government study that one in five South Africans is infected with the HIV virus.

ECONOMY

South Africa is Africa's most developed country. However, most of the black people—rural and urban—are poor with low standards of living. Natural resources include diamonds and gold, which formed the basis of its economy from the late 19th century. Today, South Africa ranks first in the world in gold production and fifth in diamond production. South Africa also produces coal, chromite, copper, iron ore, manganese, platinum, phosphate rock, silver, uranium, and vanadium. Mining and manufacturing are the most valuable economic activities and gold, metals and metal products, and gem diamonds are the chief exports.

Manufactures include chemicals, processed food, iron and steel, machinery, motor vehicles, and textiles. The main industrial areas lie in and around the cities of Cape Town, Durban, Johannesburg, Port Elizabeth and Pretoria. Investment in South African mining and manufacturing declined in the 1980s, but foreign companies began to invest again following the abolition of apartheid.

Farmland is limited by the aridity of many areas, but the country produces most of the food it needs and food products make up around 7% of South Africa's exports. Major crops include apples, grapes (for winemaking), corn, oranges, pineapples, sugarcane, tobacco and wheat. Sheep rearing is important on land that is unfit for arable farming. Other livestock products include beef, dairy products, and eggs.

CAPE TOWN

City and seaport at the foot of Table Mountain, South Africa. It is South Africa's legislative capital and the capital of Western Cape province. Founded in 1652 by the Dutch East India Company, it came under British rule in 1795. Places of interest include the Union Parliament, a 17th-century castle, the National Historic Museum, and the University of Cape Town (founded 1829). It is an important industrial and commercial center. Industries: clothing, engineering equipment, motor vehicles, and wine.

SPAIN

The colors on the Spanish flag date back to those used by the old kingdom of Aragon in the 12th century. The present design, in which the central yellow stripe is twice as wide as each of the red stripes, was adopted in 1938, during the Civil War.

The Kingdom of Spain is the second largest country in Western Europe after France. It shares the Iberian Peninsula with Portugal. A plateau, called the Meseta, covers most of Spain. Much of it is flat, but crossed by several mountain ranges (*sierras*).

The northern highlands include the Cantabrian Mountains (Cordillera Cantabrica) and the high Pyrenees, which form Spain's border with France. Mulhacén, the highest peak on the Spanish mainland, is in the Sierra Nevada in the southeast. Spain also contains fertile coastal plains. Other lowlands are the Ebro River Basin in the northeast and the Guadalquivir River Basin in the southwest.

Spain also includes the Balearic Islands (Islas Baleares) in the Mediterranean Sea and the Canary Islands off the northwest coast of Africa. Tenerife in the Canary Islands contains Pico de Teide, Spain's highest peak (12,918 ft [3,718 m]).

Forests lie to the rainier north and northwest, with beech and deciduous oak being common. Toward the drier south and east, Mediterranean pines and evergreen oaks take over, and the forests resemble open parkland. Large areas are *matorral*, a Mediterranean scrub. Where soils are thin and drought is prevalent, *matorral* gives way to steppe.

CLIMATE

Spain has the widest range of climate in Western Europe. One of the most striking contrasts is between the humid north and northwest, where winds from the Atlantic bring mild, wet weather throughout the year, and the mainly arid

Area 192,103 sq mi [497,548 sq km]
Population 40,281,000
Capital (population) Madrid (2,939,000)
Government Constitutional monarchy
Ethnic groups Mediterranean and Nordic types
Languages Castillian Spanish (official) 74%, Catalan 17%, Galician 7%, Basque 2%
Religions Roman Catholic 94%, others
Currency Euro = 100 cents
Website www.spain.info

remainder of the country. Droughts are common in much of Spain, though these are occasionally interrupted by thunderstorms.

The Meseta, removed from the influence of the sea, has a continental climate, with hot summers and cold winters, when frosts often occur and snow blankets the mountain ranges that rise above the plateau surface. By contrast, the Mediterranean coastlands and the Balearic Islands have mild, moist winters. Summers along the Mediterranean coast are hot and dry. The Canary Islands have mild to warm weather throughout the year.

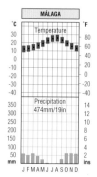

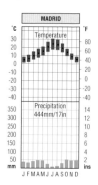

334

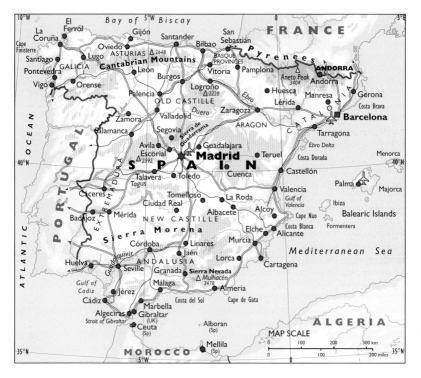

HISTORY

5,000 years ago, Spain was inhabited by farming people called Iberians. Some historians believe the Basques in northern Spain may be descendants of these people. Around 3,000 years ago, Phoenicians from the eastern Mediterranean reached the Iberian Peninsula and began to establish trading colonies, some on the sites of modern cities, such as Cádiz and Málaga. Celtic peoples arrived later from the north, while Greeks reached the east coast of Spain around 600 BC.

In the 5th century BC, Carthaginians conquered much of Spain, but after the Second Punic War (218–201 BC), the Iberian Peninsula gradually came under Roman rule. The Romans made Iberia a Roman province called Hispania.

By 573, the Visigoths had conquered the entire peninsula, including what is now Portugal, and they ruled until the early 8th century when the Muslim Moors invaded from North Africa. They introduced their culture and scholarship, far ahead of that of Europe, building superb mosques and palaces, some of which still stand. In the 11th century, the country began to divide into many small Moorish kingdoms, leaving them open to attack by the Christian kingdoms in the north. Portugal broke away from Spain in the 11th–12th centuries. By the late 13th century, Muslim power was confined to the southern Kingdom of Granada.

The rest of Spain was ruled by the Christian kingdoms of Aragon, Navarre, and, the most powerful of all, Castile. In 1469, Prince Ferdinand of Aragon married Princess Isabella of Castile. Ferdinand and Isabella started the Spanish Inquisition which persecuted Jews, Muslims, and other non-Roman Catholics. In

1492, Ferdinand's forces captured the last Muslim stronghold of Granada and, in 1512, the Kingdom of Navarre was taken by Ferdinand. This completed the union of Spain.

By the mid-16th century, Spain was a great world power controlling much of Central and South America, parts of Africa, and the Philippines in Asia. A major disaster occurred in 1588, when King Philip II sent a fleet, the Armada, to conquer England, but the English navy and bad weather destroyed half of the Spanish ships. By the 20th century all that remained of Spain's empire were a few small African territories.

A military government was established in 1923 and King Alfonso III allowed General Miguel Primo de Rivera, the prime minister, to rule as a dictator. After Primo de Rivera was forced to resign in 1930, Alfonso called for city elections. Republican candidates scored such a major victory in these elections that he left the country, though he did not renounce his claim to the throne. The republicans took over the government.

MADRID

Capital and largest city of Spain, lying on a high plain in the center of Spain on the River Manzanares. It is Europe's highest capital city, at 655 m [2,149 ft]. Founded as a Moorish fortress in the 10th century, Alfonso VI of Castile captured Madrid in 1083. In 1561, Philip II moved the capital to Madrid. The French occupied the city during the Peninsular War (1808–14). Madrid expanded considerably in the 19th century. During the Spanish Civil War, it remained loyal to the Republican cause and was under siege for almost three years. Its capitulation in March 1939 brought the war to an end. Modern Madrid is a thriving center of commerce and industry.

In October 1936, rebel Nationalists chose General Francisco Franco (1892–1975) as their commander and, and in 1939 he became the dictator of Spain, though technically the country was a monarchy. During World War II, Spain was officially neutral.

POLITICS

The revival of Spain's shattered economy began in the 1950s through the growth of manufacturing industries and tourism. As standards of living rose, people began to demand more freedom. After Franco died in 1975, the monarchy was restored and Juan Carlos, grandson of Alfonso III, became king. The ban on political parties was lifted and, in 1977, elections were held. A new constitution making Spain a parliamentary democracy, with the king as head of state, came into effect in December 1978.

From the late 1970s, Spain began to tackle the problem of its regions. In 1980, a regional parliament was set up in the Basque Country (Euskadi in Basque and Pais Vasco in Spanish). Similar parliaments were initiated in Catalonia (Cataluña) in the northeast and Galicia in the northwest. While regional devolution was welcomed in Catalonia and Galicia, it did not end the terrorist campaign of the Basque separatist movement, Euskadi Ta Askatasuna (ETA). ETA announced an indefinite ceasefire in September 1998, but the truce was ended in December 1999 and the conflict continued. The Supreme Court voted in 2003, to ban Batasuna, the Basque separatist party deemed to be the political wing of ETA.

In March 2004 terrorist bombs exploded in Madrid killing 191 people. This was seem as the work of Al Qaeda, though the government were keen to persuade the people that it was the work of ETA. The country went to the polls three days later and voted out the right-wing Aznar. This was largely seen as a reaction to his support of the US in Iraq and the sending of Spanish troops there which was to blame for the bombing some three days earlier. The new prime minister Zapatero immediately withdrew all troops from Iraq.

ECONOMY

Spain has the fifth largest economy in the EU. By the early 2000s, agriculture employed only 6% of the workforce as compared with industry at 17% and services, including tourism, which employs 77%. Farmland makes up two-thirds of the land, and forest most of the rest. Major crops include barley, citrus fruits, grapes for winemaking, olives, potatoes, and wheat.

There is some high-grade iron ore in the north. Spain's many manufacturing industries include automobiles, chemicals, clothing, electronics, processed food, metal goods, steel, and textiles.

ANDORRA

Andorra is a tiny state sandwiched between France and Spain. It lies in the Pyrenees Mountains. Most Andorrans live in the sheltered valleys.

The winters are cold and fairly dry. The summers are a little more wet, but pleasantly cool.

Tourism is Andorra's chief activity in both winter, for winter sports, and summer.

There is some farming in the valleys and tobacco is the main crop. Cattle and sheep are grazed on the mountain slopes.

Area 175 sq mi [453 sq km]
Population 68,000
Capital (population) Andorra La Vella (22,000)
Government Coprincipality
Ethnic groups Spanish 43%, Andorran 33%, Portuguese 11%, French 7%
Languages Catalan (official)
Religions Mainly Roman Catholic
Currency Euro = 100 cents
Website www.turisme.ad

GIBRALTAR

Gibraltar is a tiny British dependency on the south coast of Spain, occupying a strategic position overlooking the narrow Strait of Gibraltar which links the Mediterranean Sea with the Atlantic Ocean. The majority of the populous works for the government or in the tourist trade.

Most of the land is a huge mass of limestone, known as the Rock of Gibraltar. Between AD 711 and 1309, and again between 1333 and 1462, Gibraltar was held by Moors from North Africa. Spaniards retook the area in 1462, but it became a British territory in 1713. Gibraltar became a vital British military base, but was still claimed by Spain.

In 1967 the Gibraltarians voted to remain British. Between 1969 and 1985

Area 2.5 sq mi [6.5 sq km]
Population 28,000
Capital (population) Gibraltar Town (28,000)
Government British dependency
Ethnic groups English, Spanish, Maltese, Italian, Portuguese
Languages English (official), Spanish, Italian, Portuguese
Religions Mainly Roman Catholic
Currency Gibraltar pound = 100 pence
Website www.gibraltar.gov.uk

Spain closed its border with Gibraltar. Britain withdrew its military forces in 1991. In 2002, proposals that Gibraltar should come under joint Anglo-Spanish sovreignty were rejected by nearly all Gibraltarians.

SRI LANKA

Sri Lanka's unusual flag was adopted in 1951, three years after the country, then called Ceylon, became independent from Britain. The lion banner represents the ancient Buddhist kingdom. The stripes symbolize the minorities—Muslims (green) and Hindus (orange).

The Democratic Socialist Republic of Sri Lanka is an island nation, often called the "pearl of the Indian Ocean." It lies on the same continental shelf as India, separated by the shallow Palk Strait. Most of the land is low lying but, in the south-central part of Sri Lanka, the land rises to a mountain massif. The nation's highest peak is Pidurutalagala (8,281 ft [2,524 m]). The nearby Adam's Peak, at 7,359 ft [2,243 m], is a place of pilgrimage. The southwest is also mountainous, with long ridges.

Around the south-central highlands are broad plains, while the Jaffna Peninsula in the far north is made of limestone. Cliffs overlook the sea in the southwest, while elsewhere lagoons line the coast. Forests cover nearly two-fifths of the land, with open grasslands in the eastern highlands. Farmland, including pasture, covers another two-fifths of the country.

Area 25,332 sq mi [65,610 sq km]
Population 19,905,000
Capital (population) Colombo (642,000)
Government Multiparty republic
Ethnic groups Sinhalese 74%, Tamil 18%, Moor 7%
Languages Sinhala and Tamil (both official)
Religions Buddhism 70%, Hinduism 15%, Christianity 8%, Islam 7%
Currency Sri Lankan rupee = 100 cents
Website www.gov.lk

CLIMATE

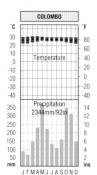

The western part of Sri Lanka has a wet equatorial climate. Temperatures are high and the rainfall is heavy. The wettest months are May and October—the advance and the retreat of the summer monsoon. Eastern Sri Lanka is drier.

HISTORY

The ancestors of the Sinhalese people settled on the island around 2,400 years ago. They pushed the Veddahs, descendants of the earliest inhabitants, into the interior. The Sinhalese founded the city of Anuradhapura, which was their center from the 3rd century BC to the 10th century AD. Tamils arrived around 2,100 years ago and the early history of Ceylon, as the island was known, was concerned with a struggle between the Sinhalese and the Tamils. Victory for the Tamils led the Sinhalese to move south.

POLITICS

From the early 16th century, Ceylon was ruled successively by the Portuguese, Dutch, and British. Independence was achieved in 1948 and the country was renamed Sri Lanka in 1972.

After independence, rivalries between the two main ethnic groups, the Sinhalese and Tamils, marred progress. In the 1950s, the government made Sinhala the official language. Following protests, the prime minister made provisions for Tamil to be used in some areas. In 1959, the prime minister was assassinated by a Sinhalese extremist and he was succeeded by Sirimavo Bandanaraike, the world's first woman prime minister.

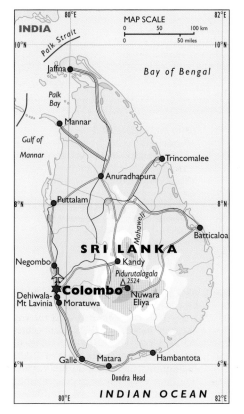

the Sinhalese Buddhists, who believe that the temple's treasured tooth belonged to Buddha.

The bombing led to rioting and provoked President Chandrika Kumaratunga to ban the LTTE. These events led to some of the fiercest fighting in the civil war, including several suicide bombings. The government lost most of the gains it had made in the mid-1990s. A longterm ceasefire agreement was signed in 2002. In December 2004, Sri Lanka was hit by a tsunami, which killed more than 30,000 people. In 2005, the election of a new president, Mahinda Rajapakse, led to hopes of a resolution to the conflict in the north.

ECONOMY

The World Bank classifies Sri Lanka as a "low-income" developing country. Agriculture employs around a third of the workforce, coconuts, rubber, and tea are the cash crops. Rice is the chief food crop. Cattle, water buffalo, and goats are the chief farm animals, while fish provide another source of protein. Manufacturing is mainly the processing of agricultural products and textile production. The leading exports are clothing and accessories, gemstones, tea, and rubber.

Conflict between Tamils and Sinhalese continued in the 1970s and 1980s. In 1987, India helped to engineer a cease-fire. Indian troops arrived to enforce the agreement. They withdrew in 1990 after failing to subdue the main guerrilla group, the Tamil Tigers, who wanted an independent Tamil homeland in northern Sri Lanka. In 1993, the country's president, Ranasinghe Premadasa, was assassinated by a suspected Tamil separatist. A ceasefire was signed in May 1993, but fighting soon broke out. In 1995, government forces captured Jaffna, the stronghold of the "Liberation Tigers of the Tamil Eelam" (LTTE). But the 1998 bombing of the Temple of the Tooth in Kandy created great outrage among

COLOMBO

Capital and chief seaport of Sri Lanka, on the southwest coast. Settled in the 6th century BC, it was taken by Portugal in the 16th century and later by the Dutch. It was captured by the British in 1796, and gained its independence in 1948. Sights include the town hall and the Aqua de Lupo Church.

339

SUDAN

Adopted in 1969, Sudan's flag uses colors associated with the Pan-Arab movement. The Islamic green triangle symbolizes prosperity and spiritual wealth. The flag is based on the one used in the Arab revolt against Turkish rule in World War I (1914–18).

The Republic of the Sudan is the largest country in Africa. It extends from the arid Sahara in the north to an equatorial swamp region (the Sudd) in the south.

Much of the land is flat, but there are mountains in the northeast and southeast; the highest point is Kinyeti, at 10,456 ft [3,187 m]. The River Nile (Bahr el Jebel) runs south to north, entering Sudan as the White Nile, converging with the Blue Nile at Khartoum, and flowing north to Egypt.

Khartoum is prone to summer dust storms (*haboobs*). From the bare deserts of the north, the land merges into dry grasslands and savanna. Dense rainforests grow in the south.

Area 967,494 sq mi [2,505, 813 sq km]
Population 39,148,000
Capital (population) Khartoum (947,000)
Government Military regime
Ethnic groups Black 52%, Arab 39%, Beja 6%, others
Languages Arabic (official), Nubian, Ta Bedawie
Religions Islam 70%, traditional beliefs
Currency Sudanese dinar = 10 Sudanese pounds
Website www.sudan.net

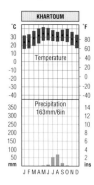

CLIMATE

Northern Sudan is hot and arid. The center of the country has an average annual rainfall of 4 to 32 in [100–510 mm], while the tropical south has between 32 and 55 in [810–1,400 mm] of rain per year.

HISTORY

One of the earliest civilizations in the Nile region of northern Sudan was Nubia, which came under Ancient Egypt around 4,000 years ago. Another Nubian civilization, called Kush, developed from about 1000 BC, finally collapsing in AD 350. Christianity was introduced to northern Sudan in the 6th century. From the 13th to 15th centuries, northern Sudan came under Muslim control, and Islam became the dominant religion.

In 1821 Muhammad Ali's forces occupied Sudan. Anglo-Egyptian forces, led by General Gordon, attempted to extend Egypt's influence into the south. Muhammad Ahmad led a Mahdi uprising, which briefly freed Sudan from Anglo-Egyptian influence. In 1898, General Kitchener's forces defeated the Mahdists, and in 1899 Sudan became Anglo-Egyptian Sudan, governed jointly by Britain and Egypt.

POLITICS

After Sudan's independence in 1952, the southern Sudanese, who are predominantly Christians or followers of traditional beliefs, revolted against the dominance of the Muslim north, and civil war broke out. In 1958, the military seized power. Civilian rule was reestablished in 1964, but overthrown again in 1969, when Gaafar Muhammad Nimeri seized control. In 1972, southern Sudan received considerable autonomy, but unrest persisted.

In 1983, the imposition of strict Islamic law sparked off further conflict

340

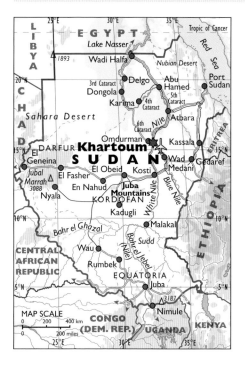

region in the west, primarily involving rebels and government-backed militias. A severe humanitarian crisis developed, and the militias were accused of ethnic cleansing. In the south, government and rebels signed a comprehensive peace deal in 2005.

ECONOMY

The World Bank classifies Sudan as a "low-income" economy. Food shortages and a refugee crisis worsened its economic plight Agriculture employs 60% of the population. The chief crop is cotton. Other crops include peanuts, gum arabic, millet, sesame, sorghum, and sugarcane, while many people raise livestock.

Minerals include chromium, gold, gypsum, and petroleum. Manufacturing industries process foods, and produce such things as cement, fertilizers, and textiles. The main exports are cotton, gum arabic, and sesame seeds, but the most valuable exports are petroleum and petroleum products.

between the government and the Sudan People's Liberation Army (SPLA) in the south. In 1985, Nimeri was deposed and a civilian government installed. In 1989, the military, led by Omar Hassan Ahmed al-Bashir, established a Revolutionary Command Council. Civil war continued in the south. In 1996, Bashir was reelected, virtually unopposed. The National Islamic Front (NIF) dominated the government and was believed to have strong links with Iranian terrorist groups.

In 1996, the UN imposed sanctions on Sudan. A South African peace initiative in 1997 led to the formation of a Southern States' Coordination Council. The US imposed sanctions on Bashir's regime and American Secretary of State Madeleine Albright met rebel leaders. In 1998, the USA bombed a pharmaceuticals factory in Khartoum in the mistaken belief that it produced chemical weapons. In 2003, conflict broke out in the Darfur

KHARTOUM

Capital of Sudan, at the junction of the Blue Nile and White Nile rivers. Khartoum was founded in the 1820s by Muhammad Ali and was besieged by the Mahdists in 1885, when General Gordon was killed. In 1898, it became the seat of government of the Anglo-Egyptian Sudan, and from 1956 the capital of independent Sudan. It was at the center of controversy in 1998 when the US bombed a pharmaceuticals plant thinking it was producing chemical weapons. Industries include cement, gum arabic, chemicals, glass, cotton, and textiles.

SURINAME

The star symbolizes national unity—each point being one of Suriname's five main ethnic groups. Yellow is for Suriname's golden future. Red stands for progress and the struggle for a better life. Green signifies hope and fertility. White symbolizes freedom and justice.

The Republic of Suriname is on the Atlantic Ocean in northeastern South America bordered by Brazil to the south, French Guiana to the east, and Guyana to the west.

Suriname is made up of the Guiana Highlands plateau, a flat coastal plain and a forested inland region. Its many rivers serve as a source of hydroelectric power. The narrow coastal plain was once swampy, but it has been drained and now consists mainly of farmland. Inland lie hills and low mountains which rise to 4,199 ft [1,280 m].

Area 63,037 sq mi [163,265 sq km]
Population 437,000
Capital (population) Paramaribo (216,000)
Government Multiparty republic
Ethnic groups Hindustani/East Indian 37%, Creole (mixed White and Black) 31%, Javanese 15%, Black 10%, Amerindian 2%, Chinese 2%, others
Languages Dutch (official), Sranang Tonga
Religions Hinduism 27%, Protestant 25%, Roman Catholic 23%, Islam 20%
Currency Surinamese dollar = 100 cents
Website www.parbo.com/tourism/

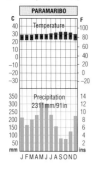

CLIMATE

Suriname has a hot, wet, and humid climate. Temperatures are high throughout the year.

HISTORY

Spanish explorer Alfonso de Ojeda discovered Suriname in 1499, but it was the British who founded the first colony in 1651. In 1667, Britain handed Suriname to the Dutch in return for New Amsterdam, an area that is now the state of New York. Slave revolts and Dutch neglect hampered development.

In the early 19th century Britain and the Netherlands disputed the ownership of the area. The British gave up their claims in 1813, and in 1815 the Congress of Vienna gave the Guyana region to Britain and reaffirmed Dutch control of "Dutch Guiana." Slavery was abolished in 1863 and soon afterward Indian and Indonesian laborers were introduced to work on plantations. It gained autonomy in 1954.

POLITICS

Suriname became fully independent from the Netherlands in 1975 and gained membership of the United Nations, but the economy was weakened when thousands of skilled people emigrated to the Netherlands.

Following a coup in 1980, Suriname was ruled by a military dictator, Dési Bouterse who banned all political parties. Guerrilla warfare disrupted the

---

BAUXITE

Rock from which most aluminum is extracted. Bauxite is a mixture of several minerals, such as diaspore, gibbsite, boehmite, and iron. It is formed by prolonged weathering and leaching of rocks containing aluminum silicates.

dent in 1988 elections, but he was overthrown by a military coup in 1990.

In 1991, Ronald Venetiaan, leader of the New Front for Democracy and Development, became president. In 1992 the government negotiated a peace agreement with the *boschneger*, descendants of African slaves, who had launched a struggle against the government. That same year, the constitution was amended in order to limit the power of the military and a peace agreement was reached with the rebels. Elections were held in 1996 and again in 2000.

In 1999, Bouterse was convicted in absentia in the Netherlands of having led a cocaine-trafficking ring during and after his tenure in office. In 2004, the government announced that he and others would face trial over the killings of 15 people in 1982.

economy. In 1987, a new constitution provided for a 51-member National Assembly, with powers to elect the president. Rameswak Shankar became presi-

ECONOMY

The World Bank classifies Suriname as an "upper-middle-income" developing country. Its economy is based on mining and metal processing. Suriname is a leading producer of bauxite, from which the metal aluminum is extracted.

The chief agricultural products are rice, bananas, sugarcane, coffee, coconuts, timber, and citrus fruits.

PARAMARIBO

Capital of Suriname, a port on the River Suriname. It was founded in the early 17th century by the French and became a British colony in 1651. It was held intermittently by the British and the Dutch until 1816, when the latter finally took control until independence. Paramaribo is the administrative and economic center. The name is derived from Paramurubo, meaning "city of parwa blossoms" after an old Arrawak village. Places of interest include Fort Zeelandia/Suriname Museum, the Palm Gardens, and the Presidential Palace. Industries include bauxite, timber, sugarcane, rice, rum, coffee, and cacao.

COPPENAME RIVER

The Coppename River is located on the Atlantic Coast of Suriname and its estuary is an important wetland region. The wide, tidal mud flats, lagoons, and swamps are typical of the coastal wetlands of the Guyanas and northern Brazil and have been protected since 1953.

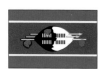

The oxhide shield with two spears and a fighting staff represent the defense of Swaziland; the tassels are symbols of the Swazi monarchy. Black and white represent racial harmony, blue denotes peace, yellow stands for the nation's mineral wealth. Red signifies blood shed in past struggles.

The Kingdom of Swaziland is a small, landlocked country in southern Africa bounded by South Africa to the north, west, and south and by Mozambique to the east. The country has four regions which run north–south.

In the west, the Highveld, with an average height of 3,937 ft [1,200 m], makes up 30% of Swaziland. The Middleveld, between 1,148 and 3,281 ft [350 –1,000 m], covers 28% of the country. The Lowveld, with an average height of 886 ft [270 m], covers another 33%, while the Lebombo Mountains reach 2,600 ft [800 m] along the eastern border.

Meadows and pasture cover 65% of Swaziland. Arable land covers 8% of the land, and forests only 6%.

Area 6,704 sq mi [17,364 sq km]
Population 1,169,000
Capital (population) Mbabane (38,000)
Government Monarchy
Ethnic groups African 97%, European 3%
Languages Siswati and English (both official)
Religions Zionist (a mixture of Christianity and traditional beliefs) 40%, Roman Catholic 20%, Islam 10%
Currency Lilangeni = 100 cents
Website www.gov.sz

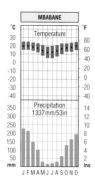

CLIMATE

The Lowveld is almost tropical, with an average temperature of 72°F [22°C] and a low rainfall of 20 in [500 mm] a year. The altitude moderates the climate in the west of the country. Mbabane has a climate typical of the Highveld with warm summers and cool winters.

HISTORY

In the 18th century, according to tradition, a group of Bantu-speaking people, under the Swazi Chief Ngwane II, crossed the Lebombo range and united with local African groups to form the Swazi nation. In the 1840s, under attack from the Zulu, the Swazi sought British protection. Gold was discovered in the 1880s, and many Europeans sought land concessions from the King, who did not realize that in acceding to their demands he lost control of the land. In 1894, Britain and the Boers of South Africa agreed to put Swaziland under the control of the South African Republic (the

---

BANTU

Group of African languages generally considered as forming part of the Benue-Congo branch of the Niger-Congo family. Swahili, Xhosa, and Zulu are among the most widely spoken of the several hundred Bantu languages spoken from the Congo Basin to South Africa, and almost all are tone languages. There are more than 200 million speakers of Bantu. The language of Swaziland, Siswati (also known as Swati or Swazi) is a member of the Bantu family of languages.

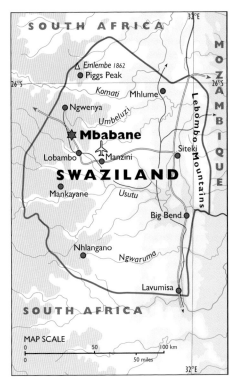

1978, he banned all political parties. Sobhuza died in 1982 after a reign of 82 years.

In 1983, his son, Prince Makhosetive, was chosen as his heir. In 1986, he became King Mswati III. Elections in 1993 and 1998, in which political parties were banned, failed to satisfy protesters who opposed the absolute monarchy.

Mswati continued to rule by decree and in 2004 he announced plans to build palaces for each of his 11 wives. At the same time the government appealed for aid in the face of a national disaster caused by the spread of HIV/AIDS and a severe drought.

ECONOMY
The World Bank classifies Swaziland as a "lower-middle-income" developing country. Agriculture employs 50% of the workforce, with many farmers living at subsistence level. Farm products and processed foods, including sugar, wood pulp, citrus fruits, and canned fruit, are the leading exports. Swaziland exhausted its high-grade iron ore reserves in 1978, while the world demand for its asbestos fell. Swaziland is heavily dependent on South Africa and the two countries are linked through a customs union.

Transvaal). Britain took control at the end of the second South African War (1899–1902).

POLITICS
In 1968, when Swaziland became fully independent as a constitutional monarchy, the head of state was King Sobhuza II. In 1973, Sobhuza suspended the constitution and assumed supreme power. In

LOBAMBO
The traditional royal capital of Swaziland, lying in the Ezulwini Valley 10 mi [16 km] from Mbabane. It is the home of the Queen Mother. The National Assembly, National Museum, and parliament are all based here.

MBABANE
Capital of Swaziland, in the northwest of the country, at the northern end of the Ezulwini Valley in the Dlangeni Hills, which are part of the Highveld region of southern Africa. It is both an administrative and commercial center, serving the surrounding agricultural region. Tin and iron ore are mined nearby.

SWEDEN

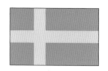

Sweden's flag was adopted in 1906, though it had been in use since the time of King Gustavus Vasa (reigned 1523–60). This king won many victories for Sweden and laid the foundations of the modern nation. The colors on the flag come from a coat of arms dating from 1364.

The Kingdom of Sweden is the largest of the countries of Scandinavia in both area and population. It shares the Scandinavian Peninsula with Norway. The western part of the country, along the border with Norway, is mountainous. The highest point is Kebnekaise, which reaches 6,946 ft [2,117 m] in the northwest.

Area 173,731 sq mi [449,964 sq km]
Population 8,986,000
Capital (population) Stockholm (744,000)
Government Constitutional monarchy
Ethnic groups Swedish 91%, Finnish, Sami
Languages Swedish (official), Finnish, Sami
Religions Lutheran 87%, Roman Catholic, Orthodox
Currency Swedish krona = 100 öre
Website www.sweden.gov.se

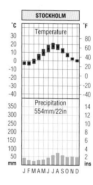

CLIMATE

The northerly latitude and high mountains and plateaus of Norway cut Sweden off from the mild influences of the Atlantic in the west. The Gulf Stream warms the southern coastlands. The February temperature in the central lowlands is just below freezing, but in the north it is 5°F [–15°C].

Precipitation is low throughout Sweden, but lies as snow for more than six months in the north. In summer there is little difference between the north and south. Most areas have an average temperature range between 59 and 68°F [15°–20°C].

HISTORY

People began to settle in Sweden around 8,000 years ago. Accounts were recorded in about AD 100.

By the seventh century, Teutonic peoples had occupied much of central Sweden. Between the 9th and 11th centuries the Swedish Vikings sailed to the east, across Russia and down to the Black and Caspian seas.

In the 11th century, Sweden, Norway and Denmark were separate kingdoms. However, in 1388, Sweden, fearing the growing influence of Germany on Sweden's affairs, turned to Queen Margaret of Denmark and Norway for help. The Germans were defeated in 1389 and, in 1397, Sweden, Denmark, and Norway were united by a treaty called the Union of Kalmar. Sweden defeated the Danes in 1523 and under Gustavus Vasa, a Swedish noble, Sweden broke away from the union. Gustavus encouraged followers of Martin Luther to spread their ideas and by 1540, Lutheranism had become the official religion.

From the late 16th century, Sweden became involved in a series of wars, during which it gained territory around the Baltic Ocean. In 1658, Sweden forced Denmark to give up its provinces on the Swedish mainland. Following defeat at the hands of Tsar Peter the Great in 1709, a coalition of Russia, Poland, and Denmark forced Sweden to give up most of its European possessions.

Sweden lost Finland to Russia in 1809, though it gained Norway from

goes on one of the widest ranging welfare programs in the world. In turn, the tax burden is the world's highest.

The elections of September 1991 saw the end of the Social Democratic government, which had been in power since 1932, with voters swinging toward parties advocating lower taxes. But the Social Democrats returned to power in 1994, advocating economic stringency.

A founder member of EFTA (European Free Trade Association), Sweden joined the European Union in 1985 following a referendum. In 2003, the government launched a referendum on replacing the krona with the euro. During the campaign Sweden's foreign minister, Anna Lindh, was murdered. Shortly afterward Swedish voters rejected the adoption of the euro.

ECONOMY
Sweden is a highly developed industrial country. Major products include iron and steel goods. Steel is used in the engineering industry to manufacture airplanes, automobiles, machinery and ships. Sweden has some of the world's richest iron ore deposits.

Denmark in 1814. By the late 19th century, Sweden was a major industrial nation. The Social Democratic Party was set up in 1889 to improve the conditions of workers. In 1905, Norway's parliament voted for independence from Sweden.

POLITICS
Sweden has a high standard of living, more than 70% of the national budget

STOCKHOLM

Port and capital of Sweden, on Lake Mälaren's outlet to the Baltic Sea. Founded in the mid-13th century, it became a trade center dominated by the Hanseatic League. Gustavus I made it the center of his kingdom, and ended the privileges of Hanseatic merchants. Stockholm became the capital of Sweden in 1436, and developed as an intellectual center in the 17th century. Industrial development dates from the mid-19th century. Industries include textiles, clothing, paper and printing, rubber, chemicals, shipbuilding, beer, and electronics.

Switzerland has used this square flag since 1848, though the white cross on the red shield has been Switzerland's emblem since the 14th century. The flag of the International Red Cross, which is based in Geneva, was derived from this flag.

The Swiss Confederation is a land-locked country in Western Europe. Much of the land is mountainous. The Jura Mountains lie along Switzerland's western border with France, while the Swiss Alps make up about 60% of the country in the south and east. Four-fifths of the people of Switzerland live on the fertile Swiss Plateau, which contains most of Switzerland's large cities.

Area 15,940 sq mi [41,284 sq km]
Population 7,451,000
Capital (population) Bern (124,000)
Government Federal republic
Ethnic groups German 65%, French 18%, Italian 10%, Romansch 1%, others
Languages French, German, Italian and Romansch (all official)
Religions Roman Catholic 46%, Protestant 40%
Currency Swiss franc = 100 centimes
Website www.vlada.hr/default.asp?ru=2

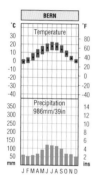

CLIMATE

The climate varies greatly according to the height of the land. The plateau region has a central European climate with warm summers, but cold and snowy winters. Rain occurs throughout the year.

POLITICS

In 1993 the Swiss voted against joining the European Union. However, in 2002, the Swiss voted by a narrow majority to end their centuries-old political isolationism and join the United Nations.

ECONOMY

Although lacking natural resources, Switzerland is a wealthy, industrialized country with many highly skilled workers. Major products include chemicals, electrical equipment, machinery and machine tools, precision instruments, processed food, watches, and textiles.

HISTORY

In 1291, three small cantons (states) united to defend their freedom against the Habsburg rulers of the Holy Roman Empire. They were Schwyz, Uri, and Unterwalden, and they called the confederation "Switzerland." In the 14th century, Switzerland defeated Austria in three wars of independence. But after a defeat by the French in 1515, the Swiss adopted a policy of neutrality, which they still follow.

In 1815, the Congress of Vienna expanded Switzerland to 22 cantons and guaranteed its neutrality. Switzerland's 23rd canton, Jura, was created in 1979 from part of the capital, Bern (Berne).

BERN (BERNE)

Capital of Switzerland, on the River Aare in Bern region. Founded in 1191 as a military post, it became part of the Swiss Confederation in 1353. Bern was occupied by France during the French Revolutionary Wars (1798). It has a Gothic cathedral, a 15th-century town hall, and is the headquarters of the Swiss National Library.

Farmers produce about three-fifths of the country's food—the rest is imported. Livestock raising, especially dairy farming, is the chief agricultural activity. Crops include fruits, potatoes, and wheat. Tourism and banking are also important. Swiss banks attract investors from all over the world.

LIECHTENSTEIN

The Principality of Liechtenstein is sandwiched between Switzerland and Austria. The River Rhine flows along its western border, while Alpine peaks rise in the east and the south. The capital, Vaduz, is situated on the Oberland Plateau above the fields and meadows of the Rhine Valley. The climate is relatively mild and the average annual precipitation is about 35 in [890 mm].

Liechtenstein, whose people speak a German dialect, has been an independent principality since 1719. Switzerland has represented Liechtenstein abroad since 1918 and Swiss currency was adopted in 1921. It has been in customs union with Switzerland since 1924.

Liechtenstein is best known abroad for its postage stamps, but is a haven for

Area 62 sq mi [160 sq km]
Population 33,000
Capital (population) Vaduz (5,000)
Government Hereditary constitutional monarchy
Ethnic groups Alemannic 86%, Italian, Turkish
Languages German (official), Alemannic dialect
Religions Roman Catholic 76%, Protestant 7%
Currency Swiss franc = 100 centimes
Website www.liechtenstein.li

international companies, attracted by the low taxation and the strictest banking codes in the world.

In 2003, the people voted to give the head of state, Prince Hans Adam III, sovereign powers. In 2004, he handed the running of the country to his son, Prince Alois, but remained titular head of state.

SYRIA

Syria has used this flag since 1980. The colors are those used by the Pan-Arab movement. This flag is the one that was used by the United Arab Republic between 1958 and 1961, when Syria was linked with Egypt and North Yemen.

The Syrian Arab Republic is in southwestern Asia. The narrow coastal plain is overlooked by a low mountain range which runs north–south. Another range, the Jabal ash Sharqi, runs along the border with Lebanon. South of this range is a region called the Golan Heights. Israel has occupied this region since 1967. East of the mountains, the bulk of Syria consists of fertile valleys, grassy plains, and large sandy deserts. This region contains the valley of the River Euphrates (Nahr al Furat).

Area 71,498 sq mi [185,180 sq km]
Population 18,017,000
Capital (population) Damascus (1,394,000)
Government Multiparty republic
Ethnic groups Arab 90%, Kurdish, Armenian, others
Languages Arabic (official), Kurdish, Armenian
Religions Sunni Muslim 74%, other Islam 16%
Currency Syrian pound = 100 piastres
Website www.syriatourism.org

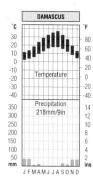

CLIMATE

The coast has a Mediterranean climate, with dry, warm summers and wet, mild winters. The low mountains cut off Damascus from the sea. It has less rainfall than the coastal areas and becomes drier to the east.

HISTORY

Syria is rich in historic sites from a wide range of periods. The earliest known settlers were Semites who arrived around 3,500 years ago. They set up city-states, such as Ebla, which existed between about 2700 and 2200 BC. The people of Ebla used clay tablets inscribed in cuneiform, an ancient system of writing developed by the Sumer people of Mesopotamia. Later conquerors of the area included the Akkadians, Canaanites, Phoenicians, Amorites, Aramaeans, and

the Hebrews, who introduced monotheism. The Assyrians occupied the area from 732 BC until 612 BC, when the Babylonians took over. The ancient Persians conquered the Babylonians in 539 BC, but the armies of Alexander the Great swept into the region in 331 BC, introducing Greek culture in their wake. The Romans took over in 64 BC, and Syria remained under Roman law for nearly 700 years.

Christianity became the state religion of Syria in the 4th century AD, but in 636 Muslims from Arabia invaded the region. Islam gradually replaced Christianity as the main religion, and Arabic became the chief language. From 661, Damascus became the capital of a vast Muslim empire which was ruled by the Ummayad Dynasty. However, the Abbasid Dynasty took over in 750 and the center of power passed to Baghdad.

From the late 11th century, Crusaders sought to win the Holy Land from the Muslims. But the Crusaders were unsuccessful in their aim because Saladin, a Muslim ruler of Egypt, defeated the Crusaders and ruled most of the area by the end of the 12th century. The Mameluke Dynasty of Egypt ruled Syria

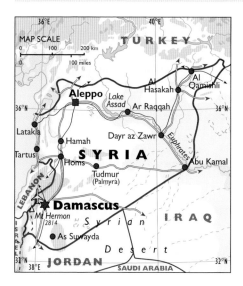

flict, greatly improved its standing in the West. In the mid-1990s, Syria had talks with Israel over the future of the Golan Heights. Negotiations were suspended after the election of Binyamin Netanyahu's right-wing government in Israel in 1996. Assad died in 2000 and was succeeded by his son, Bashar al Assad, raising hopes of a more pliable policy on the Golan Heights. Syria has been criticized for supporting Palestinian terrorist groups and keeping its troops in Lebanon. In 2005, following demonstrations against its continuing military presence in Lebanon, Syria announced the phased withdrawal of its troops.

ECONOMY

The World Bank classifies Syria as a "lower-middle-income" developing country. Its main resources are petroleum, hydroelectricity, and fertile land. Agriculture employs about 26% of the population. Petroleum is the chief mineral product, and phosphates are mined to make fertilizers.

from 1260 to 1516, when the region became part of the huge Turkish Ottoman Empire. During World War I, Syrians and other Arabs fought alongside British forces and overthrew the Turks.

POLITICS

After the collapse of the Turkish Ottoman empire in World War I, Syria was ruled by France. Syria became fully independent from France in 1946. The partition of Palestine and the creation of Israel in 1947 led to the first Arab-Israeli war, when Syria and other Arab nations failed to defeat Israeli forces. In 1949, a military coup established a military regime, starting a long period of revolts and changes of government. In 1967, in the third Arab-Israeli war (known as the Six Day War), Syria lost the strategically important Golan Heights to Israel.

In 1970, Lieutenant-General Hafez al Assad led a military revolt, becoming Syria's president in 1971. His repressive but stable regime attracted much Western criticism, and was heavily reliant on Arab aid. But Syria's anti-Iraq stance in the 1991 Gulf War, and the involvement of about 20,000 Syrian troops in the con-

DAMASCUS

Capital of Syria, on the River Barada, southwest Syria. Thought to be the oldest continuously occupied city in the world, in ancient times it belonged to the Egyptians, Persians and Greeks, and under Roman rule was a prosperous commercial center. It was held by the Ottoman Turks for 400 years, and after World War I came under French administration. It became capital of an independent Syria in 1941. Sites include the Great Mosque and the Citadel. It is Syria's administrative and financial center. Industries include damask fabric, metalware, leather goods, and sugar.

351

TAIWAN

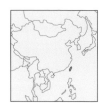

In 1928, the Chinese Nationalists adopted this design as China's national flag and used it in the long struggle against Mao Zedong's Communist army. When the nationalists were forced to retreat to Taiwan in 1949, their flag went with them.

Taiwan (formerly Formosa), is an island about 87 mi [140 km] off the south coast of mainland China. The country administers a number of islands close to the mainland. They include Quemoy (Jinmen) and Matsu (Mazu).

High mountain ranges, extending the length of the island, occupy the central and eastern regions, and only a quarter of the island's surface is used for agriculture. The highest peak is Yü Shan (Morrison Mountain), 12,966 ft [3,952 m] above sea level. Several peaks in the central ranges rise to more than 10,000 ft [3,000 m], and carry dense forests of broadleaved evergreen trees, such as camphor and Chinese cork oak. Above 5,000 ft [1,500 m], conifers, such as pine, larch, and cedar, dominate. In the east, where the mountains often drop steeply down to the sea, the short rivers have cut deep gorges. The western slopes are more gentle.

Area 36,000 sq mi [13,900 sq km]
Population 22,750,000
Capital (population) Taipei (2,619,022)
Government Unitary multiparty republic
Ethnic groups Taiwanese 84%, mainland Chinese 14%
Languages Mandarin Chinese (official), Min, Hakka
Religions Buddhism, Taoism, Confucianism
Currency New Taiwan dollar = 100 cents
Website www.roc-taiwan.org.uk

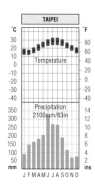

CLIMATE

Taiwan has a tropical monsoon climate. The annual rainfall exceeds 79 in [2,000 mm] in almost all areas. From July to September, the island is often hit by typhoons. When humidity is high the heat can be oppressive.

HISTORY

Chinese settlers arrived in Taiwan from the 7th century AD, displacing the Aboriginal people, but large settlements were not established until the 17th century. When the Portuguese first reached the island in 1590, they named the island Formosa (meaning "beautiful island"), but chose not to settle there. The Dutch occupied a trading port in 1624, but they were driven out in 1661 by refugees from the deposed Ming Dynasty on the mainland. A Ming official tried to use the island as a base for attacking the Manchu Dynasty, but without success as the Manchus took the island in 1683 and incorporated it into what is now Fujian province.

The Manchus settled the island in the late 18th century and, by the mid-19th century, the population had increased to about 2,500,000. The island was a major producer of sugar and rice, which were exported to the mainland. In 1886, the island became a Chinese province and Taipei became its capital in 1894. However, in 1895, Taiwan was ceded to Japan following the Chinese-Japanese War. Japan used the island as a source of food crops and, from the 1930s, they developed manufacturing industries based on hydroelectricity.

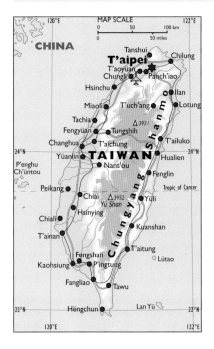

1971. It was then abandoned diplomatically by the United States in 1979, when the US switched its recognition to mainland China. However, in 1987 with continuing progress in the economy, martial law was lifted by the authoritarian regime in Taiwan. In 1988, a native Taiwanese became president and in 1991 the country's first general election was held.

China continued to regard Taiwan as a Chinese province and, in 1999, tension developed when the Taiwanese President Lee Teng-hui stated that relations between China and Taiwan should be on a "special state-by-state" basis. This angered the Chinese President Jiang Zemin, whose "one-nation" policy was based on the concept that China and Taiwan should be regarded as one country with two equal governments. Tension mounted in 2000, when Taiwan's opposition leader, Chen Shui-bian, was elected president, because Chen had adopted a pro independence stance. However, after the elections, Chen adopted a more conciliatory approach to mainland China.

POLITICS

In 1945, the Japanese army surrendered Taiwan to General Chiang Kai-shek's Nationalist Chinese government. Following victories by Mao Zedong's Communists, about 2 million Nationalists, together with their leader, fled the mainland to Taiwan in the two years before 1949, when the People's Republic of China was proclaimed. The influx met with hostility by the 8 million Taiwanese, and the new regime, the "Republic of China," was imposed with force. Boosted by help from the United States, Chiang's government set about ambitious programs for land reform and industrial expansion and, by 1980, Taiwan had become one of the top 20 industrial nations. Economic development was accompanied by a marked rise in living standards.

Nevertheless, Taiwan remained politically isolated and it lost its seat in the United Nations to Communist China in

ECONOMY

The economy depends on manufacturing and trade. Manufactures include electronic goods, footwear and clothing, ships and television sets. The western coastal plains produce large rice yields. Other products include bananas, pineapples, sugarcane, sweet potatoes, and tea.

TAIPEI

Capital and largest city of Taiwan, at the northern end of the island. A major trade center for tea in the 19th century, the city grew under Japanese rule (1895–1945), and became the seat of the Chinese Nationalist government in 1949. The city expanded from 335,000 people in 1945 to 2,619,022 in 2005. Industries include textiles, chemicals, fertilizers, metals, and machinery.

TAJIKISTAN

Tajikistan's flag was adopted in 1993. It replaced the flag used during the Communist period which showed a hammer and sickle. The new flag shows an unusual gold crown under an arc of seven stars on the central white band.

The Republic of Tajikistan is one of the five central Asian republics that formed part of the former Soviet Union. Only 7% of the land is below 3,280 ft [1,000 m], while almost all of eastern Tajikistan is above 9,840 ft [3,000 m]. The highest point is Communism Peak (Pik Kommunizma), which reaches 24,590 ft [7,495 m]. The main ranges are the westward extension of the Tian Shan Range in the north and the snowcapped Pamirs (Pamir) in the southeast. Earthquakes are common throughout the country.

Vegetation varies greatly according to altitude. Much of Tajikistan consists of desert or rocky mountain landscapes capped by snow and ice.

Area 55,521 sq mi [143,100 sq km]
Population 7,012,000
Capital (population) Dushanbe (529,000)
Government Republic
Ethnic groups Tajik 65%, Uzbek 25%, Russian
Languages Tajik (official), Russian
Religions Islam (Sunni Muslim 85%)
Currency Somoni = 100 dirams
Website www.tajiktour.taknet.com

CLIMATE

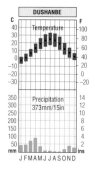

Tajikistan has an extreme continental climate. Summers are hot and dry in the lower valleys, and winters are long and bitterly cold in the mountains. Much of the country is arid, but the south east has heavy snowfalls.

HISTORY

The ancestors of the people of Tajikistan were Persians who had settled in the area about 2,500 years ago. The area was conquered many times. The Persians invaded in the 6th century BC and the Macedonian Greeks led by Alexander the Great in 331 BC. From 323 BC, the area was split into several independent states. Arab armies conquered the area in the mid-7th century and introduced Islam, which remains the chief religion today. The region was later ruled by various Turkic tribes and later by the Monguls, led by Genghis Khan. Uzbeks, a Turkic people, ruled the area as the Khanate of Bukhara from the 16th to the 19th centuries.

PAMIRS

A mountainous region in central Asia, lying mostly in Tajikistan and partly in Pakistan, Afghanistan and China. The region forms a geological structural knot from which the Tian Shan, Karakoram, Kunlun and Hindu Kush mountain ranges radiate. The climate is cold during winter and cool in the summer. The terrain includes grasslands and sparse trees. The main activity is sheep herding and some coal is mined. The highest peak is Pik Kommunizma at 24,590 ft [7,495 m].

The fragmentation of the region aided Russian conquest from 1868. Following the Russian Revolution (1917), Tajikistan rebelled against Russian rule. Although Soviet troops annexed northern Tajikistan into Turkistan in 1918, the Bukhara Emirate held out against the Red Army until 1921. In 1924 Tajikistan became an autonomous part of the Republic of Uzbekistan. In 1929 Tajikistan achieved full republic status, but Bukhara and Samarkand remained in the Republic of Uzbekistan. During the 1930s vast irrigation schemes greatly increased agricultural land. Many Russians and Uzbeks were settled in Tajikistan.

POLITICS

When the Soviet Union began to introduce reforms in the 1980s, many Tajiks demanded freedom. In 1989, the Tajik government made Tajik the official language instead of Russian and, in 1990, it stated that its local laws overruled Soviet laws. Tajikistan became fully independent in 1991, following the breakup of the Soviet Union. As the poorest of the ex-Soviet republics, Tajikistan faced many problems in trying to introduce a free-market system.

In 1992, civil war broke out between the government, which was run by former Communists, and an alliance of democrats and Islamic forces. The government maintained control, but it relied heavily on aid from the Commonwealth of Independent States, the organization through which most of the former Soviet republics kept in touch. Presidential elections in 1994 resulted in victory for Imomali Rakhmonov, though the Islamic opposition did not recognize the result.

A ceasefire was signed in December 1996. Further agreements in 1997 provided for the opposition to have 30% of the ministerial posts in the government. But many small groups excluded from the agreement continued to undermine the peace process through a series of killings and military actions. In 1999, Rakhmonov was reelected president. Changes to the constitution in 2003 enabled Rakhmonov to serve two more seven-year terms after the elections in 2006. His party won parliamentary elections in 2005.

ECONOMY

The World Bank classifies Tajikistan as a "low-income" developing country. Agriculture, mainly on irrigated land, is the main activity and cotton is the chief product. Other crops include fruits, grains and vegetables. The country has large hydroelectric power resources and it produces aluminum.

DUSHANBE

Capital of Tajikistan, at the foot of the Gissar Mountains, central Asia. Founded in the 1920s, it was known as Stalinabad from 1929–61. An industrial, trade, and transportation center, it is the site of Tajik University and Academy of Sciences. Industries include cotton milling, engineering, leather goods, and food processing.

355

TANZANIA

Tanzania's flag was adopted in 1964 when mainland Tanganyika joined with the island nation of Zanzibar to form the United Republic of Tanzania. The green represents agriculture and the yellow minerals. The black represents the people, while the blue symbolizes Zanzibar.

The United Republic of Tanzania consists of the former mainland country of Tanganyika and the island nation of Zanzibar, which also includes the island of Pemba.

Behind a narrow coastal plain, the majority of Tanzania is a plateau lying between 2,950 and 4,920 ft [900–1,500 m] above sea level. The plateau is broken by arms of the Great African Rift Valley. The western arm contains lakes Nyasa (also called Malawi) and Tanganyika, while the eastern arm contains the strongly alkaline Lake Natron, together with lakes Eyasi and Manyara. Lake Victoria occupies a shallow depression in the plateau and it is not situated within the Rift Valley.

Kilimanjaro, the highest peak, is an extinct volcano. At 19,340 ft [5,895 m], it is also Africa's highest mountain. Zanzibar and Pemba are coral islands.

Area 364,899 sq mi [945,090 sq km]
Population 36,588,000
Capital (population) Dodoma (204,000)
Government Multiparty republic
Ethnic groups Native African 99% (Bantu 95%)
Languages Swahili (Kiswahili) and English (both official)
Religions Islam 35% (99% in Zanzibar), traditional beliefs 35%, Christianity 30%
Currency Tanzanian shilling = 100 cents
Website www.tanzania.go.tz/index2E.html

its importance. Arab traders often inter-married with local people and the Arab-African people produced the distinctive Arab-Swahili culture. The Portuguese took control of coastal trade in the early 16th century, but the Arabs regained control in the 17th century.

In 1698, Arabs from Oman took control of Zanzibar. From this base, they developed inland trade, bringing gold, ivory, and slaves from the interior. During the 19th century, European explorers and missionaries were active, mapping the country and striving to stop the slave trade.

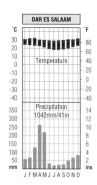

CLIMATE
The coast has a hot, humid climate. The greatest rainfall is in April and May. Inland mountains and plateaus are cooler and less humid. The Rift Valley is hot. Mount Kilimanjaro is permanently covered in snow and ice.

HISTORY
Around 2,000 years ago, Arabs, Persians and Chinese traded along the Tanzanian coast. The old cities and ruins testify to

LAKE TANGANYIKA

Second largest lake in Africa and the second deepest freshwater lake in the world. It lies in east-central Africa and borders Tanzania, DR Congo, Zambia, and Burundi in the Rift Valley. The lake's fish are a vital source of protein for the local peoples. It has an area of 12,700 sq mi [32,893 sq km] and a depth of 4,715 ft [1,437 m].

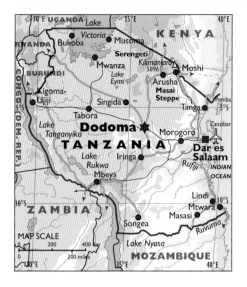

POLITICS

Mainland Tanganyika became a German territory in the 1880s, while Zanzibar (including Pemba) became a British protectorate in 1890. The Germans introduced a system of forced labor to develop plantations. This led to a major rebellion in 1905, which was put down with great brutality.

Following Germany's defeat in World War I, Britain gained control of Tanganyika and was granted a mandate to rule it by the League of Nations. Tanganyika remained a Briish territory until winning its independence in 1961, followed by Zanzibar in 1963. Tanganyika and Zanzibar united to form the United Republic of Tanzania in 1964.

The country's first president, Julius Nyerere, pursued socialist policies of self-help (called *ujamaa* in Swahili) and egalitarianism. While many of its social reforms were successful, the country failed to make economic progress. Nyerere resigned as president in 1985, though he remained influential until his death in 1999. His successors, Ali Hassan Mwinyi, who served from 1985 until 1995, and Benjamin Mkapa, who was reelected in 2000, pursued more liberal economic policies. In 2005, Mkapa was succeeded by Jakaya Kikwete, another CCM (Chama Cha Mapinduzi) candidate.

ECONOMY

Tanzania is one of the world's poorest countries. Although crops are grown on only 5% of the land, agriculture employs 85% of the people. Most farmers grow only enough to feed their families. Food crops include bananas, cassava, corn, millet, rice, and vegetables. Export crops include coffee, cotton, cashew nuts, tea and tobacco. Other crops grown for export include cloves, coconuts, and sisal. Some farmers raise animals, but sleeping sickness and drought restrict the areas for livestock farming.

Diamonds and other gems are mined, together with some coal and gold. Industry is mostly small-scale. Manufactures include processed food, fertilizers, petroleum products, and textiles.

Tourism is increasing. Tanzania has beautiful beaches, but its main attractions are its magnificent national parks and reserves, including the celebrated Serengeti and the Ngorongoro Crater. These are renowned for their wildlife and are among the world's finest.

Tanzania also contains a major archaeological site, Olduvai Gorge, west of the Serengeti. Here, in 1964, the British archaeologist and anthropologist, Louis Leakey, discovered the remains of ancient human-like creatures.

DODOMA

Capital of Tanzania, located in the center of the country. Dodoma replaced Dar es Salaam as capital in 1974. It is in an agricultural region, crops include grain, seeds, and nuts.

THAILAND

Thailand's flag was adopted in 1917. In the late 19th century, it featured a white elephant on a plain red flag. In 1916, white stripes were introduced above and below the elephant, but in 1917 the elephant was dropped and a central blue band was added.

The Kingdom of Thailand is one of ten nations in Southeast Asia. Central Thailand is a fertile plain drained mainly by the Chao Phraya. A densely populated region, it includes the capital, Bangkok. The highest land occurs in the north and includes the second largest city, Chiang Mai, and Doi Inthanon, the highest peak, which reaches 8,514 ft [2,595 m].

The Khorat Plateau, in the northeast, makes up about 30% of the country and extends to the River Mekong border with Laos. In the south, Thailand shares the finger-like Malay Peninsula with Burma and Malaysia.

The vegetation of Thailand includes many hardwood trees to the north. The south has rubber plantations. Grass, shrub, and swamp make up 20% of land. Some 33% of the land is arable, mainly comprising rice fields.

Area 198,114 sq mi [513,115 sq km]
Population 64,866,000
Capital (population) Bangkok (6,320,000)
Government Constitutional monarchy
Ethnic groups Thai 75%, Chinese 14%, others
Languages Thai (official), English, ethnic and regional dialects
Religions Buddhism 95%, Islam, Christianity
Currency Baht = 100 satang
Website www.thaigov.go.th

CLIMATE

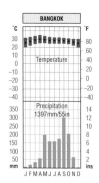

Thailand has a tropical climate. Monsoon winds from the southwest bring heavy rains between May and October. Bangkok is drier than many parts of Southeast Asia because mountains shelter the central plains from the rain-bearing winds.

HISTORY

The Mongol capture in 1253 of a Thai kingdom in southwest China forced the Thai people to move south. A new kingdom was established around Sukothai. In the 14th century the kingdom expanded and the capital moved to Ayutthaya.

European contact began in the early 16th century. However, in the late 17th century, the Thais, fearing interference in their affairs, forced all Europeans to leave. This policy continued for 150 years. Thailand remained the only Southeast Asian nation to resist colonization.

In 1782, a Thai General, Chao Phraya Chakkri, became king, founding a dynasty which continues today. The country became known as Siam, and Bangkok became its capital. From the mid-19th century, contacts with the West were restored. In World War I, Siam supported the Allies against Germany and Austria-Hungary. In 1932, Thailand became a constitutional monarchy.

POLITICS

In 1938, Pibul Songkhram became premier and changed the country's name to Thailand. In 1941, Pibul, despite opposition, invited Japanese forces into Thai-

south where the majority of the population is Muslim, many of whom claim that they suffer discrimination by the central government.

ECONOMY

Despite its rapid progress, the World Bank classifies the country as a "lower-middle-income" developing country. Manufactures, including commercial vehicles, food products, machinery, timber products, and textiles, are exported.

Agriculture still employs two-thirds of the workforce. Rice is the chief crop, while other major crops include cassava, cotton, corn, pineapples, rubber, sugarcane, and tobacco. Thailand also mines tin and other minerals.

Tourism is a major source of income, though the December 2004 tsunami, which killed over 5,000 people, cast a shadow over its future growth.

land. Pibul was overthrown in a military coup in 1957. The military governed Thailand until 1973.

Since 1967, when Thailand became a member of ASEAN (Association of South East Asian Nations), its economy has grown, especially its manufacturing and service industries. However, in 1997, it suffered recession along with other eastern Asian countries and this persisted into the 21st century.

A military group seized power in 1991, but elections were held in 1992, 1995 and 2001. Then in 2004 Thailand was rocked by sectarian violence in the

BANGKOK

Capital and chief port of Thailand, located on the east bank of the River Chao Phraya. Bangkok became the capital in 1782, when King Rama I built a royal palace here. It quickly became Thailand's largest city. The Grand Palace (including the sacred Emerald Buddha) and more than 400 Buddhist temples (*wats*) are notable examples of Thai culture. It has a large Chinese minority. During World War II it was occupied by the Japanese. Today, Bangkok is a busy market center, much of the city's commerce taking place on the numerous canals (*klongs*) which are found on the Thonburi (original site of the capital) side of the river and connect the city with the suburbs. The port handles most of Thailand's imports and exports. Industries include tourism, building materials, rice processing, textiles, and jewelry.

TOGO

The five stripes represent action and the five regions of Togo, the alternation of colors is for unity in diversity. Red represents the blood shed in the struggle for independence, the star is for life, liberty and labor. Green is for hope and agriculture. Yellow is for Togo's mineral wealth.

The Republic of Togo is a long, narrow country in West Africa. From north to south, it extends about 311 mi [500 km]. Its coastline on the Gulf of Guinea is only 64 km [40 mi] long, and it is only 145 km [90 mi] at its widest point. The coastal plain is sandy. North of the coast is an area of fertile, clay soil. North again is the Mono Tableland which reaches an altitude of 1,500 ft [450 m], and is drained by the River Mono. The Atakora Mountains are the fourth region. The vegetation is mainly open grassland.

Area 21,925 sq mi [56,785 sq km]
Population 5,557,000
Capital (population) Lomé (658,000)
Government Multiparty republic
Ethnic groups Native African 99% (largest tribes are Ewe, Mina and Kabre)
Languages French (official), African languages
Religions Traditional beliefs 51%, Christianity 29%, Islam 20%
Currency CFA franc = 100 centimes
Website www.republicoftogo.com

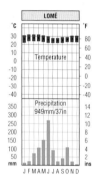

CLIMATE

Togo has year-round high temperatures. The main wet season runs between the months of March and July, and there is a further minor wet season in October and November.

HISTORY

The historic region of Togoland comprised what is now the Republic of Togo and West Ghana. From the 17th to the 19th century, the Ashanti raided Togoland and sold the indigenous inhabitants, the Ewe, to Europeans as slaves.

Togo became a German protectorate in 1884; it developed economically and Lomé was built. At the start of World War I, Britain and France captured Togoland from Germany. In 1922, it divided into two mandates, which in 1942 became UN Trust Territories. In 1956, the people of British Togoland voted to join Ghana, while French Togoland gained independence as the Republic of Togo in 1960.

In 1961 Sylvanus Olympio became Togo's first president. He was assassinated in 1963. Nicolas Grunitzky became president, but he was overthrown in the military coup of 1967, led by the head of the armed forces, General Gnassingbé Eyadéma, who then became president. In 1969 a new constitution confirmed Togo as a single-party state, the sole legal party being the Rassemblement du Peuple Togolais (RPT).

Reelected in 1972 and 1986, Eyadéma was forced to resign in 1991 after prodemocracy riots. Kokou Koffigoh led an interim government. Unrest continued with troops loyal to Eyadéma attempting to overthrow Koffigoh.

POLITICS

A new constitution was adopted in 1992. In 1993 Eyadéma won rigged elections. Multiparty elections were held in 1994 and these were won by an opposition alliance, but Eyadéma formed a coalition government. In 1998, paramilitary police

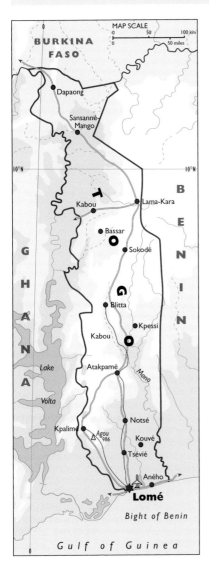

Eyadéma to stand for reelection. He won the subsequent elections in 2003.

Eyadéma died in 2005 and his son, Faure, became president. After international pressure he stepped down and called elections. Faure won these two months later amid claims by the opposition that the vote was rigged. In addition, the political violence surrounding the presidential poll was such that around 40,000 Togolese fled to neighboring countries.

These events called into question Togo's commitment to democracy which had been declared in 2004 when trying to normalize ties with the EU. The EU had cut off aid to Togo in 1993 over the country's human rights record.

ECONOMY

Togo is a poor developing country. Farming employs 65% of the people, but most farmers grow little more than they need to feed their families. Major food crops include cassava, corn, millet, and tropical yams. The chief cash crops are cocoa, coffee, and cotton. The leading exports are phosphate rock, which is used to make fertilizers, and palm oil.

Togo's small-scale manufacturing and mining industries employ about 6% of the people.

prevented the completion of the count in presidential elections when it became clear that Eyadéma had been defeated. Eyadéma continued in office and the main opposition parties boycotted the general elections in 1999. In late 2002 the constitution was changed to allow

LOMÉ

Capital and largest city of the Republic of Togo, on the Gulf of Guinea. Made capital of German Togoland in 1897, it later became an important commercial center. It was the site of two conferences (1975, 1979) that produced a trade agreement (known as the Lomé Convention) between Europe and 46 African, Caribbean, and Pacific states. Its main exports are coffee, cocoa, palm nuts, copra, and phosphates.

TRINIDAD & TOBAGO

The colors represent earth, water, and fire. The black stripe stands for abundant petroleum and natural gas resources, plus the unity and determination of the people. The white stripes symbolize the Caribbean Sea, as well as purity and equality. The red field represents the warm Caribbean sun.

The Republic of Trinidad and Tobago consists of two main islands and is the most southerly in the Lesser Antilles. The largest island, Trinidad, is just 10 mi [16 km] off Venezuela's Orinoco delta. A detached extension of Trinidad's hilly Northern Range, lies 21 mi [34 km] to the north. The country's highest point is Mount Aripo (3,085 ft [940 m]) in Trinidad's rugged and forested Northern Range. Fertile plains cover much of the country.

Area 1,981 sq mi [5,130 sq km]
Population 1,097,000
Capital (population) Port of Spain (51,000)
Government Multiparty republic
Ethnic groups Indian (South Asian) 40%, African 38%, mixed 21%, others
Languages English (official), Hindi, French, Spanish, Chinese
Religions Roman Catholic 26%, Hindu 23%, Anglican 8%, Baptist 7%, Pentecostal 7%, others
Currency Trinidad and Tobago dollar = 100 cents
Website www.visittnt.com

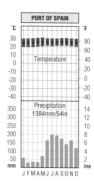

PORT OF SPAIN

Temperature

Precipitation
1384mm/54in

CLIMATE
Temperatures are high throughout the year, ranging from 64° to 92°F [18°–33°C]. Rainfall is heavy; the wettest months are from June to November. Annual rainfall ranges from 50 in [1,270 mm] on southwestern Trinidad to more than 100 in [2,540 mm] on the highlands of Tobago.

HISTORY
Christopher Columbus visited the islands, then populated by Arawak and Carib Amerindians, in 1498. He named Trinidad after three peaks at its southern tip, and Tobago after a local tobacco pipe. Spain colonized Trinidad in 1532, while Dutch settlers planted sugar on plantations in Tobago in the 1630s. In 1781, France colonized Tobago and further developed its plantation economy. The British captured Trinidad from Spain in 1797 and, in 1802, Spain formally ceded the island to Britain. In 1814, France also ceded Tobago to Britain and, in 1869, the two islands were combined into one colony. Black slaves worked on the plantations until slavery was abolished in 1834. To meet the problem of labor shortages, the British recruited Indian and Chinese indentured laborers. The presence of people of African, Asian, and European origin has resulted in a complex cultural mix in present day Trinidad and Tobago.

POLITICS
Independence was achieved in 1962. Eric Williams, moderate leader of the People's National Movement (PNP) which he had founded in 1956, became prime minister. In 1970, the government declared a state of emergency following violence by black power supporters, who called for an end to foreign influence and unemployment. The emergency was lifted in 1972, but strikes caused problems in 1975. Trinidad and Tobago became a republic in 1976, with Williams continuing as prime minister. In 1986, after 30

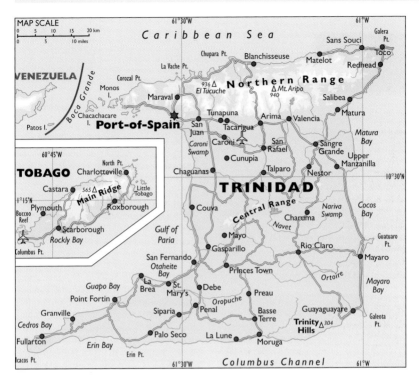

MAP SCALE
0 5 10 15 20 km
0 5 10 miles

C a r i b b e a n   S e a

61°30'W · 61°W

Galera Pt.
Sans Souci
Toco
Chupara Pt. · Blanchisseuse · Matelot · Redhead
La Vache Pt.
VENEZUELA
Corozal Pt.
Monos I.
Maraval
936△ El Tucuche · **N o r t h e r n   R a n g e** · △Mt. Aripo 940 · Salibea · Matura
Chacachacare I.
Patos I.
Port-of-Spain
San Juan · Tunapuna · Arima · Valencia
Tacarigua
Caroni · Caroni Swamp · San Rafael · Sangre Grande · Upper Manzanilla
Cunupia · Matura Bay
Talparo · Nestor

TOBAGO
60°45'W
North Pt.
Charlotteville
Castara · 565△ · **Main Ridge** · Little Tobago
1°15'N
Plymouth · Roxborough
Buccoo Reef
Scarborough
Rockly Bay
Columbus Pt.

Chaguanas · **TRINIDAD**
Couva · **Central Range** · Nariva Swamp · Cocos Bay
Charuma · Navet
Gulf of Paria
Mayo · Guataro Pt.
Gasparillo · Rio Claro · Mayaro
San Fernando
Otaheite Bay · Princes Town · Ortoire · Mayaro Bay
Guapo Bay · La Brea · St. Mary's · Debe · Preau · Guayaguayare · Galeota Pt.
Point Fortin · Oropuche
Granville · Siparia · Penal · Basse Terre · **Trinity Hills** △304
Cedros Bay
Fullarton · Palo Seco · La Lune · Moruga
Icacos Pt. · Erin Bay · Erin Pt.
10°30'N
C o l u m b u s   C h a n n e l
61°30'W · 61°W

years in office, the PNP was defeated in elections. The National Alliance for Reconstruction (NAR) coalition took office under Arthur Robinson. In 1990, Islamists seized parliament and held Robinson and other officials hostage for several days. In 1991, Patrick Manning became prime minister following an election victory for the PNP, but in 1994 Baseo Panday, leader of the Indian-based United National Congress (UNC), became prime minister, leading a coalition with the NAR. In the 2002 elections the PNP was victorious and Patrick Manning returned as prime minister.

Trinidad and Tobago is a major transshipment point for cocaine being moved from South America to North America and Europe. Cannabis is also produced in the country. The drug trade has fueled gang violence and corruption. The death penalty was reintroduced in 1999, despite strong international pressure. In 2005, a Caribbean Court of Justice was set up in Trinidad as a final court of appeal to replace the British Privy Council.

ECONOMY

Petroleum is vital to the economy. Chief exports include refined and petroleum, anhydrous ammonia, and iron and steel.

PORT OF SPAIN

Capital of Trinidad and Tobago, on the northwest coast of Trinidad. Founded by the Spanish in the late 16th century, it was seized by Britain in 1797. From 1958 to 1962 it was the capital of the Federation of the West Indies. It is a major tourist and shipping center.

Tunisia's flag originated in about 1835 when the country was officially under Turkish rule. It became the national flag in 1956, when Tunisia became independent from France. The flag contains two traditional symbols of Islam, the crescent and the star.

The Republic of Tunisia is the smallest country in North Africa. The mountains in the north are an eastward and comparatively low extension of the Atlas Mountains.

To the north and east of the mountains lie fertile plains, especially between Sfax, Tunis and Bizerte. South of the mountains lie broad plateaus which descend toward the south. This low-lying region contains a large salt pan, called the Chott Djerid, and part of the Sahara.

Area 63,170 sq mi [163,610 sq km]
Population 9,975,000
Capital (population) Tunis (702,000)
Government Multiparty republic
Ethnic groups Arab 98%, European 1%
Languages Arabic (official), French
Religions Islam 98%, Christianity 1%, others
Currency Tunisian dinar = 100 1,000 millimes
Website www.tourismtunisia.com

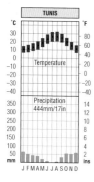

CLIMATE

Northern Tunisia has a Mediterranean climate, with dry summers, and mild winters with a moderate rainfall. The average yearly rainfall decreases toward the south, which forms part of the Sahara.

HISTORY

Tunisia has come under the influence of a succession of cultures, each of which has left its mark on the country, giving Tunisia a distinct identity and a long tradition of urban life. The Phoenicians began the Carthaginian Empire in Tunisia around 1100 BC and, according to legend, the colony of Carthage was established in 814 BC on a site near present-day Tunis. At its peak, Carthage controlled large areas in the eastern Mediterranean but, following the three Punic Wars with Rome, Carthage was

destroyed in 146 BC. The Romans ruled the area for 600 years until the Vandals defeated them in AD 439. The Vandals were finally conquered by the Byzantines. Arabs reached the area in the mid-7th century, introducing Islam and the Arabic language. In 1547, Tunisia came under the rule of the Turkish Ottoman Empire.

In 1881, France established a protectorate over Tunisia and ruled the country until 1956. Tunisian aspirations were felt before World War I, but it was not until 1934 that Habib Bourguiba founded the first effective opposition group, the Neo-Destour (New Constitution) Party, which was renamed the Socialist Destour Party in 1964, and is now known as the Constitutional Assembly.

Tunisia supported the Allies during World War II and it was the scene of much fierce fighting. Following independence, the new parliament abolished the monarchy and declared Tunisia to be a republic in 1957. The nationalist leader, Habib Bourguiba, became president.

POLITICS

In 1975, Bourguiba was elected president for life. His government introduced

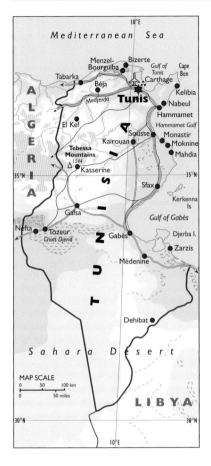

president in 1989, 1994, 1999, and 2004 with his party dominating the Chamber of Deputies, though some seats were reserved for opposition parties whatever their proportion of the popular vote. However, he faced opposition from Islamic fundamentalists. Occasional violence and suppression of human rights, including the banning of al-Nahda, the main Islamic party, marred his presidency. However, Islamic fundamentalism in Tunisia did not prove to be anything like as effective as in Algeria.

ECONOMY

The World Bank classifies Tunisia as a "middle-income" developing country. Its main natural resources are petroleum and phosphates. Agriculture employs 22% of the people. Chief crops are barley, citrus fruits, dates, grapes, olives, sugarbeet, tomatoes, and wheat. Sheep are the most important livestock, but goats and cattle are also raised. Tourism has grown considerably.

Since independence, new industries and tourism have transformed a number of coastal towns. Major manufactures include cement, flour, phosphoric acid, processed food, and steel. An important stimulus was the signing of a free-trade agreement with the EU in 1995. In doing so, Tunisia became the first Arab country on the Mediterranean to sign such an agreement.

many reforms, including votes for women. But problems arose from the government's successes. For example, the establishment of a national school system led to a very rapid increase in the number of educated people who were unable to find jobs that measured up to their qualifications. The growth of tourism, which provided a valuable source of foreign currency, also led to fears that Western influences might undermine traditional Muslim values. Finally, the prime minister, Zine el Abidine Ben Ali, removed Bourguiba from office in 1987 and succeeded him as president. He was elected

TUNIS

Capital and largest city of Tunisia. Tunis became the capital in the 13th century under the Hafsid dynasty. Seized by Barbarossa in 1534 and controlled by Turkey, it attained infamy as a haven for pirates. The French assumed control in 1881. Tunis gained independence in 1956. Products include olive oil, carpets, textiles, and handcrafts. The ruins of Carthage are nearby.

TURKEY

Turkey's flag was adopted when the Republic of Turkey was established in 1923. The crescent moon and the five-pointed star are traditional symbols of Islam. They were used on earlier Turkish flags used by the Turkish Ottoman Empire.

The Republic of Turkey lies in two continents. The European section (Thrace) lies west of a waterway between the Black and Mediterranean seas. This waterway consists of the Bosphorus, on which the city of Istanbul stands, the Sea of Marmara (Marmara Denizi) and a narrow strait called the Dardanelles.

Most of the Asian part of Turkey consists of plateaus and mountains, which rise to 16,945 ft [5,165 m] at Mount Ararat (Agri Dagi) near the border with Armenia. Earthquakes are common.

Deciduous forest grow inland with conifers on the mountains. The plateau is mainly dry steppe.

Area 299,156 sq mi [774,815 sq km]
Population 68,894,000
Capital (population) Ankara (2,984,000)
Government Multiparty republic
Ethnic groups Turkish 80%, Kurdish 20%
Languages Turkish (official), Kurdish, Arabic
Religions Islam (mainly Sunni Muslim) 99%
Currency New Turkish lira = 100 kurus
Website www.kultur.gov.tr/EN

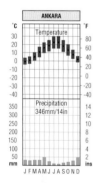

CLIMATE

Central Turkey has a dry climate, with hot, sunny summers and cold winters. The driest part of the central plateau lies south of Ankara, around Lake Tuz. Western Turkey has a Mediterranean climate, while the Black Sea coast has cooler summers.

HISTORY

In AD 330, the Roman Empire moved its capital to Byzantium, renaming it Constantinople. Constantinople became the capital of the East Roman (or Byzantine) Empire in 395. Muslim Seljuk Turks from central Asia invaded Anatolia in the 11th century. In the 14th century, another group of Turks, the Ottomans, conquered the area. In 1453, the Ottoman Turks took Constantinople, which they called Istanbul. The Ottoman Turks built up a large empire, which finally collapsed during World War I (1914–18). In 1923, Turkey became a republic. Its leader Mustafa Kemal, or Atatürk ("father of the Turks"), launched policies to modernize and secularize the country.

Turkey joined NATO in 1951 and applied to join the European Economic Community in 1987. But Turkey's conflict with Greece, together with its invasion of northern Cyprus in 1974, have led many Europeans to treat Turkey's aspirations with caution. Political instability, military coups, conflict with Kurdish nationalists in eastern Turkey and concern about the country's record on human rights are other problems.

POLITICS

Turkey has enjoyed democracy since 1983, though in 1998 the government banned the Islamist Welfare Party, accusing it of violating secular principles. In 1999, the largest numbers of parliamentary seats were won by the ruling Democratic Left Party and the far-right Nationalist Action Party. In 2001, the

Turkish parliament adopted reforms to ease the country's entry into the European Union. One reform formally recognized men and women as equals—the former code designated the man as the head of the family.

In the elections of 2002 the moderate Islamic Justice and Development Party (AKP) won 362 of the 500 seats in parliament. None of the parties in the former ruling coalition won even 10%.

Turkey finally agreed to recognize Cyprus as an EU member and this led to EU membership talks being formally launched in October 2005 with negotiations expected to take about 10 years.

In the 1980s and 1990s civil war was a problem in the east and southeast of Turkey. Fighting took place between Turkish forces and those of the secessionist Kurdistan Workers' Party (PKK). Over 30,000 people died. The PKK seeks greater political and cultural rights for the Kurdish community. A five-year ceasefire was called off in 2004 by Kurdish secessionists after what they called annihilation operations against their fighters by the Turkish authorities. There have been subsequent clashes between Kurdish fighters and Turkish forces in the southeast causing many deaths.

ECONOMY

Turkey is a "lower-middle-income" developing country. Agriculture employs 40% of the people, and barley, cotton, fruits, corn, tobacco, and wheat are major crops. Livestock farming is important and wool is a leading product. Manufacturing is the chief activity, including processed farm products and textiles, automobiles, fertilizers, iron and steel, machinery, metal products, and paper products. Turkey receives more than 9 million tourists a year.

ANKARA

Capital of Turkey, at the confluence of the Cubuk and Ankara rivers. In ancient times it was known as Ancyra, and was an important commercial center as early as the 8th century BC. It flourished under Augustus as a Roman provincial capital. Tamerlane took the city in 1402. Mustafa Kemal set up a provisional government here in 1920. It replaced Istanbul as the capital in 1923, changing its name to Ankara in 1930. It is noted for its angora wool and mohair.

TURKMENISTAN

Turkmenistan's flag was adopted in 1992. It incorporates a typical Turkmen carpet design. The crescent is a symbol of Islam, while the five stars and the five elements in the carpet represent the traditional tribal groups of Turkmenistan.

The Republic of Turkmenistan is one of five central Asian republics which once formed part of the Soviet Union. Most of the land is low-lying, with mountains on the southern and southwestern borders.

In the west lies the salty Caspian Sea. A depression called the Kara Bogaz Gol Bay contains the country's lowest point. Most of the country is arid and Asia's largest sand desert, the Garagum, covers 80% of the country, though parts of it are irrigated by the Garagum Canal.

Area 188,455 sq mi [488,100 sq km]
Population 4,863,000
Capital (population) Ashkhabad (521,000)
Government Single-party republic
Ethnic groups Turkmen 85%, Uzbek 5%, Russian 4%, others
Languages Turkmen (official), Russian, Uzbek
Religions Islam 89%, Eastern Orthodox 9%
Currency Turkmen manat = 100 tenesi
Website
www.turkmenistan.gov.tm/index_eng.html

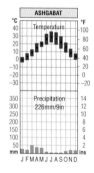

CLIMATE

Turkmenistan has a continental climate, with average annual rainfall varying from 3 in [80 mm] in the desert to 12 in [300 mm] in the mountains. Summers are very hot, but temperatures during winter drop below freezing.

HISTORY

Just over 1,000 years ago, Turkic people settled in the lands east of the Caspian Sea and the name "Turkmen" comes from this time. Genghis Khan and his Mongul armies conquered the area in the 13th century and it subsequently became part of Tamerlane's vast empire. With the breakup of the Timurid dynasty, Turkmenistan came under Uzbek control. Islam was introduced in the 14th century.

Russia took over the region during the 1870s and 1880s. In 1899, despite resistance, Turkmenistan became part of Russian Turkistan. After the Russian Revolution of 1917, the area came under Communist rule and, in 1924, as part of the Turkistan Autonomous Soviet Socialist Republic, it joined the Soviet Union. The Communists strictly controlled all aspects of life and, in particular, they discouraged religious

CASPIAN SEA

Shallow salt lake, the world's largest inland body of water. The Caspian Sea is enclosed on three sides by Russia, Kazakhstan, Turkmenistan and Azerbaijan. The southern shore forms the northern border of Iran. It has been a valuable trade route for centuries. It is fed mainly by the River Volga; there is no outlet. The chief ports are Baku and Astrakhan. It has important fisheries and covers an area of 143,000 sq mi [371,000 sq km].

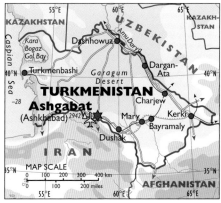

expected to take spiritual guidance from his book, *Ruhnama*, a collection of thoughts on Turkmen culture and history. After giving up smoking due to major heart surgery in 1997, he ordered all his ministers to give up and then banned smoking in public places. Subsequent bans include one on young men having beards and long hair, opera, ballet, and the playing of recorded music on television, at public events, and at weddings.

worship. But they also improved such services as education, health, housing, and transportation.

POLITICS
During the 1980s, the Soviet Union introduced reforms, and the Turkmen began to demand more freedom. In 1990, the Turkmen government stated that its laws overruled Soviet laws. In 1991, Turkmenistan became fully independent after the breakup of the Soviet Union, but kept ties with Russia through the Commonwealth of Independent States (CIS).

In 1992, Turkmenistan adopted a new constitution, allowing for political parties, providing that they were not ethnic or religious in character. But effectively Turkmenistan remained a one-party state and, in 1992, Saparmurad Niyazov, the former Communist and now Democratic leader, was the only candidate. In 1994 a referendum prolonged Niyazov's term of office to 2002, while in 1999 the parliament declared him president for life. In 2004, parliamentary elections were described as a "sham" because all the candidates supported the president. In 2005 he surprised observers by calling for contested presidential elections to take place in 2009.

Niyazov seeks to influence every aspect of his people's lives. Turkmens are

ECONOMY
Turkmenistan joined the Economic Cooperation Organization, which was set up in 1985 by Iran, Pakistan and Turkey. In 1996, the completion of a railroad link from Turkmenistan to the Iranian coast was seen as an important step in the development of Central Asia. The World Bank classifies Turkmenistan as a "lower-middle-income" country. The chief resources are petroleum and natural gas, but agriculture is important. The chief crop, grown on irrigated land, is cotton. Grains and vegetables are also important. Manufactures include cement, glass, petrochemicals, and textiles. Turkmenistan has extensive hydrocarbon and natural gas reserves that could be of major economic assistance.

ASHGABAT (ASHKHABAD)
Capital of the central Asian republic of Turkmenistan, located 25 mi [40 km] from the Iranian border. Founded in 1881 as a Russian fortress between the Garagum Desert and the Kopet Dagh Mountains, it was largely rebuilt after a severe earthquake in 1948. The city was known as Poltaratsk from 1919 to 1927. Its present name was adopted after the republic attained independence from the former Soviet Union in 1991.

UGANDA

The flag used by the party that won the first national election was adopted as the national flag when Uganda became independent from Britain in 1962. The black represents the people, the yellow the sun, and the red brotherhood. The crested crane is the country's emblem.

The Republic of Uganda is a land-locked country on the East African Plateau. It contains part of Lake Victoria, Africa's largest lake and a source of the River Nile, which occupies a shallow depression in the plateau.

The plateau varies in height from about 4,921 ft [1,500 m] in the south to 2,953 ft [900 m] in the north. The highest mountain is Margherita Peak, which reaches 16,762 ft [5,109 m] in the Ruwenzori Range in the southwest. Other mountains, including Mount Elgon at 14,177 ft [4,321 m], rise along Uganda's eastern border.

Part of the Great African Rift Valley, which contains lakes Edward and Albert, lies in western Uganda. The landscapes range from rainforests in the south, through savanna in the center, to semi-desert in the north.

Area 93,065 sq mi [241,038 sq km]
Population 26,405,000
Capital (population) Kampala (774,000)
Government Republic
Ethnic groups Baganda 17%, Ankole 8%, Basogo 8%, Iteso 8%, Bakiga 7%, Langi 6%, Rwanda 6%, Bagisu 5%, Acholi 4%, Lugbara 4% and others
Languages English and Swahili (both official), Ganda
Religions Roman Catholic 33%, Protestant 33%, traditional beliefs 18%, Islam 16%
Currency Ugandan shilling = 100 cents
Website www.statehouse.go.ug

CLIMATE

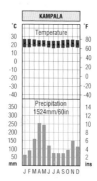

The Equator runs through Uganda and the country is warm throughout the year, though the high altitude moderates the temperature. The lands to the north of Lake Victoria, and the western mountains, especially the high Ruwenzori Range are the wettest regions. Much of Uganda has two rainy seasons from Apri to May and from October to December. In the center and the north, these merge into one, with a distinct dry season.

HISTORY

In around 1500, the Nilotic-speaking Lwo people formced various kingdoms in southwestern Uganda, including Buganda (kingdom of the Ganda) and Bunyoro. During the 18th century, the Buganda kingdom expanded and trade flourished. In 1862, a British explorer, John Speke became the first European to reach Buganda. He was closely followed in 1875 by Sir Henry Stanley. The conversion activities of Christian missionaries led to conflict with Muslims. The Kabaka (king) came to depend on Christian support. In 1894, Uganda became a British Protectorate.

In 1962, Uganda gained independence with Buganda's Kabaka, Sir Edward Mutesa II, as president and Apollo Milton Obote as prime minister. In 1966, Mutesa II was forced into exile. Obote also abolished the traditional kingdoms, including Buganda. Obote was overthrown in 1971 in a military coup led by General Idi Amin Dada. Amin quickly established a personal dictatorship and launched a war against for-

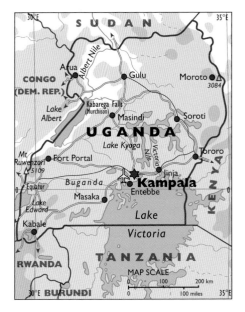

the surprise of his opponents.

Strife continued until 1986, when the NRA captured Kampala and Yoweri Museveni became president. Musveni began to rebuild the domestic economy and improve foreign relations. In 1993 the Kabaka of Uganda returned as monarch. Museveni won Uganda's first direct presidential elections in 1996 and was reelected in 2001.

In 2005, the people voted in favor of restoring multiparty politics after years of nonparty politics. Parliament also voted to remove presidential time limits, enabling Museveni to contest the elections in 2006 and beyond.

ECONOMY
Stability was restored to the economy under President Museveni and it finally expanded. Agriculture dominates, employing 80% of the people. Food crops include bananas, cassava, corn, millet, sorghum, and sweet potatoes, while the chief cash crops are coffee, cotton, sugarcane, and tea. The only important metal is copper. The Owen Falls Dam at Jinja, on the outlet of Lake Victoria, produces cheap electricity.

eign interference that resulted in the mass expulsion of Asians. Amin's regime was responsible for the murder of more than 250,000 Ugandans. Obote loyalists resisted the regime from neighboring Tanzania. In 1976, Amin declared himself president for life and Israel launched a successful raid on Entebbe Airport to end the hijack of one of its passenger planes. In 1978 Uganda annexed the Kagera region of north-west Tanzania.

POLITICS
In 1979 Tanzanian troops helped the Uganda National Liberation Front (UNLF) to overthrow Amin and capture Kampala. In 1980, Apollo Milton Obote led his party to victory in the national elections. But after charges of fraud, Obote's opponents the National Resistance Movement (NRA) began guerrilla warfare. More than 200,000 Ugandans sought refuge in Rwanda and Zaïre. A military group overthrew Obote in 1985 and he lived in exile in Zambia for the last 20 years of his life. Upon his death in 2005 he was granted a state funeral, to

KAMPALA
Capital and largest city in Uganda, on the northern shore of Lake Victoria. Founded in the late 19th century on the remains of a royal palace of the Kings of Buganda, it replaced Entebbe as capital when Uganda attained independence in 1962. It is the trading center for the agricultural goods and livestock produced in Uganda. Industries include textiles, food processing, tea blending, coffee, and brewing.

UKRAINE

Ukraine's flag was first used between 1918 and 1922. It was readopted in September 1991. The colors were first used in 1848. They are heraldic in origin and were first used on the coat of arms of one of the Ukrainian kingdoms in the Middle Ages.

Ukraine is the second largest country in Europe after Russia. This mostly flat country faces the Black Sea in the south. The Crimean Peninsula includes a highland region overlooking Yalta. The highest point of the country is in the eastern Carpathian Mountains. The most extensive land region is the central plateau which descends in the north to the Dnipro-Pripet Lowlands. A low plateau occupies the northeast.

Area 233,089 sq mi [603,700 sq km]
Population 47,732,000
Capital (population) Kiev (2,590,000)
Government Multiparty republic
Ethnic groups Ukranian 78%, Russian 17%, Belarusian, Moldovan, Bulgarian, Hungarian, Polish
Languages Ukranian (official), Russian
Religions Mainly Ukranian Orthodox
Currency Hryvnia = 100 kopiykas
Website www.president.gov.ua/en

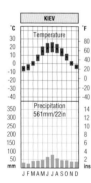

CLIMATE

Ukraine has warm summers, but the winters are cold, becoming more severe from west to east. In the summer, the east of the country is often warmer than the west. The heaviest rainfall occurs in the summer.

HISTORY

In the 9th century AD, a civilization called Kievan Rus was founded, with its capital at Kiev. Russians took over the area in 980 and the region prospered. In the 13th century, Mongol armies ravaged the area. Later, the region was split into small kingdoms and large areas fell under foreign rule. In the 17th and 18th centuries, parts of Ukraine came under Polish and Russian rule. But Russia gained most of Ukraine in the late 18th century, although Austria held an area in the west, called Galicia. After the Bolshevik Revolution of 1917, the Ukrainians set up an independent, non-Communist republic. Austrian Ukraine declared itself a republic in 1918 and the two parts joined together, but in 1919, Ukrainian Communists set up a second government and proclaimed the country a Soviet Socialist Republic. The Communists ultimately triumphed and, during 1922, Ukraine became one of the four founding republics of the Soviet Union.

Millions of people died in the 1930s as the result of Soviet policies. Millions more died during the Nazi occupation between 1941 and 1944. In 1945, areas that were formerly in Czechoslovakia, Poland, and Romania were added to Ukraine by the Soviet Union.

POLITICS

In the 1980s, the people demanded more say over their affairs. The country finally became independent when the Soviet Union broke up in 1991. Ukraine continued to work with Russia through the Commonwealth of Independent States. But Ukraine differed with Russia on some issues, including control over Crimea. In 1999, a treaty ratifying Ukraine's present boundaries failed to get the approval of Russia's upper house.

Leonid Kuchma, who became president in 1994, came under fire in the early 2000s for maladministration and for his alleged involvement in the murder of a journalist. In 2004, the prime minister, a supporter of Kuchma, was declared the winner in presidential elections, but after massive demonstrations, the election was declared invalid. The opposition and pro-Western leader Victor Yuschenko was elected president. This led to tensions with Russia. Russia feared that Ukraine might become aligned with the West. A dispute with Russia over the price of the natural gas it supplies to Ukraine in 2005-6 was said to be politically motivated.

ECONOMY

The World Bank classifies Ukraine as a "lower-middle-income" economy. Agriculture is important, the major export crops are wheat and sugarbeet. Livestock rearing and fishing are also important. Manufacturing is the chief economic activity and includes iron and steel, machinery, and vehicles. The country has large coalfields and hydroelectric and nuclear powerplants, but it imports petroleum and natural gas. In 1986, an accident at the Chernobyl nuclear power plant caused widespread nuclear radiation. The plant was finally closed in 2000.

KIEV (KYYIV)

Capital of Ukraine and a seaport on the Dnieper River. Founded in the 6th century AD, Kiev was the capital of Kievan Russia. It later came under Lithuanian, then Polish, rule before being absorbed into Russia. It became the capital of the Ukrainian Soviet Socialist Republic in 1934, and of an independent Ukraine in 1991.

UNITED ARAB EMIRATES

The flag of the United Arab Emirates was adopted on December 2, 1971 when the country was formed by a union of seven sheikdoms. Red, white, black, and green are the Pan-Arab colors, historically linked to the Arab people and Islamic faith. They stand for Arab unity and independence.

The United Arab Emirates (UAE) is a union of seven small Arab emirates (or sheikhdoms). Swamps and salt marshes border much of the coast in the north. The land is a flat, stony desert, with occasional oases. Sand dunes occur in the east. Highlands rise in the east, near the border with Oman. In the south, the land merges into the bleak Rub' al Khali (Empty Quarter) of Saudi Arabia.

Area 32,278 sq mi [83,600 sq km]
Population 2,524,000
Capital (population) Abu Dhabi (363,000)
Government Federation of Sheikdoms
Ethnic groups South Asian 50%, other Arab and Iranian 23%, Emirati 19%
Languages Arabic (official), Persian, English, Hindi, Urdu
Religions Muslim 96% (Shi'a 16%), others
Currency Emirati dirham = 100 fils
Website www.government.ae/gov/en/index.jsp

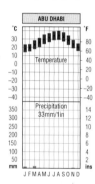

CLIMATE

In most of the country, the average annual rainfall is less than 5 in [130 mm], most of it occurring between November and March. In summer (May to September), temperatures can soar to 120°F [49°C], with high humidity along the coast where conditions can become unpleasant. Winters are warm to mild. The eastern highlands are generally cooler and rainier than the rest of the country. Sandstorms and duststorms are common.

HISTORY

The area has its roots as a trading center between the Mesopotamian and Indus Valley civilizations, later coming under Persian control and, in the 7th century AD, embracing Islam. In the 16th century various European nations set up coastal trading posts. The emirates of today began to develop in the 18th century. Their economies were based on pearl fishing and trading. In 1820, conflict between local rulers and piracy along the coast led Britain to force the states to sign a series of truces. Britain took control of the foreign affairs of the states, while promising protection from attack by outsiders. The states retained control over internal affairs. Because of these truces, the region became known as the Trucial States. In 1952, the emirates set up a Trucial Council to increase cooper-

ABU DHABI (ABU ZABY)

Largest and wealthiest of the United Arab Emirates, lying on the south coast of the Persian Gulf. Also the name of its capital city, the federal capital of the UAE. Ruled since the 18th century by the Al-bu-Falah clan of the Bani Yas tribe. There are long-standing frontier disputes with Saudi Arabia and Oman. Abu Dhabi's economy is based almost entirely on petroleum production.

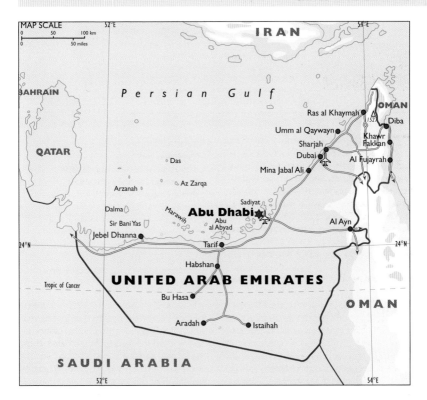

ation between them. Petroleum was discovered in 1958 and first exported in 1962. In 1968, Britain announced the withdrawal of its forces.

POLITICS

The country became independent in 1971, when six of its seven states, Abu Zaby (Abu Dhabi), Ajman, Dubayy (Dubai), Al Fujayrah, Ash Shariqah (Sharjah), and Umm-al-Qaywayn, agreed to form a single country, the United Arab Emirates. A seventh state, Ras al Khaymah, joined the country in 1972. Each of the seven emirates has its own emir, who controls internal affairs. The federal government controls foreign affairs and defense and plays a leading role in the social and economic development of the country. The seven emirs form a Federal Supreme Council, which

elects the federation's president and vice-president who serve five-year terms. The president appoints the prime minister. The country also has a Federal National Council with 40 members appointed by the rulers of the states. There are no elections and the role of the National Council is to review legislation; it cannot change or veto it. The country is one of the most liberal and tolerant of the Persian Gulf countries, but it is the only one without elected bodies. The UAE joined the allied force against Iraq in 1991 following the invasion of Kuwait, and the United States stationed forces there during the 2003 invasion of Iraq.

ECONOMY

The economy is based on petroleum production and the country is the world's sixth largest petroleum exporter.

375

UNITED KINGDOM

The flag of the United Kingdom was officially adopted in 1801. The first Union flag, combining the cross of St. George (England) and St. Andrew (Scotland), dates back to 1603. In 1801, the cross of St. Patrick, Ireland's emblem, was added to form the present flag.

The United Kingdom of Great Britain and Northern Ireland is a union of four countries. Three of them – England, Scotland and Wales – make up Great Britain. The Isle of Man and the Channel Islands, including Jersey and Guernsey, are not part of the UK, but are instead self-governing British dependencies.

Much of Scotland and Wales is mountainous, the highest peak is Scotland's Ben Nevis at 4,404 ft [1,342 m]. England has some highland areas, including the Cumbrian Mountains (Lake District) and the Pennine range in the north. England also has large areas of fertile lowland. Northern Ireland is a mixture of lowlands and uplands and contains the UK's largest lake, Lough Neagh.

Area 93,381 sq mi [241,857 sq km]
Population 60,271,000
Capital (population) London (8,089,000)
Government Constitutional monarchy
Ethnic groups English 82%, Scottish 10%, Irish 2%, Welsh 2%, Ulster 2%, West Indian, Indian, Pakistani, others
Languages English (official), Welsh, Gaelic
Religions Christianity, Islam, Sikhism, Hinduism, Judaism
Currency Pound sterling = 100 pence
Website www.pm.gov.uk

CLIMATE

The UK has a mild climate, influenced by the warm Gulf Stream flowing across the Atlantic from the Gulf of Mexico, then past the British Isles. Moist winds from the southwest bring rain, which diminishes west to east. Winds from the east and north bring cold conditions in winter. The weather is markedly changeable, because of the common occurrence of depressions with their associated fronts.

HISTORY

The isolation of the United Kingdom from mainland Europe has made a major impact on its history. Despite insularity, Britons are of mixed stock.

In ancient times, Britain was invaded by many peoples, including Iberians, Celts, Romans, Angles, Saxons, Jutes, Norsemen, Danes, and Normans, who arrived in 1066 and were the last people successfully to invade Britain. The Normans finally overcame Welsh resistance in 1282, when King Edward I annexed Wales and united it with England. Union with Scotland was

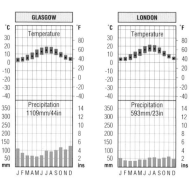

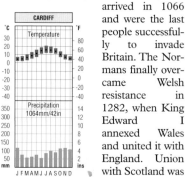

achieved by the Act of Union of 1707. This created a country known as the United Kingdom of Great Britain.

Ireland came under Norman rule in the 11th century, and much of its later history was concerned with a struggle

377

against English domination. In 1801, Ireland became part of the United Kingdom of Great Britain and Ireland, but in 1921 southern Ireland broke away to become Irish Free State. Most of the people in Irish Free State were Roman Catholics. In Northern Ireland, where the majority were Protestants, most wanted to remain citizens of the United Kingdom, and as a result the country's official name changed to the United Kingdom of Great Britain and Northern Ireland.

The British empire began to develop in the 18th century, despite the loss in 1783 of its North American colonies. In the late 18th century the UK was the first country to industrialize its economy.

The British Empire broke up after World War II (1939–45), though the UK still administers many small territories around the world. The empire was transformed into the Commonwealth of Nations, a free association of independent countries, numbering 53 in 2005.

POLITICS

A welfare state was set up in 1945, with a social security system that provided welfare for people "from the cradle to the grave." In 1960, the UK helped to set up the European Free Trade Association-with six other nations. In 1963, Britain's request to join the EEC was rejected: the UK finally joined the EEC in 1973, though a strong body of opinion still feared that the development of a federal Europe would jeopardize British sovereignty. Membership was endorsed by a referendum in 1975, but, at the turn of the century, Britons were still debating whether it was advisable for Britain to adopt the euro, the single European currency adopted by 12 of the 15 European Union members in 1999.

Since the 1960s, Northern Ireland has been the scene of conflict between the Protestant majority, who favor continuing union with the UK, and the Roman Catholic minority, many of whom are republicans who would like to see Ireland reunified. British troops were sent to the province in 1969 to control violence between the communities and, at various times, Britain has imposed direct rule. In 1998, the "Good Friday" agreement held out hope for the future, when unionists and nationalists agreed that Northern Ireland would remain part of the United Kingdom, until a majority of its people voted in favor of a change. The agreement also allowed Ireland to play a part in the affairs of the

LONDON

Capital of the United Kingdom, second-largest city in Europe, after Moscow. Located on the River Thames, 40mi [65km] from its mouth in the North Sea. Greater London, comprises the City of London "square mile" plus 13 inner and 19 outer boroughs, covering a total of 610sq mi [1,580 sq km]. The Romans called it Londinium. In the 9th century, Alfred the Great made it the seat of government. Edward the Confessor built Westminster Abbey and made Westminster his capital in 1042. The Plague of 1665 killed 75,000 Londoners, and the following year the Fire of London destroyed many buildings. Sir Christopher Wren designed many churches, including St. Paul's. The 19th century saw the population reach 4 million, and London become the world's biggest city. Further growth between the World Wars was accompanied by extensions to the transportation system. Much of London was rebuilt after bomb damage during World War II, and the docklands were regenerated in the late 1980s. London is one of the world's most important administrative, financial, and commercial cities.

north, while the republic amended its constitution to remove all claims to Northern Ireland. A Northern Ireland Assembly was set up to handle local affairs. In July 2005 the Irish Republican Army (IRA) issued a statement of full disarmament.

Before 1999, Scotland and Wales were directly ruled by the British parliament in London. In 1997, following the landslide victory of the Labor Party under Tony Blair, 74% of voters in Scotland and 50.3% of voters in Wales opted for the setting up local assemblies.

The Scottish parliament is responsible for local affairs with limited powers to raise or reduce taxes. The Welsh Assembly has no powers over taxation. Both met for the first time in 1999. Devolution has caused concern among those who fear that it might lead to the breakup of the UK, also focusing attention on the question of nationality.

The high cost of welfare services is a matter of political controversy. There is also concern about the changing economy, with a decline in traditional manufacturing and the growth of service industries, both of which affect employment. Another issue is immigration and the fear that economic migrants entering the UK will lessen the job opportunities of the indigenous workforce.

After the terrorist attacks on the United States on September 11, 2001, Britain was prominent in its support for the US, helping to create the broad alliance that launched the attack on the Taliban government of Afghanistan. However, others are concerned at the cost and morality of British military operations especially the war with Iraq. In July 2005 four suicide bombers struck in central London, killing 52 and injuring hundreds.

ECONOMY

The UK is a major industrial and trading nation. Its natural resources are coal, iron ore, petroleum, and natural gas, but it has to import most of the materials it needs for industry. It also has to import food. Service industries are vital financial and insurance services bring in much-needed foreign exchange, while tourism is a major earner.

Agriculture employs only 1% of the workforce. Production is high. Major crops include barley, potatoes, sugarbeet and wheat. Sheep are the leading livestock, beef and dairy cattle, pigs and poultry are also important, as is fishing.

CHANNEL ISLANDS

Group of islands at the southwest end of the English Channel, 10 mi [16 km] off the west coast of France. The main islands are Jersey, Guernsey, Alderney, and Sark; the chief towns are St. Helier (Jersey) and St. Peter Port (Guernsey). A dependency of the British crown since the Norman Conquest, they were under German occupation during World War II. They are divided into the administrative bailiwicks of Guernsey and Jersey, each with its own legislative assembly. The islands have a warm, sunny climate and fertile soil. The major industries are tourism and agriculture. They cover an area of 75 sq mi (194 sq km).

ISLE OF MAN

Island off the northwest coast of England, in the Irish Sea; the capital is Douglas. In the Middle Ages, it was a Norwegian dependency, subsequently coming under Scottish, then English, rule. It has been a British crown possession since 1828, but has its own government (the Tynwald). The basis of the economy is tourism although agriculture is important, the chief products being oats, fruit, and vegetables. It covers an area of 221 sq mi [572 sq km].

UNITED STATES OF AMERICA

This flag, known as the "Stars and Stripes," has had the same basic design since 1777, during the War of Independence. The 13 stripes represent the 13 original colonies in the eastern United States. The 50 stars represent the 50 states of the Union.

Area 3,717,792 sq mi [9,629,091 sq km]
Population 293,028,000
Capital (population) Washington, DC (572,000)
Government Federal republic
Ethnic groups White 77%, African American 13%, Asian 4%, Amerindian 2%, others
Languages English (official), Spanish, more than 30 others
Religions Protestant 56%, Roman Catholic 28%, Islam 2%, Judaism 2%
Currency US dollar = 100 cents
Website www.firstgov.gov

The highest mountain in the USA is Mount McKinley (6194 m) in Alaska.

MAP SCALE
0   250   500 km
0   250 miles

The United States of America is the world's fourth largest country in area and the third largest in population. It contains 50 states, 48 of which lie between Canada and Mexico, plus Alaska in northwestern North America and Hawaii, a group of volcanic islands in the North Pacific Ocean.

Densely populated coastal plains lie to the east and south of the Appalachian Mountains. The central lowlands drained by the Mississippi-Missouri rivers stretch from the Appalachians to the Rocky Mountains in the west. The Pacific region contains fertile valleys separated by mountain ranges.

CLIMATE

The climate of the United States varies greatly, ranging from the Arctic conditions in northern Alaska, with average temperatures of 9°F [–13°C], to the intense heat of Death Valley, which holds the record for the highest shade tempera-

ture recorded in the United States (134°F [57°C]).

Winters are cold and summers warm in New England, the Midwest, and the Middle Atlantic States. The southern states have long, hot summers and mild, wet winters. In the central United States, a lack of topographical features bars the northward movement of hot, moist air from the Gulf of Mexico, and in winter the southward movement of dry, cold air from the Arctic. Parts of California have a pleasant Mediterranean-type climate, but the mountains of the west are much

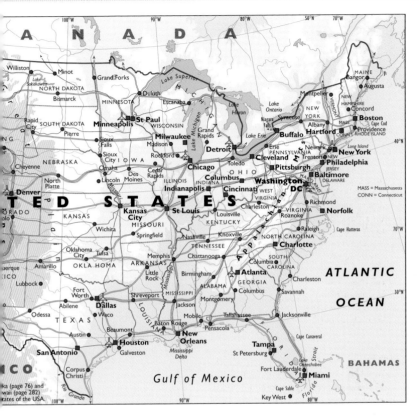

cooler and wetter. The central plains are arid, while deserts occur in parts of the west and southwest.

HISTORY

The first people in North America, the ancestors of the Native Americans arrived around 40,000 years ago from Asia. Although Vikings probably reached North America 1,000 years ago, European exploration proper did not begin until the late 15th century.

The first Europeans to settle in large numbers were the British, who founded settlements on the eastern coast in the early 17th century. British rule ended with the War of Independence (1775–83). The country expanded in 1803 when a vast territory in the south

and west was acquired through the Louisiana Purchase, while the border with Mexico was fixed in the mid-19th century. The Civil War (1861–5) ended slavery and the serious threat that the nation might split into two parts. In the late 19th century the West was opened up, while immigrants flooded in from Europe and elsewhere.

POLITICS

The United States has long played a leading role in industrial, economic, social, and technological innovation, creating problems through sheer ebullience, and solving them more or less through inventiveness, enterprise, and with huge, wealth-bringing resources of energy and raw materials. The majority of Americans

UNITED STATES OF AMERICA

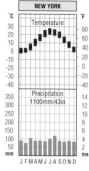

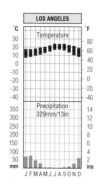

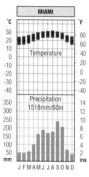

continue to enjoy one of the world's highest material standards of living, but the country faces many problems. One being the maintaining of social cohesion as the composition of American society changes, another being the issue of poverty. A sizeable underclass of poor and inadequately educated people, many of whom are members of ethnic minorities, suffers from low standards of living. Other associated problems include crime, drug addiction, and racial conflict.

The United States has one of the most diverse populations of any country in the world. Until about 1860, the population, with the exception of the Native Americans and the southern African Americans, was made up largely of immigrants of British and Irish origin, with small numbers of Spaniards and French. After the Civil War, increasing numbers of immigrants arrived from the countries of central and southeastern Europe, including Italy, the Balkans, Poland, Scandinavia, and Russia. This influx of Europeans (about 30 million between 1860 and 1920), was vastly different in culture and language from the established population. More recently, the country has received influxes of Japanese, Chinese, Filipinos, Cubans, Puerto Ricans, and Mexicans, many of them illegal immigrants. Although strong influences and pressures toward Americanization still exist, members of these groups have tended to maintain their own culture, establishing social and cultural enclaves within American society. The nation has never adopted an official language, but English was readily adopted by most immigrants in the late 19th and early 20th centuries, because they sought acceptance. However, many of the recent Hispanic immigrants speak only Spanish, which has become the country's second language. Many Americans are concerned about this trend toward "cultural pluralism" rather than integration. They argue that Hispanics who do not speak English are at a disadvantage in American society and believe that everyone should speak English, either as a first or second language. According to some population forecasts, today's white majority will be outnumbered by other ethnic groups in 2050. With a total pro-

WASHINGTON, DC

Capital of the USA, on the east bank of the Potomac River, covering the District of Columbia and extending into the neighboring states of Maryland and Virginia. The site was chosen as the seat of government in 1790. Construction of the White House began in 1793, and the building of the Capitol the following year. Congress moved from Philadelphia to Washington in 1800. It is the legislative, judicial, and administrative center of the USA. Despite its role, Washington has severe social problems; many of the large African-American population live in slum housing.

jected population of 380 million, Hispanics are expected to number around 80 million by 2050, while African Americans will account for another 62 million. Such rapid growth is seen by some as a threat to the majority.

The United States developed into a world power from the 1890s, and played a leading role in international affairs throughout the 20th century. After World Wars II it became one of the world's two superpowers—the other being the Soviet Union. It assumed leadership of the West during the Cold War. Since the end of the Cold War, the United States has faced new threats from terrorists and rogue states. Its vulnerability was demonstrated by the terrorist attacks on New York City and Washington, DC, on September 11, 2001. The United States responded vigorously, creating an international alliance to combat terrorism and the nations which shelter or aid terrorists. In 2001 it led a coalition force against the Taliban regime in Afghanistan which was protecting al Qaida terrorists. Then in 2003 the US led another coalition force to overthrow the repressive regime of Saddam Hussein in Iraq. However, despite early military successes, the conflict continued.

George W. Bush was reelected in 2004.

ECONOMY

The US is a leading producer of meat, dairy products, cereals, cotton, sugar, and many other crops, despite agriculture employing only 2.4% of the labor force. The western plains are the main centers of production. Large-scale industrial production was pioneered in the US, and as almost every raw material is available within its own boundaries, its mining and extractive industries have been heavily exploited. Anthracite from eastern Pennsylvania, bituminous and coking coals from the Appalachians, Indiana, Illinois, Colorado, and Utah are still in demand, and vast reserves remain.

Petroleum is found in major fields underlying the Midwest, the eastern and central mountain states, the Gulf of Mexico, California, and Alaska. Although the US is a major producer, it is the world's greatest consumer and has long imported petroleum. The exploitation of petroleum in Oklahoma, Texas, and Louisiana has shifted the dependence on agriculture to the refining and petrochemical industries. Natural gas is also found in abundance, usually associated with petroleum.

PUERTO RICO

The Commonwealth of Puerto Rico is the easternmost island in the Greater Antilles. The land is mountainous, with a narrow coastal plain. the highest point is Cerro de Punta (4,389 ft [1,338 m]). The climate is hot and wet.

Ceded by Spain to the United States in 1898, Puerto Rico became a self-governing commonwealth in free association with the United States after a referendum in 1952. Puerto Ricans are US citizens, but pay no federal taxes, nor do they vote in US congressional or presidential elections.

The island is the most industrialized and urbanized in the Caribbean, and

Area 3,459 sq mi [8,959 sq km]
Population 3,916,632
Capital (population) San Juan (433,733)
Government Commonwealth
Ethnic groups White 80%, black 8%, others
Languages Spanish, English
Religions Roman Catholic 85%, others 15%
Currency US dollar = 100 cents
Website www.gotopuertorico.com

manufacturing and tourism are growing industries. Cash crops include bananas, coffee, sugar, tobacco, tropical fruits, vegetables and spices. Chief exports are chemicals and chemical products, machinery and food.

URUGUAY

Uruguay has used this flag since 1830. The nine stripes represent the nine provinces which formed the country when it became an independent republic in 1828. The colors and the May Sun had originally been used by Argentina during its struggle against Spanish rule.

The Oriental Republic of Uruguay, as Uruguay is officially known, is South America's second smallest independent nation after Suriname. The River Uruguay, which forms the country's western border, flows into the Río de la Plata (River Plate), a large estuary fringed with lagoons and sand dunes, which leads into the South Atlantic Ocean.

The land consists mainly of low-lying plains and hills. The highest point lies south of Minas and is only 1,644 ft [501 m] above sea level. The main river in the interior is the Rio Negro.

Area 67,574 sq mi [175,016 sq km]
Population 3,399,000
Capital (population) Montevideo (1,303,000)
Government Multiparty republic
Ethnic groups White 88%, Mestizo 8%, Mulatto or Black 4%
Languages Spanish (official)
Religions Roman Catholic 66%, Protestant 2%, Judaism 1%
Currency Uruguayan peso = 100 centésimos
Website www.turismo.gub.uy

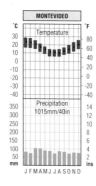

CLIMATE

Uruguay has a mild climate, with rain throughout the year, though droughts sometimes occur. The summer is pleasantly warm, especially near the coast. The weather remains relatively mild in winter.

HISTORY

The first people of Uruguay were Amerindians. But the Amerindian population has largely disappeared. Many were killed by Europeans, some died of European diseases, while others fled into the interior. The majority of Uruguayans today are of European origin, though there are some *mestizos* (of mixed European and Amerindian descent). The first European to arrive in Uruguay was a

Spanish navigator, Juan Diaz de Solis, in 1516. But he and part of his crew were killed by the local Charrúa Amerindians when they went ashore.

Few Europeans settled until the late 17th century. Spanish settlers founded Montevideo in order to prevent the Portuguese from gaining influence in the area. Uruguay was then little more than a buffer zone between the Portuguese territory to the north and Spanish territories to the west. By the late 18th century, Spaniards had settled in most of the country. Uruguay became part of a colony called the Viceroyalty of La Plata, which included Argentina, Paraguay, and parts of Bolivia, Brazil, and Chile.

Uruguay was annexed by Brazil in 1820, bringing about an end to Spanish rule. In 1825, Uruguayans, supported by Argentina, began a struggle for independence.

POLITICS

Uruguay was recognized as an independent republic by Brazil and Argentina in 1828. Social and economic developments were slow in the 19th century, but, from 1903, governments made

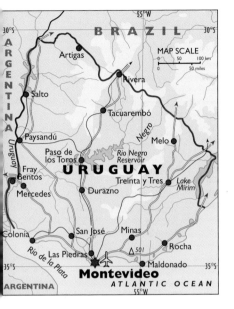

lems in trying to rebuild its weakened economy and shoring up its democratic traditions. In 1991, Uruguay joined with Argentina, Brazil and Paraguay to form Mercosur, which aimed to create a common market. Mercosur's secretariat is in Montevideo. The early 21st century brought economic problems, many of which were the result of the economic crisis in Argentina, and its imposition of banking controls. Uruguay elected its first leftist president, Tabare Vasquez, in 2004.

Uruguay a democratic and stable country. Since 1828, two political parties—the Colorados (Liberals) and the Blancos (Conservatives)—have dominated.

During World War II, Uruguay prospered because of its export trade, especially in meat and wool. However, from the 1950s, economic problems caused unrest. Terrorist groups, notably the Tupumaros (Marxist urban guerrillas), carried out murders and kidnappings in the 1960s and early 1970s. In 1972, President Juan Maria Bordaberry declared war on the Tupumaros and the army crushed them. In 1973, the military seized power, suspended the constitution and ruled with great severity, committing major human rights abuses.

Military rule continued until 1984, when elections were held. General Gregorio Alvarez, who had been president since 1981, resigned and Julio Maria Sanguinetti, leader of the Colorado Party, became president in February 1985, leading a government of National Unity. He ordered the release of all political prisoners. In the 1990s, Uruguay faced prob-

ECONOMY

Uruguay is classed by the World Bank as an "upper-middle-income" developing country. Although 90% of the population live in urban areas and agriculture employs 3% of the population, the economy depends on the exports of hides and leather goods, beef, and wool. Main crops include corn, potatoes, rice, sugarbeet, and wheat.

Manufacturing concentrates on food processing and packing. The economy has diversified into cement, chemicals, leather goods, textiles, and steel. Uruguay depends largely on hydroelectric power for energy and exports electricity to Argentina.

MONTEVIDEO

Capital of Uruguay, in the south of the country, on the River Plate. Once a Portuguese fort (1717), the city was captured by the Spanish in 1726, and became the capital of Uruguay in 1828. One of South America's major ports, it is the base of a large fishing fleet and handles most of the country's exports. Products include textiles, dairy goods, wine, and packaged meat.

UZBEKISTAN

The white crescent moon in the upper hoist is a traditional symbol of Islam and represents the rebirth of the nation. The white stars recall the 12 signs of the zodiac. Blue stands for water and the eternal sky. Red represents life. White symbolizes peace and green denotes nature.

The Republic of Uzbekistan is one of five republics in Central Asia which were once part of the Soviet Union. Plains cover most of western Uzbekistan, with highlands in the east. The main rivers, the Amu (or Amu Darya) and Syr (or Syr Darya), drain into the Aral Sea. So much water has been taken from these rivers for irrigation that the Aral Sea is now only a quarter of its size in 1960. Much of the former sea is now desert.

Area 172,741sq mi [447,400 sq km]
Population 26,410,000
Capital (population) Tashkent (2,143,000)
Government Socialist republic
Ethnic groups Uzbek 80%, Russian 5%, Tajik 5%, Kazakh 3%, Tatar 2%, Kara-Kalpak 2%
Languages Uzbek (official), Russian
Religions Islam 88%, Eastern Orthodox 9%
Currency Uzbekistani sum = 100 tyiyn
Website www.gov.uz

CLIMATE

Uzbekistan has a continental climate. Winters are cold, but temperatures soar in the summer. In the west. conditions are extremely arid with an average annual rainfall of about 8 in [200 mm].

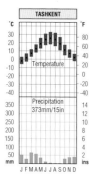

HISTORY

Uzbekistan lies on the ancient Silk Road between Europe and Asia. Great cities such as Samarkand and Bukhara, famed for their architectural opulence, were important trade and cultural centers. Russia took the area in the 19th century. After the Russian Revolution of 1917, Communists took over, setting up the Uzbek Soviet Socialist Republic in 1924. Under Communism, all aspects of Uzbek life were controlled and worship were discouraged. The country did benefit, though, and health, housing, education, and transportation were all improved. In the late 1980s, people demanded more freedom and, in 1990, the government stated that its laws overruled those of the Soviet Union.

POLITICS

Uzbekistan became independent in 1991 with the breakup of the Soviet

SAMARKAND

City in the fertile Zeravshan Valley, southeast Uzbekistan. One of the oldest cities in Asia, it was conquered by Alexander the Great in 329 BC. A vital trading center on the Silk Road, it flourished in the 8th century as part of the Umayyad Empire. Samarkand was destroyed in 1220 by Genghis Khan but became capital of the Mongol empire of Tamerlane in 1370. Ruled by the Uzbeks from the 16th century, it was captured by Russia in 1868, though it remained a center of Muslim culture. It is now a major scientific research center and has a population of 361,800.

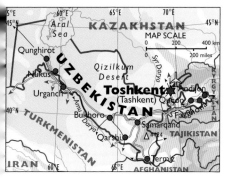

Karimov blamed fundamentalists out to destabilize the country; his opponents blamed the determination of those in power to crush all dissent and maintain a repressive state. There were calls for an international enquiry which the government rejected, as a result of which the US threatened to withold aid. Uzbekistan's reaction to this was to order US forces to leave their base.

Union, but retained links with Russia through the Commonwealth of Independent States, but it subsequently pulled out due to the leader's opposition to closer integration on post-Soviet territory. Islam Karimov, leader of the People's Democratic Party (formerly the Communist Party), was elected president in December 1991. In 1992–3, many opposition leaders were arrested because the government said that they threatened national stability. In 1994-5, the PDP was victorious in national elections and in 1995 a referendum extended Karimov's term in office until 2000, when he was reelected.

In 2001, Karimov declared Uzbekistan's support for the United States in its campaign against the terrorist al Qaida bases in Afghanistan and indeed allowed the US forces to have a base on Uzbek territory.

Due to the country's poor record on human rights the European Bank for Reconstruction and Development announced in 2004 that it would cut aid to Uzbekistan.

Recent years have seen bombings and shootings, for which the authorities have blamed Islamic extremists. In 2005 protests against the jailing of several people charged with Islamic extremism in the city of Andijan turned to violence with troops opening fire. Several hundred civilians were killed, though the government claimed a toll of 180.

ECONOMY

The World Bank classifies Uzbekistan as a "lower-middle-income" developing country. The government still controls most economic activity and economic reform has been very slow. Uzbekistan produces coal, copper, gold, petroleum, and natual gas, while manufacturing industries include agricultural machinery, chemicals, and textiles. Agriculture is important with cotton the main crop. Other crops include fruits, rice, and vegetables; cattle, sheep, and goats are raised.

TASHKENT

Largest city and capital of Uzbekistan, in the Tashkent oasis in the foothills of the Tian Shan (Celestial Range) mountains (a 1,500 mi [2,400 km] long mountain range). Tashkent is watered by the River Chirchik. It was ruled by the Arabs from the 8th until the 11th century. The city was captured by Tamerlane in 1361, and by the Russians in 1865. The modern city is the transportation and economic center of the region. Industries include textiles, chemicals, food processing, mining machinery, paper, porcelain, clothing, leather, and furniture. It has a population of 2,143,000.

VENEZUELA

Venezuela's flag has the same tricolor as the flags of Colombia and Ecuador, colors used by the Venezuelan patriot Francisco de Miranda. The stars represent the provinces. In 2006 the eighth star (Guayana) was added and the horse on the coat of arms was turned to face left.

The Bolivarian Republic of Venezuela, in northern South America, contains the Maracaibo Lowlands in the west. The lowlands surround the petroleum-rich Lake Maracaibo (Lago de Maracaibo). Arms of the Andes Mountains enclose the lowlands and extend across most of northern Venezuela. Between the northern mountains and the scenic Guiana Highlands in the southeast, where the Angel Falls are found, lie the *llanos* (tropical grasslands), a lowlying region drained by the River Orinoco and its tributaries. The Orinoco is Venezuela's longest river.

Area 352,143 sq mi [912,050 sq km]
Population 25,017,000
Capital (population) Caracas (1,823,000)
Government Federal republic
Ethnic groups Spanish, Italian, Portuguese, Arab, German, African, indigenous people
Languages Spanish (official), indigenous dialects
Religions Roman Catholic 96%
Currency Bolivar = 100 céntimos
Website www.venezlon.co.uk

CLIMATE

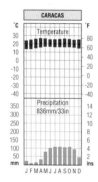

Venezuela has a tropical climate. Temperatures are high throughout the year on the lowlands, though far cooler in the mountains. There is a marked dry season in much of the country that falls between December and April. Most rainfall is in the mountains.

HISTORY

Arawak and Carib Amerindians were the main inhabitants of Venezuela before the arrival of Europeans. The first European to arrive was Christopher Columbus, who sighted the area in 1498. Spaniards began to settle in the early 16th century, but economic development was slow.

In the early 19th century, Spain's colonies in South America began their struggle for independence. The Venezuelan patriots Simón Bolívar and Francisco Miranda were prominent in the struggle. Venezuela was the first South American country to demand freedom and, in July 1811, it declared its independence, though Spaniards still held most of the country. In 1819, Venezuela became part of Gran Colombia, a republic led by Simón Bolívar that also included Colombia, Ecuador and Panama.

The country became fully independent in 1821, after the Venezuelans had defeated the Spanish in a battle at Carabobo, near Valencia. Venezuela broke away from Gran Colombia in 1829 and in 1830 a new constitution was drafted. The country's first president

ANGEL FALLS

The world's highest uninterrupted waterfall, in La Gran Sabrana, eastern Venezuela. Part of the River Caroni, it was discovered in 1935 and named after Jimmy Angel, a US aviator who died in a crash near the falls. A total drop of 3,212 ft [980 m].

Republic of Venezuela and held a referendum on a new constitution. This gave the president increased power over military and civilian institutions. Chávez argued that these powers were needed to counter corruption. In 2002, Chávez himself survived a coup and then in 2004 he won a majority in a referendum that had been intended by the opposition to remove him from office.

ECONOMY
The World Bank classifies Venezuela as an "upper-middle-income" developing country. Oil accounts for 80% of the exports. Other exports include bauxite and aluminum, iron ore, and farm products. Agriculture employs 9% of the people. Cattle ranching is important and dairy cattle and poultry are also raised. Major crops include bananas, cassava, citrus fruits, coffee, corn, plantains, rice, and sorghum. Most commercial crops are grown on large farms, but many people in remote areas farm small plots and produce barely enough to feed their families.

Manufacturing industries now employ 21% of the population. The leading industry is petroleum refining, centered on Maracaibo. Other manufactures include aluminum, cement, processed food, steel, and textiles.

was General José Antonio Páez, one of the leaders of Venezuela's independence movement.

POLITICS
The development of Venezuela in the 19th century and the first half of the 20th century was marred by instability, violence, and periods of harsh dictatorial rule. However, the country has had elected governments since 1958.

Venezuela has greatly benefited from its petroleum resources, which were first exploited in 1917. In 1960, Venezuela helped to form OPEC (the Organization of Petroleum Exporting Countries) and, in 1976, the government of Venezuela took control of the entire petroleum industry. Money from its export has helped Venezuela to raise living standards and diversify the economy.

Financial problems in the late 1990s led to the election of Hugo Chávez as president. Chávez, leader of the Patriotic Pole, a left-wing coalition, who had led an abortive military uprising in 1992, became president in February 1999. He announced that the country's official name would be changed to the Bolivarian

CARACAS

Capital of Venezuela, on the River Guaire. Caracas was under Spanish rule until 1821. It was the birthplace of Venezuelan patriot Simón Bolívar. The city grew after 1930, encouraged by the exploitation of petroleum. It has the Central University of Venezuela (1725) and a cathedral (1614). Industries: motor vehicles, petroleum, brewing, chemicals, and rubber.

VIETNAM

Vietnam's flag was first used by forces led by the Communist Ho Chi Minh during the liberation struggle against Japan in World War II (1939–45). It became the flag of North Vietnam in 1945 and it was retained when North and South Vietnam were reunited in 1975.

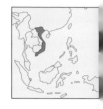

The Socialist Republic of Vietnam occupies an S-shaped strip of land facing the South China Sea in Southeast Asia. The coastal plains include two densely populated, fertile river delta areas. The Red (Hong) Delta faces the Gulf of Tonkin in the north, while the Mekong Delta is in the south.

Inland are thinly populated highland regions, including the Annam Cordillera (Chaîne Annamitique), which forms much of the boundary with Cambodia. The highlands in the northwest extend into Laos and China.

Area 128,065 sq mi [331,689 sq km]
Population 82,690,000
Capital (population) Hanoi (1,074,000)
Government Socialist republic
Ethnic groups Vietnamese 87%, Chinese, Hmong, Thai, Khmer, Cham, mountain groups
Languages Vietnamese (official), English, French, Chinese, Khmer, mountain languages
Religions Buddhism, Christianity, indigenous beliefs
Currency Dong = 10 hao - 100 xu
Website www.vietnamtourism.com

CLIMATE

Vietnam has a tropical climate, though the drier months of January to March are cooler than the wet, hot summer months, when monsoon winds blow from the southwest. Typhoons sometimes hit the coast, causing much damage.

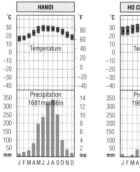

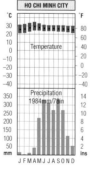

HISTORY

In 111 BC, China seized Vietnam, naming it Annam. In 939 AD, it became independent. In 1558, it split into two parts: Tonkin in the north, ruled from Hanoi, and Annam in the south, ruled from Hué. In 1802, with French support, Vietnam was united as the Empire of Vietnam, under Nguyen Anh. In 1859, the French seized Saigon, and by 1887 had formed Indochina from the union of Tonkin, Annam, and Cochin China.

Japan conquered Vietnam during World War II, and established a Vietnamese state under Emperor Bao Dai. After the war, Bao Dai's government collapsed, and the nationalist Viet Minh, led by Ho Chi Minh, set up a

MEKONG RIVER

River in southeast Asia that rises in Tibet as the Lancang Jiang and flows south through Yünnan province, China. It forms the Burma-Laos border and part of the Laos-Thailand border and then flows south through Cambodia and Vietnam creating a vast river delta that is one of the most important rice-producing regions in Asia. It is 2,600 mi [4,180 km] long.

extended its influence into South Vietnam, mainly through the Viet Cong. The USA became increasingly involved in what they perceived to be a fight against communism. The conflict escalated into the Vietnam War (1954–75). In 1975, after the withdrawal of US troops, Ho Chi Minh's forces overran South Vietnam and it surrendered.

In 1976, the reunited Vietnam became a socialist republic. In 1979, Vietnam helped overthrow the Khmer Rouge government in Cambodia, withdrawing only in 1989. The United States opened an embassy in Hanoi in 1995 and in 2002 it implemented a trade agreement which normalized the trade status between the two countries.

ECONOMY
The World Bank classifies Vietnam as a "low-income" developing country. Agriculture employs 67% of the workforce. The main food crop is rice.Other products include corn and sweet potatoes; commercial crops include bananas, coffee, peanuts, rubber, soybeans, and tea. Fishing is also important. Northern Vietnam has most of the country's natural resources, including coal. The country also produces chromium, petroleum, phosphates and tin. Manufactures include cement, fertilizers, processed food, machinery, steel, and textiles.

Vietnamese republic. In 1946, the French tried to reassert control and war broke out. Despite aid from the USA, the Viet Minh defeated the French at Dien Bien Phu.

POLITICS
In 1954, Vietnam divided along the 17th Parallel—with North Vietnam under the Communist government of Ho Chi Minh, and South Vietnam under the French-supported Bao Dai. In 1955, Bao Dai was deposed and Ngo Dinh Diem was elected president. Despite his authoritarian regime, many western countries recognized Diem as the legal ruler of Vietnam. North Vietnam, supported by China and the Soviet Union,

HANOI
Capital of Vietnam and its second largest city, on the Red River. In the 7th century the Chinese ruled Vietnam from Hanoi; it later became capital of the Vietnamese empire. Taken by the French in 1883, the city became the capital of French Indochina (1887–1945). In 1946–54, it was the scene of fierce fighting between the French and the Viet Minh. Hanoi was heavily bombed by the US during the Vietnam War.

YEMEN

Yemen's flag was adopted in 1990 when the Yemen Arab Republic (or North Yemen) united with the People's Democratic Republic of Yemen (or South Yemen). This simple flag is a tricolor of red, white, and black—colours associated with the Pan-Arab movement.

The Republic of Yemen faces the Red Sea and the Gulf of Aden in the southwestern corner of the Arabian Peninsula. Behind the narrow coastal plain along the Red Sea, the land rises to a mountain region called High Yemen. Beyond the mountains, the land slopes down towards the Rub' al Khali Desert. Other mountains rise behind the coastal plain along the Gulf of Aden. To the east lies a fertile valley called the Hadramaut and also the deserts of the Arabian Empty Quarter.

Palm trees grow along the coast. Plants such as acacia and eucalyptus flourish in the interior. Thorn shrubs and mountain pasture are found in the highlands.

Area 203,848 sq mi [527,968 sq km]
Population 20,025,000
Capital (population) Sana'a (954,000)
Government Multiparty republic
Ethnic groups Predominantly Arab
Languages Arabic (official)
Religions Islam
Currency Yemeni rial = 100 fils
Website
www.yemeninfo.gov.ye/ENGLISH/home.htm

CLIMATE

The climate in San'a is moderated by its altitude. Temperatures are much lower than in Aden (Al' Adan), which is at sea level. In summer, southwest monsoon winds bring thunderstorms. But most of Yemen is arid. The south coasts are particularly hot and humid, especially from June to September. There are two seasonal rainfalls, during March to May and from July to September. The average rainfall is about 2 in [50 mm] on most parts of the plateaus, but may rise to 40 in [1,000 mm] in the highlands, while the coastal lowlands may have no more than 0.5 in [12 mm].

HISTORY

From around 1400 BC, Yemen lay on an important trading route, with frankin-

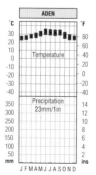

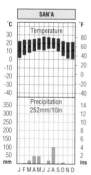

ADEN (AL'ADAN)

Commercial capital, and historic capital of the Aden Protectorate (1937–67) and the former People's Democratic Republic of Yemen (1967–90). A sea port city on the Gulf of Aden, 100 mi [160 km] east of the Red Sea, Aden was an important Roman trading port. With the opening of the Suez Canal in 1869 its importance increased. It was made a Crown Colony in 1937 and the surrounding territory became the Aden Protectorate. Industries: salt refining, cigarettes, and petroleum.

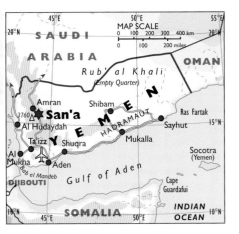

Aden by government forces. In 1995, Yemen resolved border disputes with Oman and Saudi Arabia, but clashed with Eritrea over uninhabited islands in the Red Sea. In 1998 and 1999, militants in the Aden-Abyan Islamic Army sought to destabilize the country. In 2000, a suicide bomb attack on the USS *Cole* in Aden killed 17 US personnel. In 2001, President Salih offered support to the USA in its "war on terrorism."

President Saleh has said that although the constitution entitles him to run for president again in 2006, he has chosen not to.

cense, pearls, and spices being the major commodities. But its prosperity declined in the 4th century AD, when it became divided between warring groups.

Islam was introduced during the 7th century by the son-in-law of the Prophet Muhammad. From 897, the country was ruled by a Muslim leader. In 1517, the area was taken over by the Turkish Ottoman Empire and remained under Turkish rule for the next 400 years.

POLITICS

After World War I, northern Yemen, which had been ruled by Turkey, began to evolve into a separate state from the south, where Britain was in control. Britain withdrew in 1967 and a left-wing regime took power in the south. North Yemen became a republic in 1962 when the monarchy was abolished.

Clashes occurred between the traditionalist Yemen Arab Republic in the north and the formerly British Marxist People's Democratic Republic of Yemen. But, in 1990, the two Yemens merged to form one country. The marrying of the needs of the two parts of Yemen has proved difficult. In May 1994, civil war erupted, with President Saleh, a northerner, attempting to remove the vice-president (a southerner). The war ended in July 1994, following the capture of

ECONOMY

The World Bank classifies Yemen as a "low-income" developing country. Agriculture employs up to 63% of the people. Herders raise sheep and other animals, while farmers grow such crops as barley, fruits, wheat, and vegetables in highland valleys and around oases. Cash crops include coffee and cotton.

Imported petroleum is refined at Aden and petroleum extraction began in the northwest in the 1980s. Handcrafts, leather goods, and textiles are made.

SAN'A

Capital and largest city of Yemen, 40 mi [65 km] northeast of the Red Sea port of Hodeida. Situated on a high plateau at 7,500 ft [2,286 m], it claims to be the world's oldest city, founded by Shem, eldest son of Noah. During the 17th century and from 1872 to 1918, it was part of the Ottoman Empire. In 1918, it became capital of an independent Yemen Arab Republic, and in 1990 capital of the new, unified Yemen. It is noted for its handcrafts. Agriculture (grapes) and industry (iron) are also important.

ZAMBIA

Zambia's flag was adopted when the country became independent from Britain in 1964. The colors are those of the United Nationalist Independence Party, which led the struggle against Britain and ruled until 1991. The flying eagle represents freedom.

The Republic of Zambia is a landlocked country in southern Africa. The country lies on the plateau that makes up most of southern Africa. Much of the land is between 2,950–4,920 ft [900–1,500 m] above sea level. The Muchinga Mountains in the northeast rise above this flat land.

Lakes include Bangweulu, which is entirely within Zambia, together with parts of lakes Mweru and Tanganyika in the north. Most of the land is drained by the Zambezi (from which the country takes its name) and its two main tributaries, the Kafue and Luangwa. Occupying part of the Zambezi Vally and stretching along the southern border Lake Kariba, which was dammed in 1961, is the largest artificial lake in Africa and the second largest in the world 175 mi [280 km] long and 25 mi [40 km] across at its widest point). Zambia shares Lake Kariba and the Victoria Falls with Zimbabwe.

Grassland and wooded savanna cover much of Zambia. There are also swamps. Evergreen forests exist in the drier southwest.

Area 290,586 sq mi [752,618 sq km]
Population 10,462,000
Capital (population) Lusaka (1,270,000)
Government Multiparty republic
Ethnic groups Native African (Bemba, Tonga, Maravi/Nyanja)
Languages English (official), Bemba, Kaonda
Religions Christianity 70%, Islam, IHinduism
Currency Zambian kwacha = 100 ngwee
Website www.zambiatourism.com

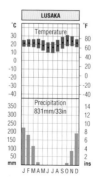

CLIMATE

Zambia lies in the tropics, although temperatures are moderated by the altitude. The rainy season runs between November and March, when the rivers sometimes flood. Northern Zambia is the wettest region of the country. The average annual rainfall ranges from about 51 in [1,300 mm] in the north down to between 20 and 30 in [510–760 mm] in the south.

HISTORY

European contact with Zambia began in the 19th century, when the explorer David Livingstone crossed the River Zambezi. In the 1890s, the British South Africa Company, set up by Cecil Rhodes (1853–1902), the British financier and statesman, made treaties with local chiefs and gradually took over the area. In 1911, the company named the area Northern Rhodesia. In 1924, Britain took over the government of the country and the discovery of copper led to a large influx of Europeans in the late 1920s.

Following World War II, the majority of Europeans living in Zambia wanted greater control of their government and some favoured a merger with their southern neighbor, Southern Rhodesia (now Zimbabwe). In 1953, Britain set up a federation of Northern Rhodesia, Southern Rhodesia, and Nyasaland (now Malawi). Local Africans opposed the setting up of the federation arguing that it concentrated power in the hands

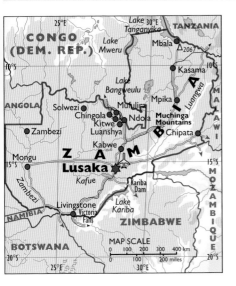

with substantial popular and parliamentary opposition. In the 2001 elections, the MMD candidate, Levy Mwanawasa, was elected president. In 2005 the Supreme Court rejected a challenge to his election, but stated that the 2001 ballot had been flawed.

ECONOMY

Zambia holds 6% of the world's copper reserves and copper is the leading export, accounting for 49% of Zambia's total exports. Zambia also produces cobalt, lead, zinc, and gemstones, but the country's dependence on minerals has created problems, especially when prices fluctuate. Agriculture employs 69% of the workforce, compared with 4% in mining and manufacturing. Major food crops include cassava, fruits, and vegetables, corn, millet, and sorghum, while cash crops include coffee, sugarcane, and tobacco.

The Copperbelt, centered on Kitwe, is the main urban region, while Lusaka, provides the other major growth pole. Rural to urban migration has increased since 1964, but work is scarce. The production of copper products is the leading industrial activity. Other manufactures include beverages, processed food, iron and steel, textiles, and tobacco.

of the white minority in Southern Rhodesia. Their opposition proved effective and the federation was dissolved in 1963. In 1964, Northern Rhodesia became an independent nation called Zambia.

POLITICS

The leading opponent of British rule, Kenneth Kaunda, became president in 1964. His government enjoyed reasonable income until copper prices crashed in the mid-1970s, but his collectivist policies failed to diversify the economy and neglected agriculture. In 1972, he declared the United Nationalist Independence Party (UNIP) the only legal party, and it was nearly 20 years before the country returned to democracy.

Under a new constitution, adopted in 1990, elections were held in 1991 in which Kaunda was trounced by Frederick Chiluba of the Movement for Multiparty Democracy (MMD)—Kaunda's first challenger in the post-colonial period. Chiluba was reelected in 1996, but he stood down in 2001 after an MMD proposal to amend the constitution to allow Chiluba to stand for a third term met

LUSAKA

Capital and largest city of Zambia, in the south-central part of the country, at an altitude of 4,200 ft [1,280 m]. Founded by Europeans in 1905 to service the local leadmining, it replaced Livingstone as the capital of Northern Rhodesia (later Zambia) in 1935. A vital road and rail junction, Lusaka is the center of a fertile agricultural region, and is a major financial and commercial city.

ZIMBABWE

Zimbabwe's flag, adopted in 1980, is based on the colors used by the ruling Zimbabwe African National Union Patriotic Front. Within the white triangle is the Great Zimbabwe soapstone bird, the national emblem. The red star symbolizes the party's socialist policies.

The Republic of Zimbabwe is a landlocked country in southern Africa. Most of the country lies on a high plateau between the Zambezi and Limpopo Rivers between 2,950 and 4,920 ft [900–1,500 m] above sea level.

The principal land feature is the High Veld, a ridge that crosses Zimbabwe from northeast to southwest. Harare lies on the northeast edge, Bulawayo on the southwest edge. Bordering the High Veld is the Middle Veld, the country's largest region and the site of many large ranches. Below 2,950 ft [900 m] is the Low Veld.

The country's highest point is Mount Inyangani, which reaches 8,507 ft [2,593 m] near the Mozambique border. Zimbabwe's best-known physical feature, Victoria Falls, is in the northeast. The Falls are shared with Zambia, as too is the artificial Lake Kariba, which is also on the River Zambezi.

Wooded savanna covers much of Zimbabwe. The Eastern Highlands and river valleys are forested. There are many tobacco plantations.

Area 150,871 sq mi [390,757 sq km]
Population 12,672,000
Capital (population) Harare (1,189,000)
Government Multiparty republic
Ethnic groups Shona 82%, Ndebele 14%, other African groups 2%, mixed and Asian 1%
Languages English (official), Shona, Ndebele
Religions Christianity, traditional belief
Currency Zimbabwean dollar= 100 cents
Website www.zim.gov.zw

CLIMATE

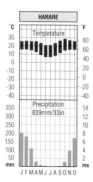

The subtropical climate varies greatly according to altitude. The Low Veld is much warmer and drier than the High Veld. November to March is mainly hot and wet. Winter in Harare is dry but cold. Frosts have been recorded between June and August.

HISTORY

The Shona people were dominant in the region about 1,000 years ago. They built Great Zimbabwe, a city of stone buildings. Under the statesman Cecil Rhodes (1853–1902), the British South Africa Company occupied the area in the 1890s, after obtaining mineral rights from local chiefs. The area was named Rhodesia and later Southern Rhodesia. It became a self-governing British colony in 1923. Between 1953 and 1963, Southern and Northern Rhodesia (now Zambia) were joined to Nyasaland (Malawi) in the Central African Federation.

POLITICS

In 1965, the European government of Southern Rhodesia (then known as Rhodesia) declared their country independent. However, Britain refused to accept this declaration. Finally, after a civil war, the country became legally independent in 1980.

After independence, rivalries between the Shona and Ndebele people threatened its stability. But order was restored when the Shona prime minister, Robert Mugabe, brought his Ndebele rivals into

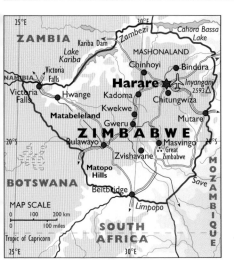

one of the world's six "outposts of tyranny," an accusation that was rejected by Zimbabwe.

ECONOMY

The World Bank classifies Zimbabwe as a "low-income" economy. Its economy has become significantly more diverse since the 1960s, having evolved to virtual self-sufficiency during the days of international sanctions between 1965 and 1980. After independence, the economy underwent a surge in most sectors, with successful agrarian policies and the exploitation of the country's mineral resources. However, a fast-growing population continues to exert pressure both on land and resources of all kinds.

Agriculture employs approximately 30% of the people. Corn is the chief food crop, while cash crops include cotton, sugar, and tobacco. Cattle ranching is another important activity. Gold, asbestos, chromium, and nickel are mined and the country also has some coal and iron ore. Manufactures include beverages, chemicals, iron and steel, metal products, processed food, textiles, and tobacco. The principal exports include tobacco, gold, other metals, cotton, and asbestos.

his government. In 1987, Mugabe became the country's executive president and, in 1991, the government renounced its Marxist ideology. In 1990, the state of emergency that had lasted since 1965 was allowed to lapse—three months after Mugabe had secured a landslide election victory. Mugabe was reelected in 1996. In the late 1990s, Mugabe threatened to seize white-owned farms without paying compensation to owners. His announcement caused much disquiet among white farmers. The situation worsened in the early 2000s, when landless "war veterans" began to occupy white-owned farms, resulting in violence and deaths.

Food shortages have become a major problem with aid agencies blaming the land reform programme while the government blames drought.

In 2002, amid accusations of electoral irregularities, Mugabe was reelected president. Mounting criticism of Mugabe led the Commonwealth to suspend Zimbabwe's membership. Later Zimbabwe confirmed that it had pulled out of the Commonwealth permanently. In 2004 the European Union renewed sanctions against the country. Zimbabwe was named by the United States, in 2005, as

HARARE

Capital of Zimbabwe, in the north east of the country. Settled by Europeans in 1890 as Fort Salisbury, it became capital of Southern Rhodesia in 1902. The city served as capital of the Federation of Rhodesia and Nyasaland (1953–63) and of Rhodesia (1965–79). It has a university (1957) and two cathedrals. Industries include gold mining, textiles, steel, tobacco, chemicals and furniture.

INDEX